Classic Texts Series

Problems in ANALYTIC GEOMETRY

D. Kletenik

arihant

ARIHANT PRAKASHAN (Series), MEERUT

arihant

ARIHANT PRAKASHAN (Series), MEERUT

Administrative & Production Offices

Regd. Office

'Ramchhaya' 4577/15, Agarwal Road, Darya Ganj, New Delhi -110002

Tele: 011- 47630600, 43518550

Head Office

Kalindi, TP Nagar, Meerut (UP) - 250002

Tel: 0121-7156203, 7156204

Sales & Support Offices

Agra, Ahmedabad, Bengaluru, Bareilly, Chennai, Delhi, Guwahati, Hyderabad, Jaipur, Jhansi, Kolkata, Lucknow, Nagpur & Pune.

ISBN 978-93-24191-75-5

Published by Arihant Publications (India) Ltd.

For further information about the books published by Arihant, log on to www.arihantbooks.com or e-mail at info@arihantbooks.com

Follow us on

Problems in

ANALYTIC GEOMETRY

Contents

Part Two : Solid Analytic Geometry

PART ONE

PLANE ANALYTIC GEOMETRY

1

ELEMENTARY PROBLEMS OF PLANE ANALYTIC GEOMETRY

1. An Axis and Segments of an Axis. Coordinates on a Straight Line

A straight line on which a positive direction has been chosen is called an axis. A segment of an axis bounded by arbitrary points A and B is called a directed segment. If one of these points has been designated as the initial point, and the other as the terminal point of the segment. A directed segment with A as its initial point and B as its terminal point is denoted by the symbol $\overline{AB}$. The value of a directed segment of an axis is defined as the length of the segment taken with a plus or minus sign according as the direction of the segment (that is, the direction from its initial to its terminal point) agrees with the positive or negative direction of the axis. The value of a segment $\overline{AB}$ is denoted by the symbol AB, whereas its length is denoted by the symbol $|AB|$. If the points A and B coincide, the segment determined by them is called a zero segment, clearly, $AB = BA = 0$ in this case (a zero segment is considered to have no definite direction).

Let there be given an arbitrary straight line a. We next choose a segment as the unit for measurement of lengths, assign a positive direction to the line a * (thereby making it an axis), and mark some point on this line by the letter O. We have thus established a coordinate system on the line a.

The coordinate of any point M of the line a (in the chosen coordinate system) is defined as the number x equal to the value of the segment OM:

$$x = OM.$$

The point O is called the origin of coordinates; the coordinate of the origin itself is equal to zero. We shall henceforth use the notation $M(x)$ to indicate that the point M has x as its coordinate.

---

* In diagrams, horizontal axes usually have their positive direction from left to right.

If $M_1(x_1)$ and $M_2(x_2)$ are two arbitrary points of the line a, then the formula

$$M_1M_2 = x_2 - x_1$$

expresses the value of the segment $\overline{M_1M_2}$ and the formula

$$|M_1M_2| = |x_2 - x_1|$$

expresses the length of this segment.

1. Plot the points:

$A(3), B(5), C(-1), D\left(\frac{2}{3}\right), E\left(-\frac{3}{7}\right), F(\sqrt{2})$, and $H(-\sqrt{5})$.

2. Plot the points whose coordinates satisfy the equations:

1) $|x| = 2$; 2) $|x - 1| = 3$; 3) $|1 - x| = 2$; 4) $|2 + x| = 2$.

3. Characterize geometrically the location of the points whose coordinates satisfy the inequalities:

1) $x > 2$; 2) $x - 3 \le 0$; 3) $12 - x < 0$; 4) $2x - 3 \le 0$;

5) $3x - 5 > 0$; 6) $1 < x < 3$; 7) $-2 \le x \le 3$; 8) $\frac{2-x}{x-1} > 0$;

9) $\frac{2x-1}{x-2} > 1$; 10) $\frac{2-x}{x-1} < 0$; 11) $\frac{2x-1}{x-2} < 1$;

12) $x^2 - 8x + 15 \le 0$; 13) $x^2 - 8x + 15 > 0$;

14) $x^2 + x - 12 > 0$; 15) $x^2 + x - 12 \le 0$.

4. Find the value AB and the length $|AB|$ of the segment determined by the points:

1) $A(3)$ and $B(11)$; 2) $A(5)$ and $B\,2)$;

3) $A(-1)$ and $B\,3)$; 4) $A(-5)$ and $B(-3)$;

5) $A(-1)$ and $B(-3)$; 6) $A(-7)$ and $B(-5)$;

5. Calculate the coordinate of a point A, given:

1) $B(3)$ and $AB = 5$ 2) $B(2)$ and $AB = -3$;

3) $B(-1)$ and $BA = 2$ 4) $B(-5)$ and $BA = -3$;

5) $B(0)$ and $|AB| = 2$ 6) $B(2)$ and $|AB| = 3$;

7) $B(-1)$ and $|AB| = 5$ 8) $B(-5)$ and $|AB| = 2$.

6. Characterize geometrically the location of the points whose coordinates satisfy the following inequalities:

1) $|x| < 1$; 2) $|x| > 2$;

3) $|x| \le 2$; 4) $|x| \ge 3$;

5) $|x - 2| < 3$; 6) $|x - 5| \le 1$;

7) $|x - 1| \ge 2$; 8) $|x - 3| \ge 1$;

9) $|x + 1| < 3$; 10) $|x + 2| > 1$;

11) $|x + 5| \le 1$; 12) $|x + 1| \ge 2$.

7. Determine the ratio $\lambda = \frac{AC}{CB}$ in which the point C divides the segment $\overline{AB}$, given:

1) $A(2)$, $B(6)$ and $C(4)$;
2) $A(2)$, $B(4)$ and $C(7)$;
3) $A(-1)$, $B(5)$ and $C(3)$;
4) $A(1)$, $B(13)$ and $C(5)$;
5) $A(5)$, $B(-2)$ and $C(-5)$;

8. Given the three points $A(-7)$, $B(-1)$ and $C(1)$. Determine the ratio λ in which each of these divides the segment bounded by the other two points.

9. Determine the ratio $\lambda = \frac{M_1M}{MM_2}$ in which a given point $M(x)$ divides the segment $\overline{M_1M_2}$ bounded by given points $M_1(x_1)$ and $M_2(x_2)$.

10. Determine the coordinate x of a point M which divides the segment $\overline{M_1M_2}$ bounded by given points $M_1(x_1)$ and $M_2(x_2)$, in a given ratio λ $\left(\lambda = \frac{M_1M}{MM_2}\right)$.

11. Find the coordinate x of the midpoint of the segment bounded by two given points $M_1(x_1)$ and $M_2(x_2)$.

12. In each of the following, find the coordinate x of the midpoint of the segment bounded by the two given points:

1) $A(3)$ and $B(5)$;
2) $C(-1)$ and $D(5)$;
3) $M_1(-1)$ and $M_2(-3)$;
4) $P_1(-5)$ and $P_2(1)$;
5) $Q_1(3)$ and $Q_2(-4)$.

13. Find the coordinate of a point M, given:

1) $M_1(3)$, $M_2(7)$ and $\lambda = \frac{M_1M}{MM_2} = 2$;

2) $A(2)$, $B(-5)$ and $\lambda = \frac{AM}{MB} = 3$;

3) $C(-1)$, $D(3)$ and $\lambda = \frac{CM}{MD} = \frac{1}{2}$;

4) $A(-1)$, $B(3)$ and $\lambda = \frac{AM}{MB} = -2$;

5) $A(1)$, $B(-3)$ and $\lambda = \frac{BM}{MA} = -3$;

6) $A(-2)$, $B(-1)$ and $\lambda = \frac{BM}{MA} = -\frac{1}{2}$.

14. Given the two points $A(5)$ and $B(-3)$. Find:

1) the coordinate of the point M symmetric to the point A with respect to the point B;
2) the coordinate of the point N symmetric to the point B with respect to the point A.

15. The segment bounded by the points. $A(-2)$ and $B(19)$ is divided into three equal parts. Determine the coordinates of the trisection points.

16. Determine the coordinates of the end points A and B of the segment whose trisection points are $P(-25)$ and $Q(-9)$.

2. Rectangular Cartesian Coordinates in a Plane

A rectangular cartesian system of coordinates is determined by the choice of a linear unit (for measurement of lengths) and of two mutually perpendicular axes numbered in any order.

The point of intersection of the axes is called the origin of coordinates, and the axes themselves are called the coordinate axes. The first of the coordinate axes is termed the x-axis or axis of abscissas, and the second, the y-axis or axis of ordinates.

The origin is denoted by the letter O, the x-axis by the symbol Ox, and the y-axis by Oy.

The coordinates of an arbitrary point M in a given system are defined as the numbers

$$x = OM_x, y = OM_y$$

(Fig. 1), where M_x and M_y are the respective projections of the point M on the axis Ox and Oy, OM_x is the value of the segment $\overline{OM_x}$ of the x-axis, and OM_y is the value of the segment $\overline{OM_y}$ of the $y-$ axis. The number x is called the abscissa of the point M, and the number y, the ordinate of M. The notation $M(x, y)$ means that the point M has the number x as its abscissa, and the number y as its ordinate.

The axis Oy divides the entire plane into two half-planes, of which the one containing the positive half of the axis Ox is called the right half-plane, and the other, the left half-plane. In like manner, the asix Ox divides the plane into two half-planes, of which the one containing the positive half of the axis Oy is called the upper half-plane, and the other, the lower half-plane.

The two coordinate axes jointly divide the plane into four quadrants, which are numbered according to the following rule: the first quadrant is the one lying simultaneously in the right and the upper half-planes; the second quadrant lies in the left and the upper half-planes; the third quadrant lies in the left and the lower half-planes; and the fourth quadrant lies in the right and the lower half-planes.

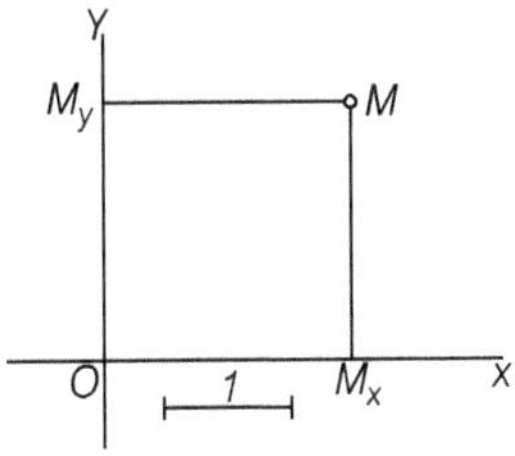

Fig. 1.

17. Plot the points:

$A(2,3), B(-5,1), C(-2,-3), D(0,3),\ E(-5,0), F\left(-\frac{1}{3},\frac{2}{3}\right)$.

18. For each of the following points, find the coordinates of its projection on the x-axis:

$A(2,-3), B(3,-1), C(-5,1), D(-3,-2), E(-5,-1)$.

19. For each of the following points, find the coordinates of its projection on the y-axis:

$A(3,2), B(-5,1), C(-3,-2), D(-1,1), E(-6,-2)$.

20. Find the coordinates of the points symmetric, with respect to the axis Ox, to the following points:

1) $A(2,3)$; 2) $B(-3,2)$; 3) $C(-1,-1)$; 4) $D(-3,-5)$;
5) $E(-4,6)$; 6) $F(a,b)$.

21. Find the coordinates of the points symmetric, with respect to the axis Oy, to the following points:

1) $A(-1,2)$; 2) $B(3,-1)$; 3) $C(-2,-2)$; 4) $D(-2,5)$;
5) $E(3,-5)$; 6) $F(a,b)$.

22. Find the coordinates of the points symmetric, with respect to the origin, to the points:

1) $A(3, 3)$; 2) $B(2, -4)$; 3) $C(-2, 1)$; 4) $D(5, -3)$;
5) $E(-5, -4)$; 6) $F(a, b)$

23. Find the coordinates of the points symmetric, with respect to the line bisecting the first quadrant, to the following points:

1) $A(2, 3)$; 2) $B(5, -2)$; 3) $C(-3, 4)$.

24. Find the coordinates of the points symmetric, with respect to the line bisecting the second quadrant, to the points:

1) $A(3, 5)$; 2) $B(-4, 3)$ $C(7, -2)$.

25. Determine the quadrants in which a point $M(x, y)$ can be situated if:

1) $xy > 0$; 2) $xy < 0$; 3) $x - y = 0$; 4) $x + y = 0$;
5) $x + y > 0$; 6) $x + y < 0$; 7) $x - y > 0$; 8) $x - y < 0$.

3. Polar Coordinates

A polar coordinate system is determined by choosing a point O, called the pole, a ray OA drawn from that point and called the polar axis, and a scale for measurement of lengths. When determining a polar system, it must also be specified which direction of rotation about the point O is to be considered positive (in diagrams, counterclockwise rotation is usually taken as positive).

The numbers $\rho = OM$ and $0 = \angle AOM$ (Fig. 2) are called the polar coordinates of the arbitrary point M (in reference to the chosen system); the angle 0 is here understood as in trigonometry. the number ρ is called the first coordinate or polarv radius *, and the number 0. the second coordinate or polar angle (0 is also termed the amplitude).

The notation $M(\rho, 0)$ means that the point M has polar coordinates ρ and 0.

The polar angle 0 has an infinite number of possible values (differing from one another by a quantity of the form $\pm 2n\pi$, where n is a positive integer). That value of the polar angle which satisfies the inequalities $-\pi < 0 < +\pi$ is called its principal value.

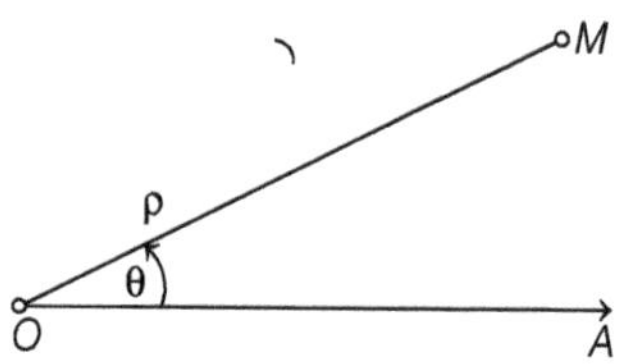

Fig. 2.

In cases where a cartesian and a polar coordinate system are to be used side by side, we shall agree: (1) to use the same scale; 2) when determining polar angles, to regard as positive the direction of the shortest rotation of the positive x-axis into the positive y-axis (thus, if the axes of the cartesian system have their usual position, i.e., with the axis Ox directed to the right and the axis Oy directed upwards, then polar angles are also measured as usual, i.e., are measured positively in the counterclockwise direction).

Under this condition, and provided that the pole of the polar coordinate system coincides with the origin of rectangular cartesian coordinates, while the polar axis coincides with the positive x-axis, the transformation from polar coordinates of an arbitrary point to its cartesian coordinates is carried out by the formulas $x = \rho\cos 0$, $y = \rho\sin 0$.

Under the same conditions, $\rho = \sqrt{x^2 + y^2}$, $\tan 0 = \dfrac{y}{x}$

---

* OM denotes here the *length* of the segment, understood as in elementary geometry (that is, the unsigned length). The more cumbersome symbol $|OM|$ need not be employed in this case since the points O and M are regarded as arbitrary points in a plane, rather than points of an axis. We shall often use this simplified notation in similar cases below.

will be the formulas for transformation from cartesian to polar coordinates.

When using two polar coordinate systems in a single problem, we shall agree to adopt the same positive direction of rotation and the same scale for both systems.

26. Plot the following points given in polar coordinates:

$A\left(3, \frac{\pi}{2}\right), B(2, \pi), C\left(3, -\frac{\pi}{4}\right), D\left(4, 3\frac{1}{7}\right),$

$E(5, 2)$ and $F(1, -1)$.

(The points D, E and F are located approximately by using a protractor.)

27. Determine the polar coordinates of the points symmetric, with respect to the polar axis, to the points

$M_1\left(3, \frac{\pi}{4}\right), M_2\left(2, -\frac{\pi}{2}\right), M_3\left(3, -\frac{\pi}{3}\right),$

$M_4(1, 2)$ and $M_5(5, -1)$, given in a polar coordinate system.

28. Determine the polar coordinates of the points symmetric, with respect to the pole, to the points

$M_1\left(1, \frac{\pi}{4}\right), M_2\left(5, \frac{\pi}{2}\right), M_3\left(2, -\frac{\pi}{3}\right),$

$M_4\left(4, \frac{5}{6}\pi\right)$ and $M_5\ (3, -2)$

given in a polar coordinate system.

29. In a polar coordinate system, $A\left(3, -\frac{4}{9}\pi\right)$ and $B\left(5, \frac{3}{14}\pi\right)$ are two given vertices of a parallelogram $ABCD$, the point of intersection of its diagonals coincides with the pole. Find the other two vertices of the parallelogram.

30. Given the points $A\left(8, -\frac{2}{3}\pi\right)$ and $B\left(6, \frac{\pi}{3}\right)$ in a polar coordinate system. Calculate the polar coordinates of the midpoint of the segment joining the points A and B.

31. Given the points

$A\left(3, \frac{\pi}{2}\right), B\left(2, -\frac{\pi}{4}\right), C(1, \pi), D\left(5, -\frac{3}{4}\pi\right), E(3, 2)$

and $F(2, -1)$

in a polar coordinate system. The positive direction of the polar axis is reversed; determine the polar coordinates of the given points in the new system.

32. Given the points

$M_1\left(3, \frac{\pi}{3}\right)$, $M_2\left(1, \frac{2}{3}\pi\right)$, $M_3(2, 0)$,

$M_4\left(5, \frac{\pi}{4}\right)$, $M_5\left(3, -\frac{2}{3}\pi\right)$

and $M_6\left(1, \frac{11}{12}\pi\right)$

in a polar coordinate system. The polar axis is turned so that, in its new position, it passes through the point M_1; determine the coordinates of the given points in the new (polar) system.

33. Given the points $M_1\left(12, \frac{4}{9}\pi\right)$, and $M_2\left(12, -\frac{2}{9}\pi\right)$ in a polar coordinate system. Calculate the polar coordinates of the midpoint of the segment joining M_1 and M_2.

34. Given the points $M_1(\rho_1, \theta_1)$ and $M_2(\rho_2, \theta_2)$ in a polar coordinate system. Compute the distance d between them.

35. Given the points $M_1\left(5, \frac{\pi}{4}\right)$ and $M_2\left(8, -\frac{\pi}{12}\right)$ in a polar coordinate system. Compute the distance d between them.

36. In a polar coordinate system, $M_1\left(12, -\frac{\pi}{10}\right)$ and $M_2\left(3, \frac{\pi}{15}\right)$ are two adjacent vertices of a square. Find its area.

37. In a polar coordinate system, $P\left(6, -\frac{7}{12}\pi\right)$ and $Q\left(4, \frac{1}{6}\pi\right)$ are two opposite vertices of a square. Find its area.

38. In a polar coordinate system, $A\left(4, -\frac{1}{12}\pi\right)$ and $B\left(8, \frac{7}{12}\pi\right)$ are two vertices of a regular triangle. Find its area.

39. One vertex of a triangle *OAB* is situated at the pole, and the other two vertices are the points $A(\rho_1, \theta_1)$ and $B(\rho_2, \theta_2)$. Calculate the area of the triangle.

40. One vertex of a triangle *OAB* is at the pole *O*, and the other two vertices are the points $A\left(5, \frac{\pi}{4}\right)$ and $B\left(4, \frac{\pi}{12}\right)$. Find the area of the triangle.

41. Calculate the area of the triangle whose vertices are $A\left(3, \frac{1}{8}\pi\right)$, $B\left(8, \frac{7}{24}\pi\right)$ and $C\left(6, \frac{5}{8}\pi\right)$ in polar coordinates.

42. The pole of a polar coordinate system coincides with the origin of rectangular cartesian coordinates, and the polar axis coincides with the positive x-axis.

$M_1\left(6, \frac{\pi}{2}\right)$, $M_2(5, 0)$, $M_3\left(2, \frac{\pi}{4}\right)$, $M_4\left(10, -\frac{\pi}{3}\right)$, $M_5\left(8, \frac{2}{3}\pi\right)$, $M_6\left(12, -\frac{\pi}{6}\right)$ are points given in the polar coordinate system. Determine the cartesian coordinates of these points.

43. The pole of a polar coordinate system coincides with the origin of rectangular cartesian coordinates, and the polar axis coincides with the positive x-axis. $M_1(0,5)$, $M_2(-3,0)$, $M_3(\sqrt{3}, 1)$, $M_4(-\sqrt{2}, -\sqrt{2})$, $M_5(1, -\sqrt{3})$ are points given in the rectangular cartesian system. Determine the polar coordinates of these points.

4. A Directed Segment. The Projection of a Segment on an Axis. The Projections of a Segment on the Coordinate Axes. The Length and the Polar Angle of a Segment. The Distance Between Two Points

A line segment is said to be directed if one of its bounding points has been designated as the initial point, and the other as the terminal point of the segment. A directed segment having A as its initial point and B as its terminal point (Fig. 3) is denoted by the symbol $\overline{AB}$ (the same as a segment of an axis; see 1). The length of a directed segment $\overline{AB}$ (in a given scale) is denoted by the symbol $|AB|$ (or AB; see the footnote on page 17).

The projection of a segment $\overline{AB}$ on an arbitrary axis u is defined as the number equal to the value of the segment $\overline{A_1B_1}$ of the axis u, where the point A_1 is the projection upon the axis u of the point A_1 and B_1 the projection of the point B.

The projection of a segment $\overline{AB}$ on an axis u is denoted by the symbol $\text{proj}\ u\overline{AB}$. *If* a rectangular cartesian system of coordinates has been attached to the plane, the projection of a segment on the x-axis is denoted by X, and its projection on the y-axis by Y.

If we know the coordinates of the points $M_1(x_1, y_1)$ and $M_2(x_2, y_2)$, then the projections X and Y of the directed segment

$\overline{M_1M_2}$ on the coordinate axes can be calculated from the formulas

$$X = x_2 - x_1,$$
$$Y = y_2 - y_1.$$

Thus, to find the projections of a directed segment of the coordinate axes, subtract the coordinates of its initial point from the corresponding coordinates of its terminal point.

The angle 0 through which the positive x-axis must be rotated to make its direction coincide with that of a segment $\overline{M_1M_2}$ is termed the polar angle of the segment $\overline{M_1M_2}$.

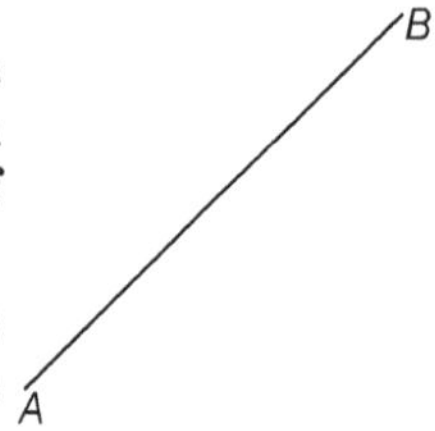

Fig. 3.

The angle 0 is understood here as in trigonometry; accordingly, 0 has an infinity of possible values, which differ from one another by a quantity of the form $\pm 2n\pi$ (where n is a positive integer). The principal value of the polar angle is defined to be that one of its values which satisfies the inequalities $-\pi < 0 \leq +\pi$.

The formulas $X = d.\cos 0, Y = d.\sin 0$

express the projections of an arbitrary segment on the coordinate axes in terms of its length and its polar angle. From these, we also have the formulas

$$d = \sqrt{X^2 + Y^2}$$

and $\cos 0 = \dfrac{X}{\sqrt{X^2 + Y^2}}, \sin 0 = \dfrac{Y}{\sqrt{x^2 + y^2}},$

which express the length and the polar angle of a segment in terms of its projections on the coordinate axes.

If $M_1(x_1, y_1)$ and $M_2(x_2, y_2)$ are two given points in the plane, the distance d between them is determined by the formula

$$d = \sqrt{(x_2 - x_1)^2 + (y_2 - y_1)^2}$$

44. Calculate the projection on an axis u of the segment whose length d and angle of inclination ϕ with respect to that axis are

1) $d = 6, \phi = \dfrac{\pi}{3}$; 2) $d = 6, \phi = \dfrac{2\pi}{3}$;

3) $d = 7, \phi = \dfrac{\pi}{2}$; 4) $d = 5, \phi = 0$;

5) $d = 5, \phi = \pi$; 6) $d = 4, \phi = -\dfrac{\pi}{3}$;

45. From the origin of coordinates, draw the segments whose projections on the coordinate axes are:

1) $X = 3, Y = 2$; 2) $X = 2, Y = -5$;

3) $X = -5, Y = 0$; 4) $X = -2, Y = 3$;

5) $X = 0, Y = 3$; 6) $X = -5, Y = -1$.

46. From the point $M(2, -1)$, draw the segments whose projections on the coordinate axes are:

1) $X = 4, Y = 3$; 2) $X = 2, Y = 0$;
3) $X = -3, Y = 1$; 4) $X = -4, Y = -2$;
5) $X = 0, Y = -3$; 6) $X = 1, Y = -3$.

47. Given the points M_1 (1, –2), M_2(2, 1), M_3 (5, 0), M_4(–1, 4), and M_5 (0, –3). Find the projections on the coordinate axes of the following segments

1) $\overline{M_1M_2}$, 2) $\overline{M_3M_1}$,
3) $\overline{M_4M_5}$, 4) $.\overline{M_5\ M_3}$,

48. $X = 5$, $Y = -4$ are the projections on the coordinate axes of the segment $\overline{M_1M_2}$ whose initial point is $M_1(-2, 3)$; find the coordinates of its terminal point.

49. $X = 4$, $Y = -5$ are the projections on the coordinate axes of the segment $\overline{AB}$ whose terminal point is at $B\,(1, -3)$, find the coordinates of the initial point of $\overline{AB}$.

50. From the origin of coordinates, draw the segment whose length d and polar angle 0 are:

1) $d = 5, 0 = \frac{\pi}{5}$; 2) $d = 3, 0 = \frac{5}{6}\pi$;
3) $d = 4, 0 = -\frac{\pi}{3}$; 4) $d = 3, 0 = -\frac{4}{3}\pi$.

51. From the point M with cartesian coordinates (2, 3), draw the segment whose length and polar angle are:

1) $d = 2, 0 = -\frac{\pi}{10}$; 2) $d = 1, 0 = \frac{\pi}{9}$;
3) $d = 5, 0 = -\frac{\pi}{2}$.

52. Find the projections on the coordinate axes of the segment whose length d and polar angle 0 are:

1) $d = 12,\ 0 = \frac{2}{3}\pi$; 2) $d = 6, 0 = -\frac{\pi}{6}$; 3) $d = 2, 0 = -\frac{\pi}{4}$;

53. Given the projections of three segments on the coordinate axes:

1) $X = 3, Y = -4$; 2) $X = 12, Y = 5$;
3) $X = -8, Y = 6$.

Calculate the length of each segment.

54. Given the projections of three segments on the coordinate axes:

1) $X = 1, Y = \sqrt{3}$; 2) $X = 3\sqrt{2}, Y = -3\sqrt{2}$;
3) $X = -2\sqrt{3}, Y = 2$.

Calculate the length d and the polar angle 0 of each segment.

55. Given the points M_1 (2, –3), M_2(1, – 4), M_3(–1, –7) and M_4(–4, 8). Calculate the length and the polar angle of the following segments:

1) $\overline{M_1M_2}$, 2) $\overline{M_1M_3}$,
3) $\overline{M_2M_4}$, 4) $\overline{M_4M_3}$.

56. The length d of a segment is 5 and its projection on the x-axis is 4. Find the projection of this segment on the y-axis if the angle which it makes with the y-axis is:

1) acute 2) obtuse

57. The length of a segment $\overline{MN}$ is 13, its initial point is M (3,–2), and its projection on the x-axis equals – 12. Find the coordinates of the terminal point of this segment if it makes with the y-axis:

1) an acute angle: 2) an obtuse angle.

58. The length of a segment $\overline{MN}$ is 17, its terminal point is N (–7, 3), and its projection on the y-axis is equal to 15. Find the coordinates of the initial point of this segment if it makes with the x-axis:

1) an acute angle; 2) an obtuse angle.

59. The projections of a segment on the coordinate axes are $X = 1$, $Y = -\sqrt{3}$; find its projection on the axis which makes an angle $\theta = \frac{2}{3}\pi$ with the axis Ox.

60. Given the two points M_1 (1, –5) and M_2 (4, –1). Find the projection of the segment $\overline{M_1M_2}$ on the axis which makes an angle $\theta = -\frac{\pi}{6}$ with the axis Ox.

61. Given the two points P (–5, 2) and Q (3, 1). Find the projection of the segment $\overline{PQ}$ on the axis which makes an angle $0 = \arctan\frac{4}{3}$ with the axis Ox.

62. Given the two points M_1 (2, –2) and M_2 (7, –3). Find the projection of the segment $\overline{M_1M_2}$ on the axis passing through the points A (5, –4), B(–7, 1) and directed:

1) from A to B; 2) from B to A.

63. Given the points A (0, 0), B (3, –4), C (–3, 4), D (–2, 2) and E(10, –3). Determine the distance d between the points:

1) A and B; 2) B and C; 3) A and C; 4) C and D; 5) A and D; 6) D and E.

64. A(3, –7) and B(–1, 4) are two adjacent vertices of a square. Compute its area.

65. P (3, 5) and Q (1, –3) are two opposite vertices of a square. Compute its area.

66. Find the area of a regular triangle, two of whose vertices are A (–3, 2) and B (1, 6).

67. A (3, –7), B(5, –7), C (–2, 5) are three vertices of a parallelogram $ABCD$; its fourth vertex D is opposite to B. Find the length of the diagonals of the parallelogram.

68. A rhombus has its side equal to $5\sqrt{10}$; two opposite vertices of the rhombus are at the points P (4, 9) and Q (–2, 1). Find the area of the rhombus.

69. A rhombus has its side equal to $5\sqrt{2}$; two opposite vertices are at the points P (3, – 4) and Q (1, 2). Find the length of the altitude of the rhombus.

70. Prove that the points A (3, –5), B (–2, –7) and C (18, 1) lie on a straight line.

71. Prove that the triangle with vertices A_1(1, 1), A_2 (2, 3) and A_3(5, –1) is a right triangle.

72. Prove that the points A (2, 2), B (–1, 6), C(–5, 3) and D(–2, –1) are the vertices of a square.

73. Determine whether the triangle with vertices M_1(1, 1), M_2(0, 2) and M_3 (2, –1) has an obtuse angle among its interior angles.

74. Prove that the interior angles of the triangle with vertices M (–1, 3), N (1, 2) and P (0, 4) are all of them acute angles.

75. The points A (5, 0), B (0, 1) and C (3, 3) are the vertices of a triangle. Calculate its interior angles.

76. The points A $(-\sqrt{3}, 1)$, B (0, 2) and C $(-2\sqrt{3}, 2)$ are the vertices of a triangle. Calculate the exterior angle at the vertex A.

77. Find a point M on the x-axis such that its distance from the point N (2, –3) will be equal to 5.

78. Find a point M on the y-axis such that its distance from the point N(–8, 13) will be equal to 17.

79. Given the two points M(2, 2) and N(5, –2); find a point P on the x-axis such that the angle MPN will be a right angle.

80. A circle tangent to both coordinate axes is drawn through the point A (4, 2). Determine the centre C and the radius R of this circle.

81. A circle of radius 5 and tangent to the axis Ox is drawn through the point M_1(1, –2). Determine the centre C of this circle.

82. Determine the coordinates of the point M_2 symmetric to the point M_1(1, 2) with respect to the straight line which passes through A (1, 0) and B (–1, –2).

83. Given two opposite vertices $A(3, 0)$ and $C(-4, 1)$ of a square. Find its other two vertices.

84. Given two adjacent vertices $A(2, -1)$ and $B(-1, 3)$ of a square. Find its other two vertices.

85. Given the vertices $M_1(-3, 6)$, $M_2(9, -10)$ and $M_3(-5, 4)$ of a triangle. Find the centre C and the radius R of the circle circumscribed about this triangle.

5. The Division of a Segment in a Given Ratio

If a point $M(x, y)$ lies on the straight line passing through two given points $M_1(x_1, y_1)$, $M_2(x_2, y_2)$ and if $\lambda = \frac{M_1M}{MM_2}$ is a given ratio in which the point M divides the line segment $\overline{M_1M_2}$, then the coordinates of M are determined by the formulas

$$x = \frac{x_1 + \lambda x_2}{1 + \lambda}, \quad y = \frac{y_1 + \lambda y_2}{1 + \lambda}.$$

If the point M is the midpoint of the segment $\overline{M_1M_2}$, its coordinates are given by the formulas

$$x = \frac{x_1 + x_2}{2}, \quad y = \frac{y_1 + y_2}{2}$$

86. Given the end points $A(3, -5)$ and $B(-1, 1)$ of a uniform rod. Determine the coordinates of its centre of gravity.

87. A uniform rod has its centre of gravity at $M(1, 4)$ and one of its end points at $P(-2, 2)$. Determine the coordinates of the other end point Q of the rod.

88. Given the vertices $A(1, -3)$, $B(3, -5)$ and $C(-5, 7)$ of a triangle. Determine the midpoints of its sides.

89. Given the two points $A(3, -1)$ and $B(2, 1)$. Find:

1) the coordinates of the point M symmetric to A with respect to B;

2) the coordinates of the point N symmetric to B with respect to A.

90. The points $M(2, -1)$, $N(-1, 4)$ and $P(-2, 2)$ are the midpoints of the sides of a triangle. Find its vertices.

91. Given three vertices $A(3, -5)$, $B(5, -3)$, $C(-1, 3)$ of a parallelogram. Find the fourth vertex D, which is opposite to B.

92. $A(-3, 5)$ and $B(1, 7)$ are two adjacent vertices of a parallelogram, and $M(1, 1)$ the point of intersection of its diagonals. Find the other two vertices.

93. Given three vertices $A(2, 3)$, $B(4, -1)$ and $C(0, 5)$ of a parallelogram $ABCD$. Find its fourth vertex D.

94. Given the vertices $A(1, 4)$, $B(3, -9)$, $C(-5, 2)$ of a triangle. Determine the length of the median drawn from the vertex B.

95. The segment bounded by the points $A(1, -3)$ and $B(4, 3)$ is divided into three equal parts. Determine the coordinates of the trisection points.

96. Given the vertices $A(2, -5)$, $B(1, -2)$, $C(4, 7)$ of a triangle. Find the point where the bisector of the interior angle at the vertex B meets the side AC.

97. Given the vertices $A(3, -5)$, $B(-3, 3)$ and $C(-1, -2)$ of a triangle. Determine the length of the bisector of the interior angle at the vertex A.

98. Given the vertices $A(-1, -1)$, $B(3, 5)$, $C(-4, 1)$ of a triangle. Find the point where the bisector of the exterior angle at the vertex A cuts the extension of the side BC.

99. Given the vertices $A(3, -5)$, $B(1, -3)$, $C(2, -2)$ of a triangle. Determine the length of the bisector of the exterior angle at the vertex B.

100. $A(1, -1)$, $B(3, 3)$ and $C(4, 5)$ are three given points lying on a straight line. Determine the ratio λ in which each of them divides the segment bounded by the other two points.

101. Determine the coordinates of the end points A and B of the segment whose trisection points are $P(2, 2)$ and $Q(1, 5)$.

102. A straight line passes through the points $M_1(-12, -13)$ and $M_2(-2, -5)$. On this line, find a point whose abscissa is equal to 3.

103. A straight line passes through the points $M(2, -3)$ and $N(-6, 5)$. On this line, find a point whose ordinate is equal to -5.

104. A straight line passes through the points $A(7, -3)$ and $B(23, -6)$. Find the point at which this line cuts the x-axis.

105. A straight line passes through the points $A(5, 2)$ and $B(-4, -7)$. Find the point at which this line cuts the y-axis.

106. Given the vertices $A(-3, 12)$, $B(3, -4)$, $C(5, -4)$ and $D(5, 8)$ of a quadrilateral. Determine the ratio in which the diagonal AC divides the diagonal BD.

107. Given the vertices $A(-2, 14)$, $B(4, -2)$, $C(6, -2)$ and $D(6, 10)$ of a quadrilateral. Determine the point of intersection of its diagonals AC and BD.

108. Given the vertices $A(x_1, y_1)$, $B(x_2, y_2)$ and $C(x_3, y_3)$ of a uniform triangular plate. Determine the coordinates of its centre of gravity.

Hint. The centre of gravity is situated at the point of intersection of the medians.

109. The point of intersection M of the medians of a triangle lies on the x-axis; two of its vertices are at $A(2, -3)$ and $B(-5, 1)$ and the third vertex C lies on the y-axis. Find the coordinates of the points M and C.

110. $A(x_1, y_1)$, $B(x_2, y_2)$ and $C(x_3, y_3)$ are the vertices of a uniform triangular plate. If the midpoints of its sides are joined, another uniform triangular plate is formed. Prove that the centres of gravity of the two plates coincide.

Hint. Use the result of Problem 108.

111. A uniform plate has the shape of a square with side equal to 12, in which a square cut is made so that the cut-off lines meet at the centre of the plate, the coordinate axes lie along the edges of the plate (Fig. 4). Determine the centre of gravity of the plate.

112. A uniform plate has the shape of a rectangle with sides a and b, in which a rectangular cut is made so that the cut-off lines meet at the centre; the coordinate axes lie along the edges of the plate (Fig. 5). Determine the centre of gravity of the plate.

113. A uniform plate has the shape of a square with side equal to $2a$, from which a triangular piece is cut so that the cut-off line joins the midpoints of two adjacent sides; the coordinate axes lie along the edges of the plate (Fig. 6). Determine the centre of gravity of the plate.

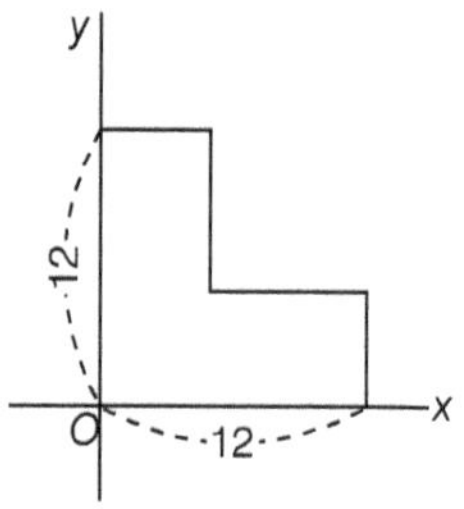

Fig. 4.

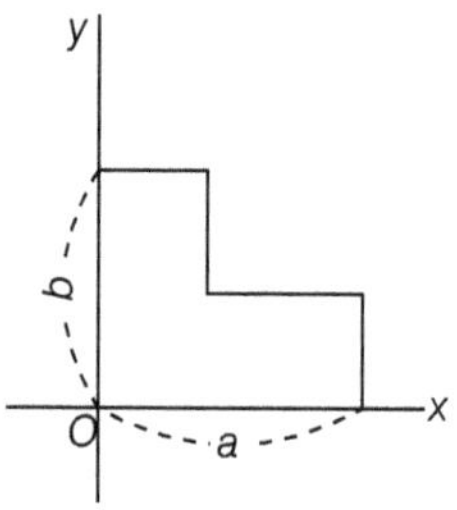

Fig. 5.

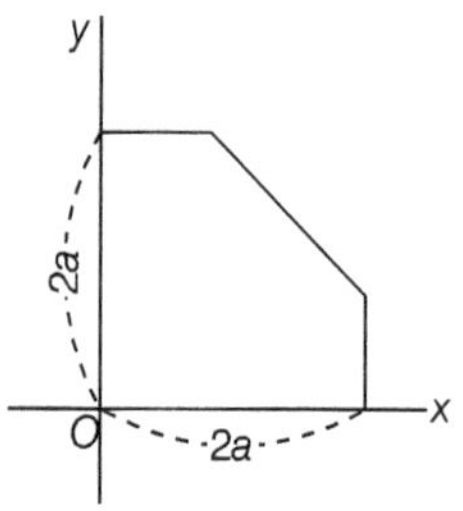

Fig. 6.

114. Three masses m, n and p are placed at the points $A(x_1, y_1)$, $B(x_2, y_2)$ and $C(x_3, y_3)$, respectively. Determine the coordinates of the centre of gravity of this system of masses.

115. The points $A(4, 2)$, $B(7, -2)$ and $C(1, 6)$ are the vertices of a triangle made of uniform wire. Find the centre of gravity of this triangle.

6. The Area of a Triangle

For any three points $A(x_1, y_1)$, $B(x_2, y_2)$, $C(x_3, y_3)$, the area S of the triangle ABC is given by the formula

$$\pm S = \frac{1}{2}\begin{vmatrix} x_2 - x_1 & y_2 - y_1 \\ x_3 - x_1 & y_3 - y_1 \end{vmatrix}.$$

The right-hand member of this formula is equal to $+S$ if the shortest rotation of the segment $\overline{AB}$ to the segment $\overline{AC}$ is in the positive direction; it is equal to $-S$ if the shortest rotation of $\overline{AB}$ to $\overline{AC}$ is in the negative direction.

116. Calculate the area of the triangle whose vertices are:

1) $A(2, -3)$, $B(3, 2)$ and $C(-2, 5)$;

2) $M_1(-3, 2)$, $M_2(5, -2)$ and $M_3(1, 3)$;

3) $M(3, -4)$, $N(-2, 3)$ and $P(4, 5)$.

117. The vertices of a triangle are the points $A(3, 6)$, $B(-1, 3)$ and $C(2, -1)$. Find the length of the altitude drawn from the vertex C.

118. Determine the area of a parallelogram, given that three of its vertices are $A(-2, 3)$, $B(4, -5)$ and $C(-3, 1)$.

119. The points $A(3, 7)$, $B(2, -3)$ and $C(-1, 4)$ are three vertices of a parallelogram. Find the length of the altitude drawn from the vertex B to the side AC.

120. Given the consecutive vertices $A(2, 1)$, $B(5, 3)$, $C(-1, 7)$ and $D(-7, 5)$ of a uniform quadrilateral plate. Find the coordinates of its centre of gravity.

121. Given the consecutive vertices $A(2, 3)$, $B(0, 6)$, $C(-1, 5)$, $D(0, 1)$ and $E(1, 1)$ of a uniform pentagonal plate. Find the coordinates of its centre of gravity.

122. The area S of a triangle is 3, two of its vertices are $A(3, 1)$ and $B(1, -3)$, and the third vertex C lies on the axis Oy. Determine the coordinates of the vertex C.

123. The area S of a triangle is 4, two of its vertices are $A(2, 1)$ and $B(3, -2)$, and the third vertex C lies on the axis Ox. Determine the coordinates of the vertex C.

124. The area S of a triangle is 3, two of its vertices are $A(3, 1)$, $B(1, -3)$ and the centre of gravity of the triangle lies on the axis Ox. Determine the coordinates of the third vertex C.

125. The area S of a parallelogram is 12 square units; two of its vertices are the points $A(-1, 3)$ and $B(-2, 4)$. Find the other two vertices of the parallelogram, if the point of intersection of its diagonals lies on the x-axis.

126. The area S of a parallelogram is 17 square units; two of its vertices are the points $A(2, 1)$ and $B(5, -3)$. Find the other two vertices of the parallelogram, if the point of intersection of its diagonals lies on the y-axis.

7. Transformation of Coordinates

The transformation of rectangular cartesian coordinates under a translation of axes is determined by the formulas

$$x = x' + a,\ y = y' + b$$

Here x, y are the coordinates of an arbitrary point M of the plane with reference to the old axes; x', y' are the coordinates of M with reference to the new axes; a, b are the coordinates of the new origin O' with reference to the old axes (a is also spoken of as the amount of shift in the direction of the x-axis, and b as the amount of shift in the direction of the y-axis).

The transformation of rectangular cartesian coordinates under a rotation of axes (through an angle α understood as in trigonometry) is determined by the formulas

$$x = x'\cos\alpha - y'\sin\alpha,$$
$$y = x'\sin\alpha + y'\cos\alpha.$$

Here x, y are the coordinates of an arbitrary point M of the plane with reference to the old axes, and x', y' are the coordinates of M with reference to the new axes.

The formulas

$$x = x'\cos\alpha - y'\sin\alpha + a,$$
$$y = x'\sin\alpha + y'\cos\alpha + b.$$

determine the transformation of coordinates under a translation of the set of axes (by an amount a in the direction of Ox and by an amount b in the direction of Oy) followed by a rotation of the axes through an angle α.

In each of the above formulas, the same scale is assumed to be used before and after the transformation of coordinates. This assumption is also made in the problems that follow.

127. Write the coordinate transformation formulas if the origin is moved (without changing the direction of the axes) to the point :

1) $A(3, 4)$; 2) $B(-2, 1)$; 3) $C(-3, 5)$.

128. The origin is moved to the point $O'(3, -4)$ without changing the direction of the axes. The coordinates of the point $A(1,3)$, $B(-3,0)$ and $C(-1,4)$ are determined with reference to the new system. Calculate the coordinates of A, B, C in the old coordinate system.

129. Given the points $A(2, 1)$, $B(-1, 3)$ and $C(-2, 5)$.

Find their coordinates in the new system when the origin is moved (without changing the direction of the axes)

1) to the point A; 2) to the point B; 3) to the point C.

130. Determine the old coordinates of the origin O' of the new system if the coordinate transformation formulas are as follows

1) $x = x'+3,\ y = y'+5$; 2) $x = x'-2,\ y = y'+1$;
3) $x = x',\ y = y'-1$; 4) $x = x'-5,\ y = y'$.

131. Write the coordinate transformation formulas if the coordinate axis are rotated through one of the following angles:

1) 60°; 2) – 45°; 3) 90°; 4) – 90° ;
5) 180°.

132. The coordinate axes are rotated through all angle $\alpha = 60°$. The coordinates of the points $A\,(2\sqrt{3}, -4)$, $B\,(\sqrt{3}, 0)$ and $C(0, -2\sqrt{3})$ are given with reference to the new system. Calculate the coordinates of A, B, C in the old coordinate system.

133. Given the points M (3, 1), N (–1,5) and P (– 3, –1). Find their new coordinates when the axes are rotated through the angle:

1) – 45°; 2) 90°; 3) – 90° ; 4) 180°.

134. Determine the angle α through which the axes have been rotated if the coordinate transformation formulas are as follows:

1) $x = \frac{1}{2}x' - \frac{\sqrt{3}}{2}y',\ y = \frac{\sqrt{3}}{2}x' + \frac{1}{2}y'$;

2) $x = \frac{\sqrt{3}}{2}x' + \frac{1}{2}y',\ y = -\frac{1}{2}x' + \frac{\sqrt{3}}{2}y'$.

135. Determine the old coordinates of the new origin O', if the point A (3,–4) lies on the new axis of abscissas, the point B(2, 3) on the new axis of ordinates, and the corresponding axes of the old and the new coordinate system have the same direction.

136. Write the coordinate transformation formulas if the point M_1 (2,–3) lies on the new axis of abscissas, the point M_2(1,–7) on the new axis of ordinates, and the corresponding axes of the old and the new coordinate system have the same direction.

137. Two systems of coordinate axes, Ox, Oy and Ox', Oy', have a common origin O; the transformation from one system to the other is accomplished by a rotation through a certain angle. The coordinates of the point $A\,(3, -4)$ are given with reference to the first system. Derive the coordinate transformation formulas if the positive direction of the axis Ox' is determined by the segment $\overline{OA}$.

138. The origin is moved to the point O' (–1, 2), and the coordinate axes are rotated through the angle $\alpha = \text{arc tan}\,\frac{5}{12}$. The coordinates of the points M_1 (3,2), M_2 (2,–3) and M_3 (13,–13) refer to the new system. Calculate the coordinates of M_1, M_2, M_3 in the old coordinate system.

139. Given the three points A (5, 5), B (2, –1) and C (12, – 6). Find their coordinates in the new system when the origin is moved to the point B and the coordinate axes are rotated through the angle $\alpha = \arc\tan \frac{3}{4}$.

140. Determine the old coordinates of the new origin and the angle α through which the axes have been turned if the coordinate transformation formulas are:

1) $x = -y' + 3,\ y = x' - 2$; 2) $x = -x' - 1,\ y = -y' + 3$;

3) $x = \frac{\sqrt{2}}{2}x' + \frac{\sqrt{2}}{2}y' + 5,\ y = -\frac{\sqrt{2}}{2}x' + \frac{\sqrt{2}}{2}y' - 3$.

141. Given the two points M_1 (9, –3) and M_2(–6, 5). The origin is moved to M_1, and the coordinate axes are rotated so that the positive direction of the new axis of abscissas agrees with the direction of the segment $\overline{M_1M_2}$. Derive the coordinate transformation formulas.

142. The polar axis of a polar coordinate system and the x-axis of a rectangular cartesian system are parallel and similarly directed. Given the rectangular cartesian coordinates of the pole O (1, 2) and the polar coordinates of the points $M_1\left(7, \frac{\pi}{2}\right)$, M_2 (3, 0), $M_3\left(5, -\frac{\pi}{2}\right)$, $M_4\left(2, \frac{2}{3}\pi\right)$ and $M_5\left(2, -\frac{\pi}{6}\right)$; determine the rectangular cartesian coordinates of these points.

143. The pole of a polar coordinate system coincides with the origin of rectangular cartesian coordinates and the polar axis lies along the bisector of the first quadrant.

Given the polar coordinates of the points $M_1\left(5, \frac{\pi}{4}\right)$, $M_2\left(3, -\frac{\pi}{4}\right)$, $M_3\left(1, \frac{3}{4}\pi\right)$, $M_4\left(6, -\frac{3}{4}\pi\right)$ and $M_5\left(2, -\frac{\pi}{12}\right)$; determine their rectangular cartesian coordinates.

144. The polar axis of a polar coordinate system and the x-axis of a rectangular cartesian system are parallel and similarly directed. Given the rectangular cartesian coordinates of the pole O (3, 2) and of the points M_1(5, 2), M_2(3, 1), M_3(3,5), M_4 $(3+\sqrt{2}, 2-\sqrt{2})$ and M_5 $(3+\sqrt{3}, 3)$. Find the polar coordinates of these points.

145. The pole of a polar coordinate system coincides with the origin of rectangular cartesian coordinates, and the polar axis goes along the bisector of the first quadrant. Given the rectangular cartesian coordinates of the points

$M_1(-1, 1)$, $M_2(\sqrt{2}, -\sqrt{2})$, $M_3(1, \sqrt{3})$, $M_4(-\sqrt{3}, 1)$ and $M_5(2\sqrt{3}, -2)$; determine their polar coordinates.

2

THE EQUATION OF A CURVE

8. A Function of Two Variables

If a rule has been given according to which a number u is associated with each point M of the plane (or of some portion of the plane), then we say that "a function of a point" has been specified for the plane (or for the portion of the plane), and we express this symbolically by a relation of the form $u = f(M)$. The number u associated with a point M is called the value of the function at the point M. For example, if A is a fixed point in the plane, and M an arbitrary point, then the distance from A to M is a function of the point M. In this case, $f(M) = AM$.

Let there be given a function $u = f(M)$, and let a coordinate system be chosen. Then an arbitrary point M will be determined by its coordinates x,y. Accordingly, the value of the given function at M will also be determined by the coordinates x, y, or (as is also said) $u = f(M)$ will be a function of two variables x and y. A function of two variables x, y is denoted by the symbol $f(x, y)$; if $f(M) = f(x, y)$, then the formula $u = f(x, y)$ is referred to as the expression for the given function in the chosen coordinate system.

Thus, $f(M) = AM$ in the above example; if we introduce a rectangular cartesian coordinate system with origin at the point A, this function will be expressed by $u = \sqrt{x^2 + y^2}$.

146. Given two points P and Q, a units apart, and the function $f(M) = d_1^2 - d_2^2$, where $d_1 = MP$ and $d_2 = MQ$.

Find the expression for this function when the point P is chosen as the origin, and the axis Ox is directed along the segment $\overline{PQ}$.

147. Solve the previous problem when: 1) the midpoint of the segment $\overline{PQ}$ is taken as the origin, and the direction of the axis Ox agrees with that of the segment $\overline{PQ}$; 2) the point P is taken as the origin, and the direction of the axis Ox agrees with that of the segment $\overline{QP}$. (Find the expression for the function $f(M)$ first directly and then by using the result of Problem 146 and transforming the coordinates.)

148. Given the square $ABCD$ with side a and the function $f(M) = d_1^2 + d_2^2 + d_3^2 + d_4^2$, where $d_1 = MA$, $d_2 = MB$, $d_3 = MC$ and $d_4 = MD$. Find the expression for this function when the diagonals of the square are taken as the coordinate axes (so that the axis Ox has the direction of the segment $\overline{AC}$, and the axis Oy the direction of $\overline{BD}$).

149. Solve the previous problem when the point A is chosen as the origin, and the sides of the square are taken as the coordinate axes so that the axis Ox has the direction of the segment $\overline{AB}$, and the axis Oy the direction of $\overline{AD}$. (Find the expression for $f(M)$ first directly and then by using the result of Problem 148 and transforming the coordinates.)

150. Given the function $f(x, y) = x^2 + y^2 - 6x + 8y$. Find the expression for this function in the new coordinate system when the origin is moved (without changing the direction of the axes) to the point $O'(3, -4)$.

151. Given the function $f(x, y) = x^2 - y^2 - 16$. Find the expression for this function in the new coordinate system when the coordinate axes are rotated through an angle of $-45°$.

152. Given the function $f(x, y) = x^2 + y^2$. Find the expression for this function in the new coordinate system when the coordinate axes are rotated through an angle α.

153. Find a point such that, if the origin is moved to that point, the transformed expression for the function $f(x, y) = x^2 - 4y^2 - 6x + 8y + 3$ will be free from first-degree terms (with respect to the new variables).

154. Find a point such that, if the origin is moved to that point, the function $f(x, y) = x^2 - 4xy + 4y^2 + 2y + y - 7$ will have an expression free from first-degree terms (with respect to the new variables).

155. What is the angle through which the coordinate axes must be rotated so that the transformed expression for the function $f(x, y) = x^2 - 2xy + y^2 - 6x + 3$ will lack the term in $x'y'$?

156. What is the angle through which the coordinate axes must be rotated so that the function $f(x, y) = 3x^2 + 2\sqrt{3}xy + y^2$ will have an expression lacking the term in $x'y'$?

9. The Concept of the Equation of a Curve. Curves Represented by Equations

An equality of the form $F(x, y) = 0$ is called an equation in two variables x, y provided that it is valid not for every pair of numbers x, y. Two numbers $x = x_0$, $y = y_0$ are said to satisfy an equation of the form $F(x, y) = 0$ if the left-hand member of the equation vanishes upon substitution of these numbers for the variables x, y.

The equation of a given curve (in a chosen coordinate system) is defined as the equation in two variables which is satisfied by the coordinates of all points lying on the curve and by the coordinates of no other point.

Throughout the rest of the book, we shall often use the expression "Given the curve $F(x, y) = 0$" instead of the longer one: "Given the curve whose equation is $F(x, y) = 0$".

If $F(x, y) = 0$ and $\phi(x, y) = 0$ are the equations of two given curves, all their points of intersection are obtained by solving the system

$$\begin{cases} F(x, y) = 0 \\ \varphi(x, y) = 0 \end{cases}$$

simultaneously. More precisely, each pair of numbers constituting a simultaneous solution of the system represents one of the points of intersection.

157. Given the points* $M_1(2,-2)$, $M_2(2, 2)$, $M_3(2, -1)$, $M_4(3, -3)$, $M_5(5, -5)$, $M_6(3, -2)$. Determine which of the given points lie on the curve represented by the equation $x + y = 0$. Identify and plot the curve.

158. On the curve represented by the equation $x^2 + y^2 = 25$, find the points whose abscissas are equal to the following numbers: a) 0, b) –3, c) 5, d) 7; on the same curve find the points whose ordinates are equal to the following numbers: e) 3, f) –5, g) –8. Identify and plot the curve.

159. Identify and plot the curves represented by the following equations:

1) $x - y = 0$; 2) $x + y = 0$;
3) $x - 2 = 0$; 4) $x + 3 = 0$;
5) $y - 5 = 0$; 6) $y + 2 = 0$;
7) $x = 0$; 8) $y = 0$;
9) $x^2 - xy = 0$; 10) $xy + y^2 = 0$;
11) $x^2 - y^2 = 0$; 12) $xy = 0$;
13) $y^2 - 9 = 0$; 14) $x^2 - 8x + 15 = 0$; 15) $y^2 + 5y + 4 = 0$;
16) $x^2y - 7xy + 10y = 0$; 17) $y = |x|$;
18) $x = |y|$; 19) $y + |x| = 0$;
20) $x + y = 0$ (21) $y = |x - 1|$; 22) $y = |x + 2|$, (23) $x^2 + y^2) = 16$;
24) $(x-2)^2 + (y-1)^2 = 16$; 25) $(x+5)^2 + (y-1)^2 = 9$;
26) $(x-1)^2 + y^2 = 4$; 27) $x^2 + (y+3)^2 = 1$;
28) $(x-3)^2 + y^2 = 0$; 29) $x^2 + 2y^2 = 0$;
30) $2x^2 + 3y^2 + 5 = 0$;
31) $(x-2)^2 + (y+3)^2 + 1 = 0$.

---

*Rectangular cartesian coordinates are used in all cases when the coordinate system is not specified.

160. Given the curves:

1) $x+y=0$; 2) $x-y=0$; 3) $x^2+y^2-36=0$;
4) $x^2+y^2-2x+y=0$; 5) $x^2+y^2+4x-6y-1=0$.

Determine which of them pass through the origin.

161. Given the curves:

1) $x^2+y^2=49$; 2) $(x-3)^2+(y+4)^2=25$;
3) $(x+6)^2+(y-3)^2=25$; 4) $(x+5)^2+(y-4)^2=9$;
5) $x^2+y^2-12x+16y=0$; 6) $x^2+y^2-2x+8y+7=0$;
7) $x^2+y^2-6x+4y+12=0$.

Find their points of intersection: a) with the axis Ox; b) with the axis Oy.

162. In each of the following, find the points of intersection of the two given curves:

1) $x^2+y^2=8,\ x-y=0$;
2) $x^2+y^2-16x+4y+18=0,\ x+y=0$;
3) $x^2+y^2-2x+4y-3=0,\ x^2+y^2=25$;
4) $x^2+y^2-8x+10y+40=0,\ x^2+y^2=4$.

163. Given the points

$$M_1\left(1, \frac{\pi}{3}\right),\ M_2(2, 0),\ M_3\left(2, \frac{\pi}{4}\right),\ M_4\left(\sqrt{3}, \frac{\pi}{6}\right) \text{ and } M_5\left(1, \frac{2}{3}\pi\right)$$

in a polar coordinate system. Determine which of these points lie on the curve represented by the polar equation $\rho = 2\cos\theta$. Identify and plot the curve.

164. On the curve represented by the equation $\rho = \frac{3}{\cos\theta}$, find the points whose polar angles are equal to the following numbers:

a) $\frac{\pi}{3}$ b) $-\frac{\pi}{3}$ c) 0 d) $\frac{\pi}{6}$.

Identify and plot the curve.

165. On the curve represented by the equation $\rho = \frac{1}{\sin 0}$, find the points whose polar radii are equal to the following numbers: a) 1, b) 2, c) $\sqrt{2}$. Identify and plot the curve.

166. Identify and plot the curves represented, in polar coordinates, by the following equations:

1) $\rho = 5$; 2) $0 = \frac{\pi}{3}$; 3) $0 = -\frac{\pi}{4}$; 4) $\rho\cos 0 = 2$;
5) $\rho\sin 0 = 1$; 6) $\rho = 6\cos 0$; 7) $\rho = 10\sin 0$;
8) $\sin\theta = \frac{1}{2}$; 9) $\sin\rho = \frac{1}{2}$.

167. Plot the following spirals of Archimedes:

1) $\rho = 20$ 2) $\rho = 50$ 3) $\rho = \frac{0}{\pi}$ 4) $\rho = -\frac{0}{\pi}$.

168. Plot the following hyperbolic spirals:

1) $\rho = \frac{1}{0}$; 2) $\rho = \frac{5}{0}$; 3) $\rho = \frac{\pi}{0}$; 4) $\rho = -\frac{\pi}{0}$.

169. Plot the following logarithmic spirals:

1) $\rho = 2^0, \rho = \left(\frac{1}{2}\right)^0$.

170. Determine the lengths of the segments into which the spiral of Archimedes $\rho = 30$ cuts the ray extending from the pole and making an angle $0 = \frac{\pi}{6}$ with the polar axis. Draw the figure.

171. A point C, whose polar radius is equal to 47, is taken on the spiral of Archimedes $\rho = \frac{5}{\pi}0$.

Determine into how many parts the spiral cuts the polar radius of C. Draw the figure.

172. On the hyperbolic spiral $\rho = \frac{6}{0}$ find the point P whose polar radius is equal to 12. Draw the figure.

173. On the logarithmic spiral $\rho = 3^4$

find the point Q whose polar radius is 81. Draw the figure.

10. Derivation of the Equation of a Given Curve

The problems of the preceding section dealt with the determination of curves from given equations. In this section, we shall have problems of an opposite character; in each of them, a curve is defined in purely geometric terms, and it is the equation of the curve that we are required to find.

Example 1. In a rectangular cartesian system of coordinates, derive the equation of the locus of points, the sum of the squares of whose distances from the two given points $A_1(-a, 0)$ and $A_2(a, 0)$ is a constant equal to $4a^2$.

Solution. Let M denote an arbitrary point of the curve, and let x and y be the coordinates of that point. Since the point M can occupy any position on the curve, it follows that x and y are variables; they are called the current coordinates.

Let us write the geometric property of the curve symbolically:

$$(MA_1)^2 + (MA_2)^2 = 4a^2. \quad (1)$$

In this relation, the lengths MA_1 and MA_2 will vary with the motion of the point M. Expressing them in terms of the current coordinates of M, we get

$MA_1 = \sqrt{(x+a)^2 + y^2}$, $MA_2 = \sqrt{(x-a)^2 + y^2}$.

Substituting these expressions in (1), we obtain the equation connecting the coordinates x, y of the point M:

$$(x+a)^2 + y^2 + (x-a)^2 + y^2 = 4a^2. \quad (2)$$

This is the equation of the given curve. For, the condition (1) is fulfilled for every point M lying on the curve, and hence the coordinates of M will satisfy equation (2); on the other hand, the condition (1) is not fulfilled for any point M not lying on the line, and hence its coordinates will not satisfy equation (2).

The problem is thus solved. But equation (2) can be simplified. Removing the parentheses and collecting like terins, we obtain the equation of the given curve in the form $x^2 + y^2 = a^2$.

It is now easy to see that the curve is a circle with centre at the origin and radius a.

Example 2. In a polar coordinate system, derive the equation of a circle with centre C (ρ_0, θ_0) and radius r (Fig. 7).

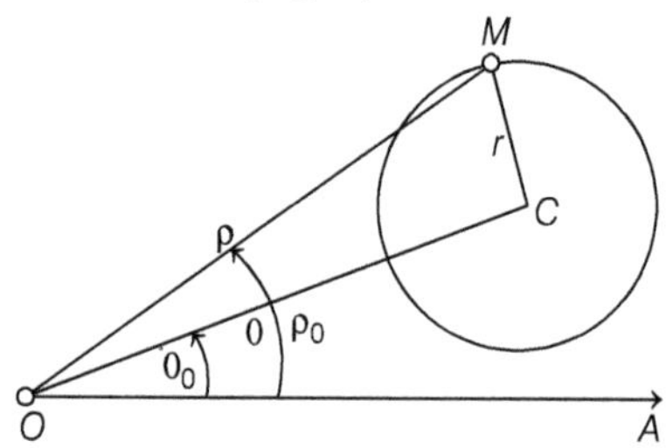

Fig. 7.

Solution. Let M denote an arbitrary point of the circle, and let ρ and θ be its polar coordinates. Since the point M may occupy any position on the circle, ρ and θ are variables, as in the case of a cartesian system, they are called the current coordinates.

All points of the circle are at a distance r from the centre; writing this condition symbolically, we have $CM = r$. (1)

Let us express CM in terms of the current coordinates of the point M (by applying the cosine theorem see Fig. 7):

$$CM = \sqrt{\rho^2 + \rho_0^2 - 2\rho_0 \rho \cos(\theta - \theta_0)}$$

Substituting this expression in (1), we obtain the equation connecting the coordinates ρ, O of the point M:

$$\sqrt{\rho^2 + \rho_0^2 - 2\rho_0 \rho \cos(\theta - \theta_0)} = r. \quad (2)$$

This is the equation of the given circle. For, the condition (1) is fulfilled for every point M lying on the circle, and hence the coordinates of M will satisfy equation (2); the condition (1) is not fulfilled for any point M not lying on the circle, so that the coordinates of all such points will not satisfy equation (2).

The problem is thus solved. By clearing radicals, the equation may be reduced to the somewhat simpler form

$$\rho^2 - 2\rho_0\rho\cos(\theta - \theta_0) = r^2 - \rho_0^2.$$

174. Derive the equation of the locus of points equidistant from the coordinate axes.

175. Derive the equation of the locus of points which are at a distance a from the axis Oy.

176. Derive the equation of the locus of points which are at a distance b from the axis Ox.

177. From the point $P(6, -8)$, all possible rays are drawn to cut the x-axis. Find the equation of the locus of their midpoints.

178. From the point $C(10, -3)$, all possible rays are drawn to cut the y-axis. Find the equation of the locus of their midpoints.

179. Derive the equation of the path of a point which moves so that it is always equidistant from the points:

1) $A(3, 2)$ and $B(2, 3)$; 2) $A(5, -1)$ and $B(1, -5)$;

3) $A(5, -2)$ and $B(-3, -2)$; 4) $A(3, -1)$ and $B(3, 5)$.

180. Write the equation of the locus of points, the difference of the squares of whose distances from the points $A(-a, 0)$ and $B(a, 0)$ is equal to c.

181. Derive the equation of the circle with centre at the origin and radius r.

182. Derive the equation of the circle with centre at $C(\alpha, \beta)$ and radius r.

183. Given the equation $x^2 + y^2 = 25$ of a circle. Write the equation of the locus of the midpoints of those chords of the circle whose length is equal to 8.

184. Find the equation of the locus of points, the sum of the squares of whose distances from the points $A(-3, 0)$ and $B(3, 0)$ is equal to 50.

185. The points $A(a, a)$, $B(-a, a)$, $C(-a -a)$ and $D(\alpha, -a)$ are the vertices of a square. Find the equation of the locus of points, the sum of the squares of whose distances from the sides of the given square is a constant equal to $6a^2$.

186. All possible chords of the circle $(x-8)^2 + y^2 = 64$ are drawn through the origin. Write the equation of the locus of the midpoints of these chords.

187. Derive the equation of the locus of points, the sum of whose distances from the two given points $F_1(-3, O)$ and $F_2(3, 0)$ is a constant equal to 10.

188. Derive the equation of the locus of points, the difference of whose distances from the two given points $F_1(-5, 0)$ and $F_2(5, 0)$ is a constant equal to 6.

189. Derive the equation of the locus of points whose distance from the given point F (3, 0) is equal to their distance from the given straight line $x + 3 = 0$.

190. Derive the equation of the locus of points, the sum of whose distances from two given points $F_1(-c, 0)$ and $F_2(c, 0)$ is a constant equal to $2a$. This locus is called an ellipse, and the points F_1, F_2 are called the foci of the ellipse.

Prove that the equation of an ellipse is of the form

$$\frac{x^2}{a^2} + \frac{y^2}{b^2} = 1,$$

where $b^2 = a^2 - c^2$.

191. Derive the equation of the locus of points, the difference of whose distances from two given points $F_1(-c, 0)$ and $F_2(c, 0)$ is a constant equal to $2a$. This locus is called a hyperbola, and the points F_1, F_2 are called the foci of the hyperbola.

Prove that the equation of a hyperbola is of the form

$$\frac{x^2}{a^2} - \frac{y^2}{b^2} = 1,$$

where $b^2 = c^2 - a^2$.

192. Derive the equation of the locus of points whose distance from a given point $F\left(\frac{p}{2}, 0\right)$ is equal to their distance from a given straight line $x = -\frac{p}{2}$. This locus is called a parabola, the point F is referred to as the focus of the parabola, and the given straight line as the directrix of the parabola.

193. Derive the equation of the locus of points, the ratio of whose distance from the given point F (– 4, 0) to their distance from the given straight line $4x + 25 = 0$ is equal to $\frac{4}{5}$.

194. Derive the equation of the locus of points, the ratio of whose distance from the given point F (– 5, 0) to their distance from the given straight line $5x + 16 = 0$ is equal to $\frac{5}{4}$.

195. Derive the equation of the locus of points whose shortest distance from the circle $(x+3)^2+y^2=1$ is equal to their shortest distance from the circle $(x-3)^2+y^2=81$.

196. Derive the equation of the locus of points whose shortest distance from the circle $(x+10)^2+y^2=289$ is equal to their shortest distance from the circle $(x-10)^2+y^2=1$.

197. Derive the equation of the locus of points whose shortest distance from the circle $(x-5)^2+y^2=9$ is equal to their shortest distance from the straight line $x+2=0$.

198. A straight line is perpendicular to the polar axis and intercepts a segment equal to 3 on that axis. Write the equation of the straight line in polar coordinates.

199. A ray is drawn from the pole at an angle $\frac{\pi}{3}$ with respect to the polar axis. Find the equation of this ray in polar coordinates.

200. A straight line passes through the pole and makes an angle of 45° with the polar axis. Find the equation of this straight line in polar coordinates.

201. In polar coordinates, write the equation of the locus of points 5 units distant from the polar axis.

202. A circle of radius $R=5$ passes through the pole and has its centre upon the polar axis. Find the equation of the circle in the polar coordinate system.

203. A circle of radius $R=3$ touches the polar axis at the pole. Find the equation of the circle in the polar coordinate system.

11. Parametric Equations of a Curve

Let us denote the coordinates of a point. M by the letters x, y and consider the two functions of an independent variable t:

$$\left.\begin{aligned} x &= \varphi(t), \\ x &= \psi(t). \end{aligned}\right\} \tag{1}$$

The quantities x and y, in general, change with t: it follows that the point M moves in the plane. Relations (1) are called the parametric equations of the curve which is the path traced by the point M; the independent variable t is called a parameter. If the parameter t can be eliminated from relations (1), we obtain the equation of the path of M in the form

$$F(x, y) = 0.$$

204. A rod AB slides with its ends A and B along the coordinate axes. A point M divides the rod into two parts $AM = a$ and $BM = b$. Derive the parametric equations

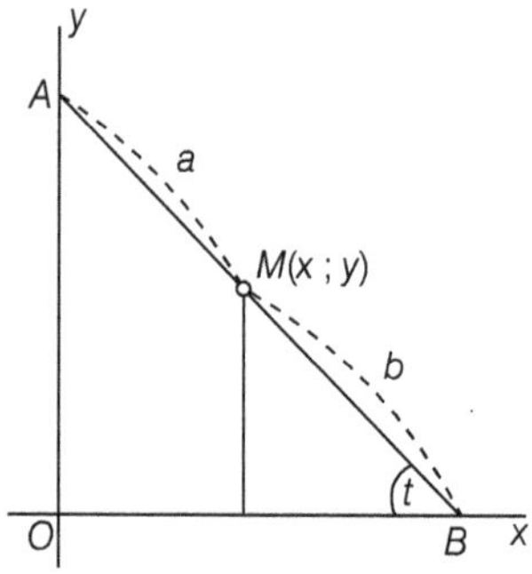

Fig. 8.

of the path traced by the point M, using the angle $t = \measuredangle OBA$ (Fig. 8) as parameter. Next, eliminate the parameter t and find the equation of the path of M in the form $F(x, y) = 0$.

205. The path of a point M is an ellipse having $\frac{x^2}{a^2} + \frac{y^2}{b^2} = 1$ as its equation (see Problem 190). Derive the parametric equations of the path of M, using the angle of inclination of the segment $\overline{OM}$ (with respect to the axis Ox) as the parameter t.

206. The path of a point M is a hyperbola having $\frac{x^2}{a^2} - \frac{y^2}{b^2} = 1$ as its equation (see Problem 191). Derive the parametric equations of the path of M, using the angle of inclination of the segment $\overline{OM}$ (with respect to the axis Ox) as the parameter t.

207. The path of a point M is a parabola having $y^2 = 2px$ as its equation (see problem 192). Derive the parametric equations of the path of M, using as the parameter t:

1) the ordinate of the point M;
2) the angle of inclination of the segment $\overline{OM}$ (with respect to the axis Ox);
3) the angle of inclination of the segment $\overline{FM}$ (with respect to the axis Ox), where the point F is the focus of the parabola.

208. Given the polar equations of the following curves.

1) $\rho = 2R\cos 0$ 2) $\rho = 2R\sin 0$

3) $\rho = 2p\dfrac{\cos 0}{\sin^2 0}$.

Find the parametric equations of these curves in rectangular cartesian coordinates, when the positive x-axis coincides with the polar axis and the polar angle is taken as parameter.

209. Given the parametric equations of the curves:

1) $\left.\begin{array}{l} x = t^2 - 2t + 1, \\ y = t - 1; \end{array}\right\}$

2) $\left.\begin{array}{l} x = a\cos t, \\ y = a\sin t, \end{array}\right\}$ 3) $\left.\begin{array}{l} x = a\sec t, \\ y = b\tan t; \end{array}\right\}$

4) $\left.\begin{array}{l} x = \dfrac{a}{2}\left(t + \dfrac{1}{t}\right), \\ y = \dfrac{b}{2}\left(t - \dfrac{1}{t}\right); \end{array}\right\}$ 5) $\left.\begin{array}{l} x = 2R\cos^2 t, \\ y = R\sin 2t; \end{array}\right\}$

6) $\left.\begin{array}{l} x = R\sin 2t, \\ y = 2R\sin^2 t; \end{array}\right\}$ 7) $\left.\begin{array}{l} x = 2p\cot^2 t, \\ y = 2p\cot t; \end{array}\right\}$

eliminate the parameter t and write the equations of these curves in the form

$F(x, y) = 0$.

3

CURVES OF THE FIRST ORDER

12. The General Equation of a Straight Line. The Slope-intercept Equation of a Straight Line. The Angle Between Two Straight Lines. The Conditions for the Parallelism and Perpendicularity of Two Straight Lines

In cartesian coordinates, every straight line is represented by an equation of the first degree and conversely, every equation of the first degree represents a straight line.

An equation of the form $Ax + By + C = 0$ (1)

is called the general equation of a straight line.

The angle α determined as shown in Fig. 9, is called the angle of inclination of a given straight line (with respect to the axis Ox).

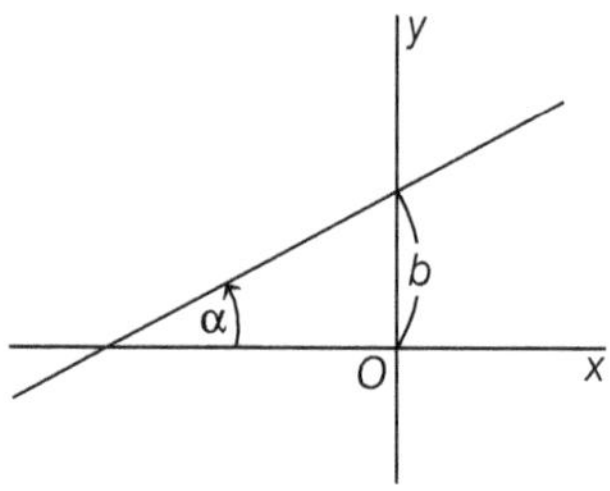

Fig. 9.

The tangent of the angle of inclination of a straight line is called the slope of that line and is usually denoted by the letter k:

$$k = \tan\alpha.$$

The equation $y = kx + b$ is called the slope-intercept equation of a straight line; k is here the slope, and b is the y-intercept (that is, the value of the segment cut off by the line on the axis Oy; see Fig. 9). If a straight line is represented by its general equation $Ax + By + C = 0$, then its slope is determined from the formula

$$k = -\frac{A}{B}.$$

The equation $y - y_0 = k(x - x_0)$ is the equation of the straight line with slope k and passing through the point $M_0(x_0, y_0)$.

If a straight line passes through the points $M_1(x_1, y_1)$ and $M_2(x_2, y_2)$, its slope is determined from the formula

$$k = \frac{y_2 - y_1}{x_2 - x_1}.$$

The equation $\frac{x - x_1}{x_2 - x_1} = \frac{y - y_1}{y_2 - y_1}$ is the equation of the straight line passing through the two points $M_1(x_1, y_1)$ and $M_2(x_2, y_2)$.

Given the slopes k_1 and k_2 of two straight lines, one of the angles φ between these lines is determined by the formula

$$\tan\varphi = \frac{k_2 - k_1}{1 + k_1 k_2}.$$

The condition for the parallelism of two straight lines is the equality of their slopes: $k_1 = k_2$. The condition for the perpendicularity of two straight lines is given by the relation $k_1 k_2 = -1$, or $k_2 = -\frac{1}{k_1}$.

In other words, the slopes of perpendicular lines are negative reciprocals

210. Determine which of the points $M_1(3, 1)$, $M_2(2, 3)$, $M_3(6, 3)$, $M_4(-3, -3)$, $M_5(3, -1)$, $M_6(-2, 1)$ lie on the line $2x - 3y - 3 = 0$.

211. The points P_1, P_2, P_3, P_4 and P_5 are situated on the line $3x - 2y - 6 = 0$; their respective abscissas are 4, 0, 2, –2 and – 6. Determine the ordinates of these points.

212. The points Q_1, Q_2, Q_3, Q_4 and Q_5 are situated on the line $x - 3y + 2 = 0$; their respective ordinates are 1, 0, 2, –1 and 3. Determine the abscissas of these points.

213. Determine the points of intersection of the straight line $2x - 3y - 12 = 0$ with the coordinate axes and plot the line.

214. Find the point of intersection of the two lines $3x - 4y - 29 = 0$, $2x + 5y + 19 = 0$.

215. The equations of the sides* *AB, BC, AC* of a triangle *ABC* are, respectively, $4x + 3y - 5 = 0$, $x - 3y + 10 = 0$, $x - 2 = 0$.

Determine the coordinates of the vertices of the triangle.

* The expression "the equations of the sides", as used in this book, means "the equations of the straight lines on which the sides lie".

216. Given the equations $8x+3y+1=0$, $2x+y-1=0$ of two sides of a parallelogram and the equation $3x+2y+3=0$ of one of its diagonals. Determine the coordinates of the vertices of the parallelogram.

217. The sides of a triangle lie on the lines $x+5y-7=0$, $3x-2y-4=0$, $7x+y+19=0$. Calculate the area S of the triangle.

218. The area S of a triangle is 8 square units; two of its vertices are the points A (1, –2) and B (2, 3), and the third vertex C lies on the line $2x+y-2=0$.
Find the coordinates of the vertex C.

219. The area S of the triangle is 1.5 square units; two of its vertices are the points A (2, –3) and B (3, –2); the centre of gravity of the triangle lies on the line $3x-y-8=0$. Determine the coordinates of the third vertex C.

220. Write the equation of the straight line whose slope k and y-intercept b are as follows:

1) $k=\frac{2}{3}, b=3$; 2) $k=3, b=0$; 3) $k=0, b=-2$;

4) $k=-\frac{3}{4}, b=3$; 5) $k=-2, b=-5$; 6) $k=-\frac{1}{3}, b=\frac{2}{3}$.

In each case, plot the line.

221. Determine the slope k and the y-intercept b for each of the lines:

1) $5x-y+3=0$; 2) $2x+3y-6=0$; 3) $5x+3y+2=0$;

4) $3x+2y=0$; 5) $y-3=0$.

222. Given the line $5x+3y-3=0$. Determine the slope k of a straight line:

1) parallel to the given line;

2) perpendicular to the given line.

223. Given the line $2x+3y+4=0$. Write the equation of the straight line passing through the point $M_0(2, 1)$ and:

1) parallel to the given line; 2) perpendicular to the given line.

224. Given the equations $2x-3y+5=0$, $3x+2y-7=0$ of two sides of a rectangle, and one of its vertices A (2, –3). Write the equations of the other two sides of the rectangle.

225. Given the equations $x-2y=0$, $x-2y+15=0$ of two sides of a rectangle and the equation $7x+y-15=0$ of one of its diagonals. Find the vertices of the rectangle.

226. Find the projection of the point P (– 6, 4) on the line $4x-5y+3=0$.

227. Find the point Q symmetric to the point P (–5, 13) with respect to the line
$2x-3y-3=0$.

228. In each of the following, find the equation of the straight line parallel to the two given lines and passing midway between them:

1) $3x-2y-1=0$, $3x-2y-13=0$;

2) $5x+y+3=0$, $5x+y-17=0$;

3) $2x+3y-6=0$, $4x+6y+17=0$;

4) $5x+7y+15=0$, $5x+7y+3=0$;

5) $3x-15y-1=0$, $x-5y-2=0$.

229. In each of the following, calculate the slope k of the straight line passing through the two given points:

1) $M_1(2, -5)$, $M_2(3, 2)$;

2) $P(-3, 1)$, $Q(7, 8)$;

3) $A(5, -3)$, $B(-1, 6)$.

230. Write the equations of the straight lines passing through the vertices $A(5, -4)$, $B(-1, 3)$, $C(-3, -2)$ of a triangle and parallel to the opposite sides.

231. Given the midpoints $M_1(2, 1)$, $M_2(5, 3)$, $M_3(3, -4)$ of the sides of a triangle. Write the equations of its sides.

232. Given the two points $P(2, 3)$ and $Q(-1, 0)$. Find the equation of the straight line through the point Q perpendicular to the segment $\overline{PQ}$.

233. The point $P(2, 3)$ is the foot of the perpendicular dropped from the origin to a straight line. Write the equation of this line.

234. Given the vertices $M_1(2, 1)$, $M_2(-1, -1)$, $M_3(3, 2)$ of a triangle. Find the equations of its altitudes.

235. The sides of a triangle are given by the equations $4x-y-7=0$, $x+3y-31=0$, $x+5y-7=0$. Determine the point of intersection of its altitudes.

236. Given the vertices $A(1, -1)$, $B(-2, 1)$ and $C(3, 5)$ of a triangle. Write the equation of the perpendicular dropped from the vertex A to the median through the vertex B.

237. Given the vertices $A(2, -2)$, $B(3, -5)$ and $C(5, 7)$ of a triangle. Find the equation of the perpendicular dropped from the vertex C to the bisector of the interior angle at the vertex A.

238. Write the equations of the sides and medians of the triangle with vertices $A(3, 2)$, $B(5, -2)$, $C(1, 0)$.

239. A straight line is drawn through the points $M_1(-1, 2)$ and $M_2(2, 3)$. Determine the points of intersection of this line with the coordinate axes.

240. Prove that the condition for three points $M_1(x_1, y_1)$, $M_2(x_2, y_2)$ and $M_3(x_3, y_3)$ to lie on a straight line can be written in the form

$$\begin{vmatrix} x_1 & y_1 & 1 \\ x_2 & y_2 & 1 \\ x_3 & y_3 & 1 \end{vmatrix} = 0.$$

241. Prove that the equation of the straight line passing through two given points $M_1(x_1, y_1)$ and $M_2(x_2, y_2)$ can be written as

$$\begin{vmatrix} x & y & 1 \\ x_1 & y_1 & 1 \\ x_2 & y_2 & 1 \end{vmatrix} = 0.$$

242. Given the consecutive vertices $A(-3, 1)$, $B(3, 9)$, $C(7, 6)$, $D(-2, -6)$ of a convex quadrilateral. Determine the point of intersection of its diagonals.

243. $A(-3, -1)$ and $B(2, 2)$ are two adjacent vertices of a parallelogram $ABCD$ and $Q(3, 0)$ is the point of intersection of its diagonals. Write the equations of the sides of the parallelogram.

244. Given the equations $5x+2y-7=0$, $5x+2y-36=0$

of two sides of a rectangle and the equation $3x+7y-10=0$

of its diagonal. Write the equations of the remaining sides and of the other diagonal of the rectangle.

245. Given the vertices $A(1, -2)$, $B(5, 4)$ and $C(-2, 0)$ of a triangle. Write the equations of the bisectors of the interior and exterior angles at the vertex A.

246. Find the equation of the straight line passing through the point $P(3,5)$ and equidistant from the points $A(-7, 3)$ and $B(11, -15)$.

247. Find the projection of the point $P(-8, 12)$ on the straight line passing through the points A $(2, -3)$ and $B(-5, 1)$.

248. Find the point M_1, symmetric to the point M_2, $(8, -9)$ with respect to the straight line which passes through the points $A(3, -4)$ and $B(-1, -2)$.

249. Find a point P on the x-axis such that the sum of its distances from the points $M(1, 2)$ and $N(3, 4)$ will have the least value.

250. Find a point P on the y-axis such that the difference of its distances from the points $M(-3, 2)$ and $N(2, 5)$ will have the greatest value.

251. On the line $2x-y-5=0$, find a point P such that the sum of its distances from the points $A(-7, 1)$ and $B(-5, 5)$ will have the least value.

252. On the line $3x-y-1=0$, find a point P such that the difference of its distances from the points $A(4, 1)$ and $B(0, 4)$ will have the greatest value.

253. Determine the angle ϕ between the two lines:

1) $5x-y+7=0$, $3x+2y=0$;
2) $3x-2y+7=0$, $2x+3y-3=0$;
3) $x-2y-4=0$, $2x-4y+3=0$;
4) $3x+2y-1=0$, $5x-2y+3=0$.

254. Given the line $2x+3y+4=0$.

Find the equation of the straight line passing through the point $M_0(2, 1)$ and making an angle of $45°$ with the given line.

255. The point $A(-4, 5)$ is a vertex of a square, whose diagonal lies on the line $7x-y+8=0$.

Write the equations of the sides and of the other diagonal of the square.

256. $A(-1, 3)$ and $C(6, 2)$ are two opposite vertices of a square. Find the equations of its sides.

257. The point E $(1, -1)$ is the centre of a square, one of whose sides lies on the line $x-2y+12=0$.

Find the equations of the straight lines which contain the remaining sides of the square.

258. From the point $M_0(-2, 3)$, a ray of light is sent at an angle α ($\tan\alpha=3$) to the axis Ox. Upon reaching the axis Ox, the ray is reflected from it. Find the equations of the straight lines which contain the incident and reflected rays.

259. A ray of light is sent along the line $x-2y+5=0$. Upon reaching the line $3x-2y+7=0$, the ray is reflected from it. Find the equation of the line containing the reflected ray.

260. Given that $3x+4y-1=0$, $x-7y-17=0$, $7x+y+31=0$

are the equations of the sides of a triangle. Prove that the triangle is isosceles. Solve the problem by comparing the angles of the triangle.

261. Prove that the equation of the straight line passing through the point M_1 (x_1, y_1) and parallel to the line $Ax+By+C=0$ can be written in the form $A(x-x_1)+B(y-y_1)=0$.

262. Find the equation of the straight line passing through the point M_1 $(2, -3)$ and parallel to the line:

1) $3x-7y+3=0$;
2) $x+9y-11=0$;
3) $16x-24y-7=0$;
4) $2x+3=0$;
5) $3y-1=0$.

Solve the problem without calculating the slopes of the given lines.

Hint. Use the results of the preceding problem.

263. Prove that the condition for the perpendicularity of the lines $A_1x+B_1y+C_1=0$, $A_2x+B_2y+C_2=0$ may be written as $A_1A_2+B_1B_2=0$.

264. Determine which of the following pairs of lines are perpendicular

1) $3x-y+5=0$, $x+3y-1=0$;
2) $3x-4y+1=0$, $4x-3y+7=0$;
3) $6x-15y+7=0$, $10x+4y-3=0$;
4) $9x-12y+5=0$, $8x+6y-13=0$;
5) $7x-2y+1=0$, $4x+6y+17=0$;
6) $5x-7y+3=0$, $3x+2y-5=0$.

Solve the problem without calculating the slopes of the given lines.

Hint. Use the perpendicularity condition derived in Problem 263.

265. Prove that the formula for determining the angle φ between the lines

$A_1x+B_1y+C_1=0$, $A_2x+B_2y+C_2=0$

may be written in the form $\tan\varphi=\dfrac{A_1B_2-A_2B_1}{A_1A_2+B_1B_2}$.

266. Determine the angle φ formed by the two lines:

1) $3x-y+5=0$, $2x+y-7=0$;
2) $x\sqrt{2}-y\sqrt{3}-5=0$, $(3+\sqrt{2})x+(\sqrt{6}-\sqrt{3})y+7=0$;
3) $x\sqrt{3}+y\sqrt{2}-2=0$, $x\sqrt{6}-3y+3=0$.

Solve the problem without calculating the slopes of the given lines.

Hint. Use the formula for determining the angle between two lines which was derived in Problem 265.

267. Given two vertices $M_1(-10, 2)$ and $M_2(6, 4)$ of a triangle whose altitudes intersect in the point $N(5, 2)$. Find the coordinates of the third vertex M_3.

268. $A(3,-1)$ and $B(5, 7)$ are two vertices of a triangle ABC whose altitudes intersect in the point $N(4, -1)$. Write the equations of the sides of the triangle.

269. In a triangle ABC, the equations of the side AB, of the altitude AN and of the altitude BN are $5x-3y+2=0$, $4x-3y+1=0$ and $7x+2y-22=0$, respectively. Write the equations of the other two sides and of the third altitude of the triangle.

270. Find the equations of the sides of a triangle ABC with $A(1, 3)$ as a vertex, if $x-2y+1=0$ and $y-1=0$ are the equations of two of its medians.

271. Find the equations of the sides of a triangle having $B(-4, -5)$ as a vertex, if $5x+3y-4=0$ and $3x+8y+13=0$ are the equations of two of its altitudes.

272. Find the equations of the sides of a triangle having $A(4,-1)$ as a vertex, if $x-1=0$ and $x-y-1=0$ are the equations of two bisectors of its angles.

273. Find the equations of the sides of a triangle having $B(2, 6)$ as a vertex, if $x-7y+15=0$ and $7x+y+5=0$ are the respective equations of an altitude and an angle bisector drawn from one and the same vertex.

274. Find the equations of the sides of a triangle having $B(2,-1)$ as a vertex, if $3x-4y+27=0$ and $x+2y-5=0$ are the respective equations of an altitude and an angle bisector drawn from different vertices.

275. Find the equations of the sides of a triangle having $C(4,-1)$ as a vertex, if $2x-3y+12=0$ and $2x+3y=0$ are the respective equations of an altitude and a median drawn from one and the same vertex.

276. Find the equations of the sides of a triangle having $B(2,-7)$ as a vertex, if $3x+y+11=0$ and $x+2y+7=0$ are the respective equations of an altitude and a median drawn from different vertices.

277. Find the equations of the sides of a triangle having $C(4, 3)$ as a vertex, if $x+2y-5=0$ and $4x+13y-10=0$ are the respective equations of an angle bisector and a median drawn from one and the same vertex.

278. Find the equations of the sides of a triangle having $A(3,-1)$ as a vertex, if $x-4y+10=0$ and $6x+10y-59=0$

are the respective equations of an angle bisector and a median drawn from different vertices.

279. Write the equation of the line passing through the origin and forming, together with the lines $x-y+12=0$, $2x+y+9=0$, a triangle of an area equal to 1.5 square units.

280. From lines passing through the point $P(3, 0)$, select the line whose segment intercepted by the lines $2x-y-2=0$, $x+y+3=0$ is bisected at the point P.

281. All possible straight lines are drawn through the point $P(-3,-1)$. Prove that the lines $x-2y-3=0$, $x-2y+5=0$

cut from each of them a segment whose midpoint is at P.

282. All possible straight lines are drawn through the point $P(0, 1)$. Prove that the segments cut from them by the lines $x-2y-3=0$, $x-2y+17=0$ are not bisected at the point P.

283. Write the equation of a straight line through the origin, if its segment intercepted by the lines $2x-y+5=0$, $2x-y+10=0$ is of length $\sqrt{10}$.

284. Write the equation of a straight line through the point $C(-5, 4)$, given that its segment intercepted by the lines
$x+2y+1=0$, $x+2y-1=0$
is of length 5.

13. Incomplete Equations of Straight Lines. Discussion of a System of Equations Representing Two or Three Straight Lines. The Intercept Equation of a Straight Line

If, in the general equation $Ax+By+C=0$ ...(1)

of a straight line, one or two of the three coefficients (counting the constant term) are zero, the equation is said to be incomplete. The following cases are possible:

1) $C=0$; the equation has the form $Ax+By=0$ and represents a straight line through the origin.
2) $B=0$ $(A\neq 0)$; the equation has the form $Ax+C=0$ and represents a straight line perpendicular to the axis Ox. This equation may be written in the form $x=a$, where $a=-\frac{C}{A}$ is the intercept cutoff by the line on the axis Ox.
3) $B=0$, $C=0$ $(A\neq 0)$; the equation, which can be written in the form $x=0$, represents the y-axis.
4) $A=0$, $(B\neq 0)$; the equation assumes the form $By+C=0$ and represents a straight line perpendicular to the axis Oy. This equation can be written in the form $y=b$, where $b=-\frac{C}{B}$ is the intercept cutoff by the line on the axis Oy.
5) $A=0$, $C=0$ $(B\neq 0)$; the equation, which can be written in the form $y=0$, represents the x-axis.

If the coefficients of equation (1) are all different from zero, then the equation is reducible to the form $\frac{x}{a}+\frac{y}{b}=1$, ...(2)

where $a=-\frac{C}{A}$ and $b=-\frac{C}{B}$ are the intercepts cutoff by the line on the x-and y-axes, respectively.

Equation (2) is called the intercept equation of a straight line.

If two straight lines are represented by the equations
$A_1x+B_1y+C_1=0$ and $A_2x+B_2y+C_2=0$, the following cases may arise:

a) $\frac{A_1}{A_2} \neq \frac{B_1}{B_2}$ — the lines have one point in common;

b) $\frac{A_1}{A_2} = \frac{B_1}{B_2} \neq \frac{C_1}{C_2}$ — the lines are parallel;

c) $\frac{A_1}{A_2} = \frac{B_1}{B_2} = \frac{C_1}{C_2}$ — the lines coincide, that is, the two equations represent the same straight line.

285. Determine the values of a for which the line
$(a+2)x+(a^2-9)y+3a^2-8a+5=0$
1) is parallel to the x-axis; 2) is parallel to the y-axis;
3) passes through the origin.
In each case write the equation of the line.

286. Determine the values of m and n for which the line $(m+2n-3)x+(2m-n+1)y+6m+9=0$ is parallel to the x-axis and has a y-intercept of -3. Write the equation of this line.

287. Determine the values of m and n for which the line $(2m-n+5)x+(m+3n-2)y+2m+7n+19=0$ is parallel to the y-axis and has an x-intercept of $+5$. Write the equation of this line.

288. In each of the following, prove that the two given straight lines intersect, and find their point of intersection:
1) $x+5y-35=0$, $3x+2y-27=0$;
2) $14x-9y-24=0$, $7x-2y-17=0$;
3) $12x+15y-8=0$, $16x+9y-7=0$;
4) $8x-33y-19=0$, $12x+55y-19=0$;
5) $3x+5=0$, $y-2=0$.

289. In each of the following, prove that the two given straight lines are parallel:
1) $3x+5y-4=0$, $6x+10y+7=0$; 2) $2x-4y+3=0$, $x-2y=0$;
3) $2x-1=0$, $x+3=0$; 4) $y+3=0$, $5y-7=0$;

290. In each of the following, prove that the two given straight lines coincide:
1) $3x+5y-4=0$, $6x+10y-8=0$; 2) $x-y\sqrt{2}=0$, $x\sqrt{2}-2y=0$;
3) $x\sqrt{3}-1=0$, $3x-\sqrt{3}=0$.

291. Find the values of a and b for which the two lines
$ax-2y-1=0$, $6x-4y-b=0$
1) have one common point; 2) are parallel;
3) coincide.

292. Find the values of m and n for which the two lines

$mx+8y+n=0,\ 2x+my-1=0.$

1) are parallel; 2) coincide; 3) are perpendicular.

293. Find the value of m for which the two lines

$(m-1)x+my-5=0,\ mx+(2m-1)y+7=0$ intersect at a point lying on the x-axis.

294. Find the values of m for which the two lines $mx+(2m+3)y+m+6=0$, $(2m+1)x+(m-1)y+m-2=0$ intersect in a point lying on the y-axis.

295. In each of the following, determine whether the three straight lines intersect in a single point:

1) $2x+3y-1=0,\ 4x-5y+5=0,\ 3x-y+2=0;$

2) $3x-y+3=0,\ 5x+3y-7=0,\ x-2y-4=0;$

3) $2x-y+1=0,\ x+2y-17=0,\ x+2y-3=0.$

296. Prove that, if the three lines $A_1x+B_1y+C_1=0$, $A_2x+B_2y+C_2=0$, $A_3x+B_3y+C_3=0$ intersect in a single point, then

$$\begin{vmatrix} A_1 & B_1 & C_1 \\ A_2 & B_2 & C_2 \\ A_3 & B_3 & C_3 \end{vmatrix}=0.$$

297. Prove that, if

$$\begin{vmatrix} A_1 & B_1 & C_1 \\ A_2 & B_2 & C_2 \\ A_3 & B_3 & C_3 \end{vmatrix}=0,$$

then the three lines

$A_1x+B_1y+C_1=0,\ A_2x+B_2y+C_2=0,\ A_3x+B_3y+C_3=0$

intersect in a single point or are parallel.

298. Determine the value of a for which the three lines $2x-y+3=0$, $x+y+3=0$, $ax+y-13=0$, intersect in a single point.

299. Given the lines

1) $2x+3y-6=0;$ 2) $4x-3y+24=0;$

3) $2x+3y-9=0;$ 4) $3x-5y-2=0;$

6) $5x+2y-1=0.$

Write their intercept equations and plot the lines.

300. Calculate the area of the triangle formed by the line $3x-4y-12=0$ and the coordinate axes.

301. Write the equation of the straight line passing through the point $M_1(3,\ -7)$ and making equal (non-zero) intercepts on the coordinate axes.

302. Write the equation of a straight line passing through the point $P(2, 3)$ and making intercepts of equal absolute value on the coordinate axes.

303. Find the equation of a straight line passing through the point $C(1, 1)$ and forming with the coordinate axes a triangle whose area is 2 square units.

304. Find the equation of a straight line passing through the point $B(5, -5)$ and forming with the coordinate axes a triangle whose area is 50 square units.

305. Find the equation of a straight line passing through the point $P(8, 6)$ and forming with the coordinate axes a triangle whose area is 12 square units.

306. Find the equation of a straight line which passes through the point $P(12, 6)$ and forms with the coordinate axes a triangle whose area is 150 square units.

307. Through the point $M(4, 3)$, a straight line is drawn so as to form with the coordinate axes a triangle whose area is 3 square units. Determine the points of intersection of the line with the coordinate axes.

308. Through the point $M_1(x_1, y_1)$, where $x_1 y_1 > 0$, the line

$\frac{x}{a} + \frac{y}{b} = 1$ is drawn so as to form with the coordinate axes a triangle of area S. Determine the relation between the quantities x_1, y_1 and S for which the intercepts a and b will be of like sign.

14. The Normal Equation of a Straight Line. The Problem of Determining the Distance of a Point from a Straight Line

Let a straight line be given in the plane xOy. Through the origin, draw a line (called the normal) perpendicular to the given line. Denote by P the point of intersection of the normal and the given line, and choose the direction from the point O to the point P as the positive direction of the normal.

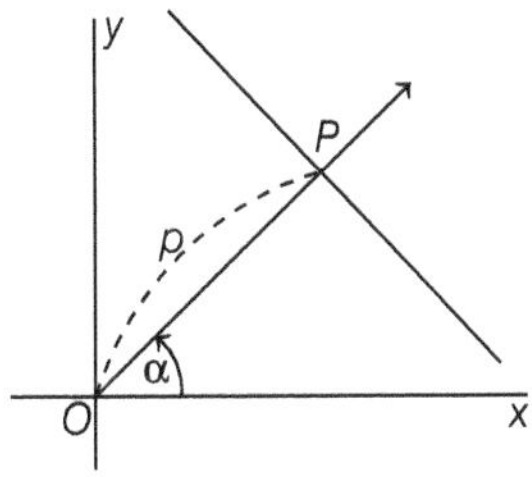

Fig. 10.

If α is the polar angle of the normal, and p is the length of the segment $\overline{OP}$ (Fig. 10), then the equation of the given straight line can be written as

$x\cos\alpha + y\cdot\sin\alpha - p = 0$; an equation of this form is called the normal equation of a straight line.

Let there be given any straight line and an arbitrary point M^*; let d denote the distance of M^* from the given line. The departure δ of the point M^* from the line is defined as the number $+d$ when M^* and the origin lie on opposite sides of the given line, and as $-d$ when M^* and the origin lie on the same side of the given line (For points lying on the line itself, $\delta = 0$.)

If we are given the coordinates x^*, y^* of the point M^* and the normal equation $x\cos\alpha + y\sin\alpha - p = 0$ of a straight line, then the departure δ of M^* from this line can be calculated from the formula

$\delta = x^*\cos\alpha + y^*\sin\alpha - p.$

Thus, to find the departure of a point M^* from a given straight line, the coordinates of the point M^* must be substituted for the current coordinates in the left-hand member of the normal equation of this straight line. The resulting number will be the required departure.

To find the distance d of a point from a straight line, we have merely to calculate the departure and to take its modulus: $d = |\delta|$.

Given the general equation $Ax + By + C = 0$ of a straight line; to reduce this equation to the normal form, all its terms must be multiplied by the normalizing factor μ which is determined from the formula

$$\mu = \pm\frac{1}{\sqrt{A^2 + B^2}}.$$

The normalizing factor must be taken with the sign opposite to that of the constant term of the equation to be normalised.

309. Determine which of the following equations of straight lines are in the normal form:

1) $\frac{3}{5}x - \frac{4}{5}y - 3 = 0$; 2) $\frac{2}{5}x - \frac{3}{5}y - 1 = 0$;

3) $\frac{5}{13}x - \frac{12}{13}y + 2 = 0$; 4) $-\frac{5}{13}x + \frac{12}{13}y - 2 = 0$;

5) $-x + 2 = 0$; 6) $x - 2 = 0$;

7) $y + 2 = 0$; 8) $-y - 2 = 0$.

310. In each of the following, reduce the general equation of the given straight line to the normal form:

1) $4x - 3y - 10 = 0$; 2) $\frac{4}{5}x - \frac{3}{5}y + 10 = 0$;

3) $12x - 5y + 13 = 0$; 4) $x + 2 = 0$;

5) $2x - y - \sqrt{5} = 0$.

311. Given the equations of straight lines:

1) $x-2=0$; 2) $x+2=0$; 3) $y-3=0$; 4) $y+3=0$;
5) $x\sqrt{3}+y-6=0$; 6) $x-y+2=0$;
7) $x+y\sqrt{3}+2=0$;
8) $x\cos\beta-y\sin\beta-q=0$, $q>0$; β is an acute angle;
9) $x\cos\beta+y\sin\beta+q=0$, $q>0$; β is an acute angle.

Determine the polar angle α of the normal and the segment p for each of the given lines; from the obtained values of the parameters α and p, plot the lines (setting $\beta=30°$ and $q=2$ in the last two cases).

312. In each of the following, calculate the value of the departure δ and the distance d of the point from the line:

1) $A(2,-1)$, $4x+3y+10=0$; 2) $B(0,-3)$, $5x-12y-23=0$;
3) $P(-2,3)$, $3x-4y-2=0$; 4) $Q(1,-2)$, $x-2y-5=0$.

313. In each of the following, determine whether the point $M(1, -3)$ and the origin lie on the same side or on opposite sides of the given line:

1) $2x-y+5=0$; 2) $x-3y-5=0$;
3) $3x+2y-1=0$; 4) $x-3y+2=0$;
5) $10x+24y+15=0$.

314. The point $A(2, -5)$ is a vertex of a square, one of whose sides lies on the line $x-2y-7=0$. Calculate the area of the square.

315. Given that $3x-2y-5=0$, $2x+3y+7=0$ are the equations of two sides of a rectangle, and that $A(-2, 1)$ is one of its vertices; calculate the area of the rectangle.

316. Prove that the line $2x+y+3=0$ cuts the segment bounded by the points $A(-5, 1)$ and $B(3, 7)$.

317. Prove that the line $2x-3y+6=0$ does not cut the segment bounded by the points $M_1(-2, -3)$ and $M_2(1, -2)$.

318. The points $A(-3, 5)$, $B(-1, -4)$, $C(7, -1)$ and $D(2, 9)$ are the consecutive vertices of a quadrilateral. Determine whether the quadrilateral is convex.

319. The points $A(-1, 6)$, $B(1, -3)$, $C(4, 10)$ and $D(9, 0)$ are the consecutive vertices of a quadrilateral. Determine whether this quadrilateral is convex.

320. The vertices of a triangle are $A(-10, -13)$, $B(-2, 3)$ and $C(2, 1)$. Calculate the length of the perpendicular dropped from the vertex B to the median through C.

321. The sides AB, BC, CA of a triangle ABC are given, respectively, by the equations $x+21y-22=0$, $5x-12y+7=0$, $4x-33y+146=0$.

Calculate the distance from the centre of gravity of this triangle to the side BC.

322. In each of the following, calculate the distance d between the given parallel lines:

1) $3x-4y-10=0$, $6x-8y+5=0$; 2) $5x-12y+26=0$, $5x-12y-13=0$;

3) $4x-3y+15=0$, $8x-6y+25=0$; 4) $24x-10y+39=0$, $12x-5y-26=0$.

323. Two sides of a square lie on the lines $5x-12y-65=0$, $5x-12y+26=0$. Calculate the area of the square.

324. Prove that the line $5x-2y-1=0$ is parallel to the lines $5x-2y+7=0$, $5x-2y-9=0$ and lies midway between them.

325. Given the three parallel lines $10x+15y-3=0$, $2x+3y+5=0$, $2x+3y-9=0$. Show that the first of them lies between the other two, and find the ratio in which the first line divides the distance between the second and the third line.

326. Prove that two lines can be drawn through the point $P(2, 7)$ so that their distances from the point Q (1, 2) will each be equal to 5. Write the equations of these lines.

327. Prove that two straight lines can be drawn through the point P (2,5) so that their distances from the point $Q(5, 1)$ will each be equal to 3. Write the equations of these lines.

328. Prove that one line only can be drawn through the point $C(7, -2)$ so that its distance from the point A (4, – 6) will be equal to 5. Write the equation of this line.

329. Prove that no line can be drawn through the point $B(4, -5)$ so that its distance from the point C (–2, 3) will be equal to 12.

330. Derive the equation of the locus of points whose departure from the line $8x-15y-25=0$ is equal to –2.

331. Find the equations of the straight lines which are parallel to the line $3x-4y-10=0$ and lie at a distance $d=3$ from that line.

332. The points $A(2, 0)$ and B (–1, 4) are two adjacent vertices of a square. Find the equations of its sides.

333. The point $A(5, -1)$ is a vertex of a square, one of whose sides lies on the line $4x-3y-7=0$. Write the equations of the lines containing the other sides of this square.

334. Given that $4x-3y+3=0$, $4x-3y-17=0$ are the equations of two sides of a square and that $A(2, -3)$ is one of its vertices find the equations of the remaining two sides of this square.

335. Given that $5x+12y-10=0$, $5x+12y+29=0$ are the equations of two sides of a square and that the point $M_1(-3, 5)$ lies on one of its sides; find the equations of the remaining two sides of this square.

336. The departures of a point M from the lines $5x-12y-13=0$ and $3x-4y-19=0$ are equal to –3 and – 5, respectively. Find the coordinates of the point M.

337. Write the equation of the straight line passing through the point $P(-2, 3)$ and equidistant from the points $A(5, -1)$ and $B(3, 7)$.

338. Find the equation of the locus of points equidistant from the two parallel lines:

1) $3x-y+7=0, 3x-y-3=0$;

2) $x-2y+3=0, x-2y+7=0$;

3) $5x-2y-6=0$; $10x-4y+3=0$.

339. Write the equations of the bisectors of the angles formed by the two intersecting lines

1) $x-3y+5=0, 3x-y-2=0$;

2) $x-2y-3=0, 2x+4y+7=0$;

3) $3x+4y-1=0$, $5x+12y-2=0$.

340. Write the equations of the straight lines which pass through the point P (2, –1) and form, together with the lines $2x-y+5=0$, $3x+6y-1=0$, isosceles triangles.

341. In each of the following, determine whether the point $M(1, -2)$ and the origin are contained by the same angle, by the supplementary angles, or by the vertical angles formed by the two intersecting lines:

1) $2x-y-5=0, 3x+y+10=0$;

2) $4x+3y-10=0, 12x-5y-5=0$;

3) $x-2y-1=0$, $3x-y-2=0$.

342. In each of the following, determine whether the points M (2, 3) and N (5, –1) are contained by the same angle, by the supplementary angles, or by the vertical angles formed by the two intersecting lines:

1) $x-3y-5=0, 2x+9y-2=0$;

2) $2x+7y-5=0, x+3y+7=0$;

3) $12x+y-1=0$, $13x+2y-5=0$.

343. Determine whether the origin lies inside or outside the triangle whose sides are given by the equations $7x-5y-11=0$, $8x+3y+31=0$, $x+8y-19=0$.

344. Determine whether the point $M(-3, 2)$ lies inside or outside the triangle whose sides are given by the equations $x+y-4=0$, $3x-7y+8=0$, $4x-y-31=0$.

345. Determine which of the angles (the acute or the obtuse one) formed by the two lines $3x-2y+5=0$, $2x+y-3=0$ contains the origin.

346. Determine which of the angles (the acute or the obtuse one) formed by the two lines $3x-5y-4=0$, $x+2y+3=0$ contains the point $M(2,-5)$.

347. Find the equation of the bisector of that angle between the lines $3x-y-4=0$ and $2x+6y+3=0$ which contains the origin.

348. Find the equation of the bisector of that 'angle between the lines $x-7y+5=0$, $5x+5y-3=0$ which is the supplement of the angle containing the origin.

349. Find the equation of the bisector of that angle between the lines $x+2y-11=0$, $3x—6y-5=0$ which contains the point $M(1,-3)$.

350. Write the equation of the bisector of that angle between the lines $2x-3y-5=0$, $6x-4y+7=0$ which is the supplement of the angle containing the point $C(2,-1)$.

351. Write the equation of the bisector of the acute angle formed by the two lines $3x+4y-5=0$, $5x-12y+3=0$.

352. Write the equation of the bisector of the obtuse angle formed by the two lines $x-3y+5=0$, $3x-y+15=0$.

15. The Equation of a Pencil of Lines

The totality of lines passing through a point S is called the pencil of lines with vertex S.

If $A_1x+B_1y+C_1=0$ and $A_2x+B_2y+C_2=0$, are the equations of two straight lines intersecting in a point S, then the equation

$$\alpha(A_1x+B_1y+C_1)+\beta(A_2x++B_2y+C_2)=0, \qquad ...(1)$$

where α, β are any numbers which are not both simultaneously equal to zero, represents a straight line also passing through the point S.

Moreover, it is always possible to choose the numbers α, β in equation (1) so as to make the equation represent any (previously assigned) line through the point S, that is, any line of the pencil with vertex S. An equation of the form (1) is therefore called the equation of a pencil of lines (with vertex S).

If $\alpha \neq 0$, then, dividing both members of (1) by α and letting $\frac{\beta}{\alpha}=\lambda$, we obtain

$$A_1x+B_1y+C_1+\lambda(A_2x+B_2y+C_2)=0. \qquad (2)$$

This equation can be made to represent every line of the pencil with vertex S except the line corresponding to $\alpha=0$, that is, except the line $A_2x+B_2y+C_2=0$.

353. Find the vertex of the pencil of lines represented by the equation $\alpha(2x+3y-1)+\beta(x-2y-4)=0$.

354. Find the equation of the line belonging to the pencil of lines $\alpha(x+2y-5)+\beta(3x-2y+1)=0$ and

1) passing through the point $A(3,-1)$; 2) passing through the origin;

3) parallel to the axis Ox; 4) parallel to the axis Oy;

5) parallel to the line $4x+3y-5=0$;

6) perpendicular to the line $2x+3y+7=0$.

355. Write the equation of the line passing through the point of intersection of the lines $3x-2y+5=0$, $4x+3y-1=0$ and making an intercept $b=-3$ on the y-axis. Solve the problem without determining the coordinates of the point of intersection of the given lines.

356. Write the equation of the line which passes through the point of intersection of the lines $2x+y-2=0$, $x-5y-23=0$ and bisects the segment bounded by the points $M_1(5,-6)$ and $M_2(-1,-4)$. Solve the problem without calculating the coordinates of the point of intersection of the given lines.

357. Given the equation $\alpha(3x-4y-3)+\beta(2x+3y-1)=0$ of a pencil of lines. Write the equation of that line of the pencil which passes through the centre of gravity of a uniform triangular plate whose vertices are the points $A(-1,2)$, $B(4,-4)$ and $C(6,-1)$.

358. Given the equation $\alpha(3x-2y-1)+\beta(4x-5y+8)=0$ of a pencil of lines. Find that line of the pencil which bisects the segment cut from the line $x+2y+4=0$ by the lines $2x+3y+5=0$, $x+7y-1=0$.

359. Given the equations $x+2y-1=0$, $5x+4y-17=0$, $x-4y+11=0$ of the sides of a triangle. Without determining the coordinates of its vertices, find the equations of the altitudes of the triangle.

360. Write the equation of the line passing through the point of intersection of the lines $2x+7y-8=0$, $3x+2y+5=0$ and making an angle of 45° with the line $2x+3y-7=0$.

Solve the problem without calculating the coordinates of the point of intersection of the given lines.

361. In a triangle ABC, the altitudes AN, BN and the side AB are represented respectively by the equations $x+5y-3=0$, $x+y-1=0$ and $x+3y-1=0$. Without determining the coordinates of the vertices and of the intersection point of the altitudes, write the equations of the other two sides and the third altitude of the triangle.

362. Find the equations of the sides of a triangle ABC, if $A(2,-1)$ is one of its vertices and if $7x-10y+1=0$ and $3x-2y+5=0$ are the respective equations of an altitude and an angle bisector drawn from one and the same vertex. Solve the problem without computing the coordinates of the vertices B and C.

363. Given that $\alpha\,(2x+y+8)+\beta\,(x+y+3)=0$ is the equation of a pencil of lines. Find those lines of the pencil whose segments intercepted by the lines $x-y-5=0$, $x-y-2=0$ are equal to $\sqrt{5}$.

364. The equation of a pencil of lines is $\alpha\,(3x+y-1)+\beta\,(2x-y-9)=0$. Prove that the line $x+3y+13=0$ belongs to this pencil.

365. The equation of a pencil of lines is $\alpha\,(5x+3y+6)+\beta\,(3x-4y-37)=0$.

Prove that the line $7x+2y-15=0$ does not belong to this pencil.

366. The equation of a pencil of lines is $\alpha\,(3x+2y-9)+\beta\,(2x+5y+5)=0$. Find the value of C for which the line $4x-3y+C=0$ will belong to this pencil.

367. The equation of a pencil of lines is $\alpha\,(5x+3y-7)+\beta\,(3x+10y+4)=0$.

Find the values of a for which the line $ax+5y+9=0$ will not belong to this pencil.

368. The vertex of the pencil of lines $\alpha\,(2x-3y+20)+\beta\,(3x+5y-27)=0$

is one of the vertices of a square whose diagonal lies along the line

$$x+7y-16=0.$$

Write the equations of the sides and of the other diagonal of the square.

369. Given the pencil of lines $\alpha\,(2x+5y+4)+\beta\,(3x-2y+25)=0$. Find that line of the pencil which makes equal (non-zero) intercepts on the coordinate axes.

370. Given the pencil of lines $\alpha\,(2x+y+1)+\beta\,(x-3y-10)=0$. Find those lines of the pencil whose intercepts on the x and y-axes are equal in absolute value.

371. Given the pencil of lines $\alpha\,(21x+8y-18)+\beta\,(11x+3y+12)=0$.

Find those lines of the pencil which form with the coordinate axes a triangle whose area is 9 square units.

372. Given the pencil of lines $\alpha\,(2x+y+4)+\beta\,(x-2y-3)=0$. Prove that, among the lines of the pencil, there exists only one line whose distance from the point $P(2,-3)$ is $d=\sqrt{10}$. Write the equation of this line.

373. Given the pencil of lines $\alpha\,(2x-y-6)+\beta\,(x-y-4)=0$. Prove that, among the lines of the pencil, there is no line situated at the distance $d=3$ from the point $P\,(3,-1)$.

374. Find the equation of the line passing through the intersection of the lines $3x+y-5=0$, $x-2y+10=0$ and situated at the distance $d=5$ from the point $C\,(-1,-2)$. Solve the problem without computing the coordinates of the point of intersection of the given lines.

375. Given the pencil of lines $\alpha(5x+2y+4)+\beta(x+9y-25)=0$.

Write the equations of those lines of the pencil which, together with the lines $2x-3y+5=0$, $12x+8y-7=0$, form isosceles triangles.

376. Find the equation of the straight line passing through the intersection of the lines $11x+3y-7=0$, $12x+y-19=0$

and equidistant from the points $A(3, -2)$ and $B(-1, 6)$. Solve the problem without determining the coordinates of the point of intersection of the given lines.

377. Given the equations $\alpha_1(5x+3y-2)+\beta_1(3x-y-4)=0$, $\alpha_2(x-y+1)+\beta_2(2x-y-2)=0$ of two pencils of lines. Without determining their vertices, find the equation of the line belonging to both pencils.

378. The sides *AB, BC, CD, DA* of a quadrilateral *ABCD* are represented, respectively, by the equations $5x+y+13=0$, $2x-7y-17=0$, $3x+2y-13=0$, $3x-4y+17=0$.

Find the equations of the diagonals *AC* and *BD* of this quadrilateral without determining the coordinates of its vertices.

379. The vertex of the pencil of lines $\alpha(2x+3y+5)+\beta(3x-y+2)=0$ is a vertex of a triangle, two of whose altitudes are given by the equations $x-4y+1=0$, $2x+y+1=0$. Write the equations of the sides of this triangle.

16. The Polar Equation of a Straight Line

The straight line drawn through the pole perpendicular to a given straight line is called the normal to the given line. Denote by P the point where the normal meets the line; take the direction from the point O to the point P as the positive direction of the normal. The angle through which the polar axis must be turned to reach coincidence with the direction of the segment $\overline{OP}$ is referred to as the polar angle of the normal.

380. Derive the polar equation of a straight line, given its distance p from the pole and the polar angle α of the normal.

Solution. First Method. On the given straight line s (Fig. 11), take an arbitrary point M with polar coordinates ρ and θ. Let P denote the point where the line s meets its normal. From the right triangle *OPM*, we find:

$$\rho = \frac{p}{\cos(\theta-\alpha)}. \tag{1}$$

We have obtained an equation in two variables, ρ and θ, which is satisfied by the coordinates of every point M lying on the line s and by the coordinates of no other point. Consequently, equation (1) is the equation of the straight line s. The problem is thus solved.

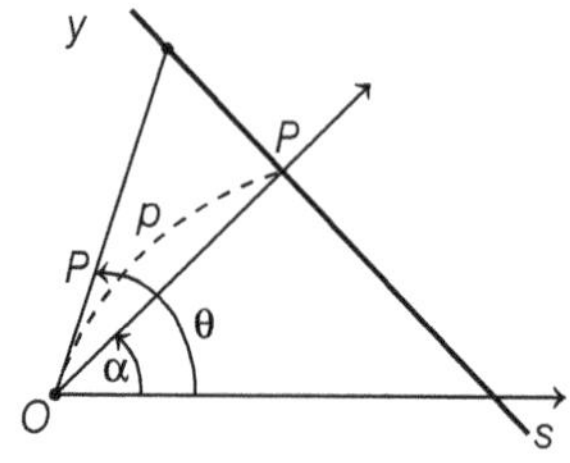

Fig. 11.

Second Method. Consider a rectangular cartesian coordinate system whose positive x-axis coincides with the polar axis of the given polar system. In this cartesian system, the normal equation of the line s is

$$x\cos\alpha + y\sin\alpha - p = 0 \qquad (2)$$

Recalling the formulas for transformation from polar to cartesian coordinates,

$$x = \rho cos\theta, \quad y = \rho\sin\,\theta. \qquad (3)$$

and substituting the expressions (3) for x and y in equation (2), we have

$$\rho\,(\cos\theta\,\cos\alpha + \sin\,\theta\,\sin\alpha) = p \text{ or } \quad \rho = \frac{p}{\cos\,(\theta-\alpha)}.$$

381. Derive the polar equation of a straight line, given:

1) the angle of inclination β of the line (with respect to the polar axis), and the length p of the perpendicular dropped from the pole to the line. Also, write the equation of the line for the case
$$\beta = \frac{\pi}{6}, \quad p = 3;$$

2) the intercept a cut-off by the line on the polar axis, and the polar angle α of the normal to the line. Also, write the equation of the line for the case
$$a = 2, \quad \alpha = -\frac{2}{3}\,\pi;$$

3) the angle of inclination β of the line (with respect to the polar axis), and the intercept a cut-off by the line on the polar axis. Also, write the equation of the line for the case
$$\beta = \frac{\pi}{6}, \qquad \alpha = 6.$$

382. Derive the polar equation of the straight line passing through the point $M_1(\rho_1, \theta_1)$ and making an angle β with the polar axis.

383. Derive the polar equation of a straight line through the point $M_1(\rho_1, \theta_1)$, if the polar angle of its normal is α.

384. Find the polar equation of the line through the points $M_1(\rho_1, \theta_1)$ and $M_2(\rho_2, \theta_2)$.

4

GEOMETRIC PROPERTIES OF CURVES OF THE SECOND ORDER

17. The Circle

The equation

$$(x-\alpha)^2+(y-\beta)^2=R^2 \qquad (1)$$

represents a circle of radius R with centre $C(\alpha, \beta)$.

If the centre of the circle coincides with the origin, that is, if $\alpha=0$, $\beta=0$, then equation (1) takes the form

$$x^2+y^2=R^2. \qquad (2)$$

385. In each of the following, find the equation of the circle determined by the stated conditions:

1) the centre of the circle is at the origin, and the radius $R=3$;
2) the centre is at the point $C(2, -3)$, and the radius $R=7$;
3) the circle passes through the origin, and the centre is at the point $C(6, -8)$;
4) the circle passes through the point $A(2, 6)$ and the centre is at $C(-1, 2)$;
5) $A(3, 2)$ and $B(-1, 6)$ are the end points of a diameter of the circle;
6) the centre is at the origin, and the line $3x-4y+20=0$ is tangent to the circle;
7) the centre is at the point $C(1, -1)$, and the line $5x-12y+9=0$ is tangent to the circle;
8) the circle passes through the points $A(3, 1)$, $B(-1, 3)$, and its centre lies on the line $3x-y-2=0$;
9) the circle passes through the three points $A(1, 1)$, $B(1, -1)$ and $C(2, 0)$;
10) the circle passes through the three points $M_1(-1, 5)$, $M_2(-2, -2)$ and $M_3(5, 5)$.

386. The point C (3, –1) is the centre of a circle which cuts off a chord of length 6 on the line $2x-5y+18=0$. Write the equation of the circle.

387. Write the equations of the circles of radius $R=\sqrt{5}$ and tangent to the line $x-2y-1=0$ at the point M_1 (3, 1).

388. Find the equation of a circle tangent to the two parallel lines $2x+y-5=0, 2x+y+15=0$, if A (2, 1) is their point of contact with one of the lines.

389. Find the equations of the circles passing through the point A (1, 0) and tangent to the two parallel lines $2x+y+2=0$, $2x+y-18=0$.

390. Find the equation of the circle with centre on the line $2x+y=0$ and tangent to the lines $4x-3y+10=0$, $4x-3y-30=0$.

391. Write the equations of circles tangent to the two intersecting lines $7x-y-5=0$, $x+y+13=0$, if $M_1(1, 2)$ is their point of contact with one of the lines.

392. Write the equations of the circles passing through the origin and tangent to the two intersecting lines $x+2y-9=0$, $2x-y+2=0$.

393. Find the equations of the circles which have their centres on the line $4x-5y-3=0$ and are tangent to the lines $2x-3y-10=0$, $3x-2y+5=0$.

394. Write the equations of the circles passing through the point A (–1, 5) and tangent to the two intersecting lines

$$3x+4y-35=0,$$
$$4x+3y+14=0.$$

395. Write the equations of the circles tangent to the three lines $4x-3y-10=0, 3x-4y-5=0$ and $3x-4y-15=0$.

396. Write the equations of the circles tangent to the three lines $3x+4y-35=0$, $3x-4y-35=0$ and $x-1=0$.

397. Determine which of the following equations represent circles and find the centre C and the radius R of each circle.

1) $(x-5)^2+(y+2)^2=25$; 2) $(x+2)^2+y^2=64$;
3) $(x-5)^2+(y+2)^2=0$; 4) $x^2+(y-5)^2=5$;
5) $x^2+y^2-2x+4y-20=0$; 6) $x^2+y^2-2x+4y+14=0$;
7) $x^2+y^2+4x-2y+5=0$; 8) $x^2+y^2+x=0$;
9) $x^2+y^2+6x-4y+14=0$; 10) $x^2+y^2+y=0$.

398. Identify and plot the lines represented by the following equations:

1) $y=+\sqrt{9-x^2}$; 2) $y=-\sqrt{25-x^2}$;
3) $x=-\sqrt{4-y^2}$; 4) $x=+\sqrt{16-y^2}$;
5) $y=15+\sqrt{64-x^2}$; 6) $y=15-\sqrt{64-x^2}$;

7) $x = -2 - \sqrt{9 - y^2}$; 8) $x = -2 + \sqrt{9 - y^2}$;

9) $y = -3 - \sqrt{21 - 4x - x^2}$; 10) $x = -5 + \sqrt{40 - 6y - y^2}$.

399. Determine whether the point $A(1, -2)$ is inside, on, or outside each of the following circles:

1) $x^2 + y^2 = 1$; 2) $x^2 + y^2 = 5$;

3) $x^2 + y^2 = 9$; 4) $x^2 + y^2 - 8x - 4y - 5 = 0$;

5) $x^2 + y^2 - 10x + 8y = 0$.

400. In each of the following, find the equation of the line of centres of the two given circles:

1) $(x-3)^2 + y^2 = 9$ and $(x+2)^2 + (y-1)^2 = 1$;

2) $(x+2)^2 + (y-1)^2 = 16$ and $(x+2)^2 + (y+5)^2 = 25$;

3) $x^2 + y^2 - 4x + 6y = 0$ and $x^2 + y^2 - 6x = 0$;

4) $x^2 + y^2 - x + 2y = 0$ and $x^2 + y^2 + 5x + 2y - 1 = 0$.

401. Find the equation of that diameter of the circle $x^2 + y^2 + 4x - 6y - 17 = 0$ which is perpendicular to the line $5x + 2y - 13 = 0$.

402. In each of the following, calculate the shortest distance from the given point to the given circle:

1) $A(6, -8), x^2 + y^2 = 9$;

2) $B(3, 9), x^2 + y^2 - 26x + 30y + 313 = 0$;

3) $C(-7, 2), x^2 + y^2 - 10x - 14y - 151 = 0$.

403. Determine the coordinates of the points of intersection of the line $7x - y + 12 = 0$ and the circle $(x-2)^2 + (y-1)^2 = 25$.

404. In each of the following, determine whether the given line cuts, touches, or fails to meet the given circle:

1) $y = 2x - 3$ and $x^2 + y^2 - 3x + 2y - 3 = 0$;

2) $y = \frac{1}{2}x - \frac{1}{2}$ and $x^2 + y^2 - 8x + 2y + 12 = 0$;

3) $y = x + 10$ and $x^2 + y^2 - 1 = 0$.

405. Determine the values of the slope k for which the line $y = kx$

1) cuts the circle $x^2 + y^2 - 10x + 16 = 0$;

2) touches this circle; 3) passes outside this circle.

406. Find the condition under which the line $y = kx + b$ touches the circle $x^2 + y^2 = R^2$.

407. Write the equation of that diameter of the circle $(x-2)^2 + (y+1)^2 = 16$ which bisects the chord cuts off by the circle on the line $x - 2y - 3 = 0$.

408. Find the equation of that chord of the circle $(x-3)^2+(y-7)^2=169$ whose midpoint is at $M(8.5, 3.5)$.

409. Determine the length of that chord of the circle $(x-2)^2+(y-4)^2=10$ whose midpoint is at $A(1, 2)$.

410. Given the pencil of lines
$\alpha\,(x-8y+30)+\beta(x+5y-22)=0.$
Find those lines of the pencil on which the circle
$x^2+y^2-2x+2y-14=0$
cuts off chords of length $2\sqrt{3}$.

411. Given the two circles
$(x-m_1)^2+(y-n_1)^2=R_1^2,\ (x-m_2)^2+(y-n_2)^2=R_2^2$
intersecting at the points $M_1(x_1, y_1)$ and $M_2(x_2, y_2)$. Prove that, by a judicious choice of the numbers α and β, an equation of the form
$\alpha[(x-m_1)^2+(y-n_1)^2-R_1^2]+\beta[(x-m_2)^2+(y-n_2)^2-R_2^2]=0$
can be made to represent every circle through the points M_1, M_2, and also to represent the straight line M_1M_2.

412. Find the equation of the circle which passes through the point $A(1, -1)$ and through the points of intersection of the two circles
$x^2+y^2+2x-2y-23=0,\ x^2+y^2-6x+12y-35=0.$

413. Find the equation of the circle which passes through the origin and through the points of intersection of the two circles
$(x+3)^2+(y+1)^2=25,\ (x-2)^2+(y+4)^2=9.$

414. Write the equation of the straignt line passing through the points of intersection of the two circles
$x^2+y^2+3x-y=0,\ 3x^2+3y^2+2x+y=0.$

415. Calculate the distance from the centre of the circle $x^2+y^2=2x$ to the straight line passing through the points of intersection of the two circles
$x^2+y^2+5x-8y+1=0,\ x^2+y^2-3x+7y-25=0.$

416. Determine the length of the common chord of the two circles
$x^2+y^2-10x-10y=0,\ x^2+y^2+6x+2y-40=0.$

417. Write the equation of the circle which has its centre on the line $x+y=0$ and passes through the points of intersection of the two circles
$(x-1)^2+(y+5)^2=50,\ (x+1)^2+(y+1)^2=10.$

418. Write the equation of the line tangent to the circle $x^2+y^2=5$ at the point $A(-1, 2)$.

419. Write the equation of the line tangent to the circle $(x+2)^2+(y-3)^2=25$ at the point $A(-5, 7)$.

420. On the circle

$16x^2 + 16y^2 + 48x - 8y - 43 = 0,$

find the point M_1 nearest to the line $8x - 4y + 73 = 0,$

and calculate the distance d between M_1 and the line.

421. The point $M_1(x_1, y_1)$ lies on the circle $x^2 + y^2 = R^2$. Write the equation of the tangent line at M_1 to the circle.

422. The point $M_1(x_1, y_1)$ lies on the circle $(x-\alpha)^2 + (y-\beta)^2 = R^2$. Find the equation of the tangent line at M_1 to the circle.

423. Determine the acute angle at which the line $3x - y - 1 = 0$ intersects the circle $(x-2)^2 + y^2 = 5$. (The angle of intersection of a straight line and a circle is defined as the angle between the line and the tangent to the circle at their point of intersection).

424. Determine the angle of intersection of the two circles

$(x-3)^2 + (y-1)^2 = 8,\ (x-2)^2 + (y+2)^2 = 2.$

(The angle of intersection of two circles is defined as the angle between their tangents at a point of intersection.)

425. Find the condition under which the two circles

$(x-\alpha_1)^2 + (y-\beta_1)^2 = R_1^2,\ (x-\alpha_2)^2 + (y-\beta_2)^2 = R_2^2$

will intersect at right angles.

426. Prove that the two circles

$x^2 + y^2 - 2mx - 2ny - m^2 + n^2 = 0,\ x^2 + y^2 - 2nx + 2my + m^2 - n^2 = 0$

intersect at right angles.

427. From the point $A\left(\frac{5}{3}, -\frac{5}{3}\right)$, tangent lines are drawn to the circle $x^2 + y^2 = 5$. Find their equations.

428. From the point $A(1, 6)$, tangent lines are drawn to the circle $x^2 + y^2 + 2x - 19 = 0$. Find their equations.

429. Given the pencil of lines

$\alpha\,(3x + 4y - 10) + \beta\,(3x - y - 5) = 0.$

Find those lines of the pencil which are tangent to the circle

$$x^2 + y^2 + 2x - 4y = 0.$$

430. Tangent lines are drawn to the circle $x^2 + y^2 = 10$, from the point $A(4, 2)$. Determine the angle between these tangent lines.

431. Tangent lines are drawn to the circle $(x-1)^2 + (y+5)^2 = 4$ from the point $P(2, -3)$. Write the equation of the chord joining the points of contact.

432. From the point $C(6,-8)$, tangent lines are drawn to the circle $x^2+y^2=25$. Calculate the distance d between the point C and the chord joining the points of contact.

433. From the point $P(-9,3)$, tangent lines are drawn to the circle

$x^2+y^2-6x+4y-78=0$.

Calculate the distance d from the centre of the circle to the chord joining the points of contact.

434. From the point $M(4,-4)$, tangent lines are drawn to the circle $x^2+y^2-6x+2y+5=0$.

Calculate the length d of the chord joining the points of contact.

435. Calculate the length of the tangent line from the point $A(1,-2)$ to the circle $x^2+y^2+x-3y-3=0$.

436. Find the equations of the lines tangent to the circle $x^2+y^2+10x-2y+6=0$ and parallel to the line $2x+y-7=0$.

437. Find the equations of the lines tangent to the circle

$x^2+y^2-2x+4y=0$ and perpendicular to the line $x-2y+9=0$.

438. Write the polar equation of a circle with radius R, if the polar coordinates of the centre are $C(R,\theta_0)$.

439. Write the polar equation of a cricle with radius R, if the polar coordinates of its centre are

1) $C(R,0)$; 2) $C(R,\pi)$;

3) $C\left(R,\frac{\pi}{2}\right)$; 4) $C\left(R,-\frac{\pi}{2}\right)$.

440. Determine the polar coordinates of the centre and find the radius of each of the following circles:

1) $\rho=4\cos\theta$; 2) $\rho=3\sin\theta$;

3) $\rho=-2\cos\theta$; 4) $\rho=-5\sin\theta$;

5) $\rho=6\cos\left(\frac{\pi}{3}-\theta\right)$; 6) $\rho=8\sin\left(\theta-\frac{\pi}{3}\right)$;

7) $\rho=8\sin\left(\frac{\pi}{3}-\theta\right)$.

441. In each of the following, a circle is represented by its polar equation:

1) $\rho=3\cos\theta$; 2) $\rho=-4\sin\theta$;

3) $\rho=\cos\theta-\sin\theta$.

Write the equation of each circle in a rectangular cartesian coordinate system whose positive x-axis coincides with the polar axis and whose origin coincides with the pole.

442. In each of the following, a circle is represented by its cartesian equation:

1) $x^2 + y^2 = x$; 2) $x^2 + y^2 = -3x$; 3) $x^2 + y^2 = 5y$;

4) $x^2 + y^2 = -y$; 5) $x^2 + y^2 = x + y$.

Write the equation of each circle in a polar coordinate system whose polar axis coincides with the positive x-axis and whose pole coincides with the origin.

443. Find the polar equation of the line tangent to the circle $\rho = R$ at the point $M_1(R, \theta_0)$.

18. The Ellipse

An ellipse is the locus of points, the sum of whose distances from two fixed points (called the foci) in the plane is a constant greater than the distance between the foci. The constant sum of the distances of an arbitrary point of an ellipse from its foci is generally denoted by $2a$. The foci of an ellipse are designated as F_1, F_2 and the distance between them as $2c$. By the definition of the ellipse, $2a > 2c$, or $a > c$.

Let there be given an ellipse. If the axes of a rectangular cartesian coordinate system are chosen so that the foci of the given ellipse are symmetrically situated on the x-axis with respect to the origin, then the equation of the given ellipse (referred to this coordinate system) has the form

$$\frac{x^2}{a^2} + \frac{y^2}{b^2} = 1, \tag{1}$$

where, $b = \sqrt{a^2 - c^2}$; obviously, $a > b$. An equation of the form (1) is called the canonical equation of an ellipse.

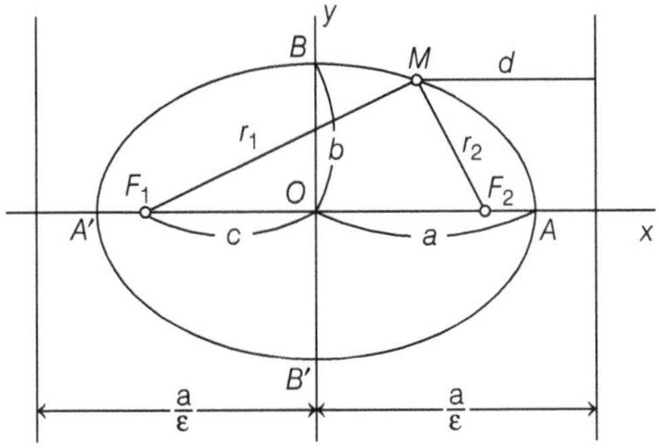

Fig. 12.

When the coordinate system is chosen as indicated above, then the coordinate axes are the axes of symmetry of the ellipse, and the origin is its centre of symmetry (Fig. 12). The axes of symmetry of an ellipse are referred to simply as its axes, and the centre of symmetry simply as the centre of the ellipse. The points where an ellipse cuts its axes are called its vertices. In Fig. 12, the vertices of the ellipse are the points A', A, B', and B. The term "axes of the ellipse" is often applied also to the segments $A'A = 2a$ and $B'B = 2b$; the segment $OA = a$ is then called the semi-major axis and the segment $OB = b$, the semi-minor axis of the ellipse.

If the foci of an ellipse are situated on the y-axis (symmetrically with respect to the origin), then the equation of the ellipse still is of the form (1), but in this case $b > a$; hence, if a is still to denote the semi-major axis, the letters a and b must be interchanged in equation (1). However, for convenient formulation of subsequent problems, we shall agree to let a always denote the semi-axis lying on the x-axis, and b the semi-axis lying on the y-axis, irrespective of whether a is greater or smaller than b. If $a = b$, equation (1) represents a circle, which is regarded as a special case of the ellipse.

The number $$\varepsilon = \frac{c}{a},$$

where a is the semi-major axis, is called the eccentricity of the ellipse. Clearly, $\varepsilon < 1$ (for a circle, $\varepsilon = 0$). Let $M(x, y)$ be an arbitrary point of an ellipse; then the segments $F_1M = r_1$ and $F_2M = r_2$ are called the focal radii of the point M (see Fig. 12). The focal radii can be calculated from the formulas

$$r_1 = a + \varepsilon x,\ r_2 = a - \varepsilon x.$$

In the case of an ellipse represented by equation (1), where $a > b$, the lines $x = -\frac{a}{\varepsilon}$, $x = \frac{a}{\varepsilon}$

(Fig. 12) are called the directrices of the ellipse. (If $b > a$, the directrices are determined by the equations $y = -\frac{b}{\varepsilon}$, $y = \frac{b}{\varepsilon}$).

Each of the directrices possesses the following property: If r is the distance from an arbitrary point of an ellipse to one of its foci, and d is the distance from the same point of the directrix.

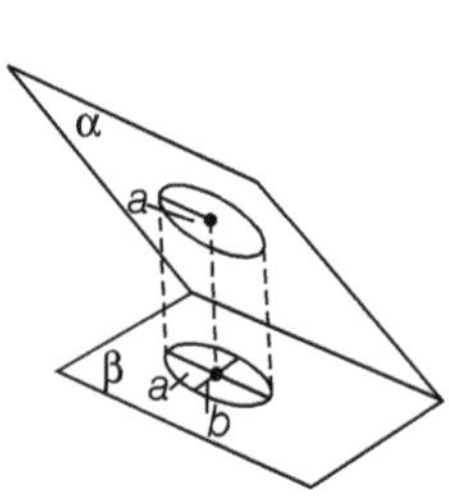

Fig. 13.

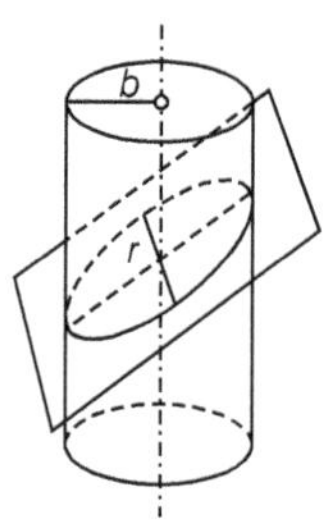

Fig.14.

associated with that focus, then the ratio $\frac{r}{d}$ is a constant equal to the eccentricity of the ellipse: $\frac{r}{d} = \varepsilon$.

Let two planes, α and β, make an acute angle ϕ, and let a circle of radius a lie in the plane α; then the projection of this circle on the plane β is an ellipse with semi-major axis a; the semi-minor axis b of the ellipse (Fig. 13) is given by the formula $b = a \cos \phi$.

Let a circular cylinder have a circle of radius b as its directing curve; then the section of this cylinder by a plane (making an acute angle ϕ with the axis of the cylinder) will be an ellipse with its semi-minor axis equal to b, the semi-major axis of the ellipse (Fig. 14) is determined by the formula

$$a = \frac{b}{\sin \phi}.$$

444. Find the equation of the ellipse whose foci are symmetrically situated on the x-axis with respect to the origin, and which satisfies the following conditions:

1) the semi-axes are equal to 5 and 2;
2) the major axis equals 10, and the distance between the foci $2c = 8$;
3) the minor axis equals 24, and the distance between the foci $2c = 10$;
4) the distance between the foci $2c = 6$, and the eccentricity $\varepsilon = \frac{3}{5}$;
5) the major axis equals 20, and the eccentricity $\varepsilon = \frac{3}{5}$;
6) the minor axis is 10, and the eccentricity $\varepsilon = \frac{12}{13}$;
7) the distance between the directrices is 5 and the distance between the foci $2c = 4$;
8) the major axis equals 8, and the distance between the directrices is 16;
9) the minor axis equals 6, and the distance between the directrices is 13;
10) the distance between the directrices is 32, and $\varepsilon = \frac{1}{2}$.

445. Write the equation of the ellipse whose foci are symmetrically situated on the y-axis with respect to the origin, and which satisfies the following conditions:

1) the semi-axes are equal to 7 and 2;
2) the major axis equals 10, and the distance between the foci $2c = 8$;
3) the distance between the foci $2c = 24$, and the eccentricity $\varepsilon = \frac{12}{13}$;

4) the minor axis is 16, and the eccentricity $\varepsilon = \frac{3}{5}$

5) the distance between the foci $2c = 6$ and the distance between the directrices is $16\frac{2}{3}$

6) the distance between the directrices is $10\frac{2}{3}$, and the eccentricity $\varepsilon = \frac{3}{4}$.

446. Determine the semi-axes of each of the following ellipses

1) $\frac{x^2}{16} + \frac{y^2}{9} = 1;$ 2) $\frac{x^2}{4} + y^2 = 1;$

3) $x^2 + 25y^2 = 25;$ 4) $x^2 + 5y^2 = 15;$

5) $4x^2 + 9y^2 = 25;$ 6) $9x^2 + 25y^2 = 1;$

7) $x^2 + 4y^2 = 1;$ 8) $16x^2 + y^2 = 16;$

9) $25x^2 + 9y^2 = 1;$ 10) $9x^2 + y^2 = 1.$

447. Given the ellipse $9x^2 + 25y^2 = 225$. Find :

1) the semi-axes; 2) the foci;

3) the eccentricity 4) the equations of the directrices.

448. Calculate the area of the quadrilateral, two of whose vertices lie at the foci of the ellipse $x^2 + 5y^2 = 20$, and whose other two vertices coincide with the ends of the minor axis of the ellipse.

449. Given the ellipse $9x^2 + 5y^2 = 45$. Find:

1) the semi-axes; 2) the foci; 3) the eccentricity;

4) the equations of the directrices.

450. Calculate the area of the quadrilateral, two of whose vertices lie at the foci of the ellipse $9x^2 + 5y^2 = 1$ and whose other two vertices coincide with the ends of the minor axis of the ellipse.

451. Calculate the distance from the focus $F(c, 0)$ of the ellipse $\frac{x^2}{a^2} + \frac{y^2}{b^2} = 1$ to the directrix associated with this focus (i.e, to the directrix lying on the same side of the centre as the focus F).

452. Using a pair of compasses alone, construct the foci of the ellipse $\frac{x^2}{16} + \frac{y^2}{9} = 1$ (assuming that the coordinate axes have been drawn and a unit segment chosen).

453. On the ellipse $\frac{x^2}{25} + \frac{y^2}{4} = 1$, find the points whose abscissa is equal to -3.

454. Determine which of the points $A_1(-2, 3)$, $A_2(2, -2)$, $A_3(2, -4)$, $A_4(-1, 3)$, $A_5(-4, -3)$, $A_6(3, -1)$, $A_7(3, -2)$, $A_8(2, 1)$, $A_9(0, 15)$ and $A_{10}(0, -16)$ lie inside, on, or outside the ellipse $8x^2 + 5y^2 = 77$.

455. Identify and plot the curves represented by the following equations:

1) $y = +\frac{3}{4}\sqrt{16 - x^2}$; 2) $y = -\frac{5}{3}\sqrt{9 - x^2}$;

3) $x = -\frac{2}{3}\sqrt{9 - y^2}$; 4) $x = +\frac{1}{7}\sqrt{49 - y^2}$.

456. The eccentricity ε of an ellipse is $\frac{2}{3}$, and one focal radius of a point M on this ellipse is equal to 10. Calculate the distance from the point M to the directrix associated with the focus in question.

457. The eccentricity ε of an ellipse is $\frac{2}{5}$, and the distance from a point M on the ellipse to one of the directrices is equal to 20. Calculate the distance from the point M to the focus associated with this directrix.

458. Given the point $M_1\left(2, -\frac{5}{3}\right)$ on the ellipse $\frac{x^2}{9} + \frac{y^2}{5} = 1$; write the equations of the straight lines along which the focal radii of the point M_1 lie.

459. Verify that the point $M_1(-4, 2.4)$ lies on the ellipse

$$\frac{x^2}{25} + \frac{y^2}{16} = 1,$$

and determine the focal radii of the point M_1.

460. The eccentricity ε of an ellipse is $\frac{1}{3}$, its centre is at the origin, and one of its foci is $F(-2, 0)$. Calculate the distance from a point M_1 of the ellipse to the directrix associated with the given focus, if the abscissa of M_1 is 2.

461. The eccentricity ε of an ellipse is $\frac{1}{2}$, its centre is at the origin and one of its directrices is given by the equation $x = 16$. Calculate the distance from a point M_1 of the ellipse to the focus associated with the given directrix, if the abscissa of M_1 is -4.

462. Determine those points of the ellipse $\frac{x^2}{100} + \frac{y^2}{36} = 1$ whose distance from the right-hand focus is equal to 14.

463. Determine those points of the ellipse $\frac{x^2}{16} + \frac{y^2}{7} = 1$ whose distance from the left-hand focus is equal to 2.5.

464. Through a focus of the ellipse $\frac{x^2}{25}+\frac{y^2}{15}=1$, a perpendicular is drawn to its major axis. Determine the distances from the foci to the points in which the perpendicualr cuts the ellipse.

465. Write the equation of the ellipse with its foci symmetrically situated on the x-axis with respect to the origin, given:

1) the point $M_1(-2\sqrt{5},2)$ of the ellipse and its semi-minor axis $b=3$;
2) the point $M_1(2,-2)$ of the ellipse and its semi-major axis $a=4$;
3) the points $M_1(4,-\sqrt{3})$ and $M_2(2\sqrt{2},3)$ of the ellipse;
4) the point $M_1(\sqrt{15},-1)$ of the ellipse and the distance between its foci $2c=8$;
5) the point $M_1\left(2,-\frac{5}{3}\right)$ of the ellipse and its eccentricity $\varepsilon=\frac{2}{3}$;
6) the point $M_1(8,12)$ of the ellipse and the distance $r_1=20$ of M_1 from the left-hand focus;
7) the point $M_1(-\sqrt{5},2)$ of the ellipse and the distance between its directrices equal to 10.

466. Determine the eccentricity ε of an ellipse, if:

1) the minor axis subtends an angle of 60° at each focus;
2) the line segment between the foci subtends a right angle at each of the vertices on the minor axis;
3) the distance between the directrices is three times the distance between the foci;
4) the segment of the perpendicular from the centre of the ellipse to a directrix is bisected at the corresponding vertex.

467. Through the focus F of an ellipse, a perpendicular is drawn to its major axis (Fig. 15). Determine that value of the eccentricity of the ellipse for which the line segments $\overline{AB}$ and $\overline{OC}$ will be parallel.

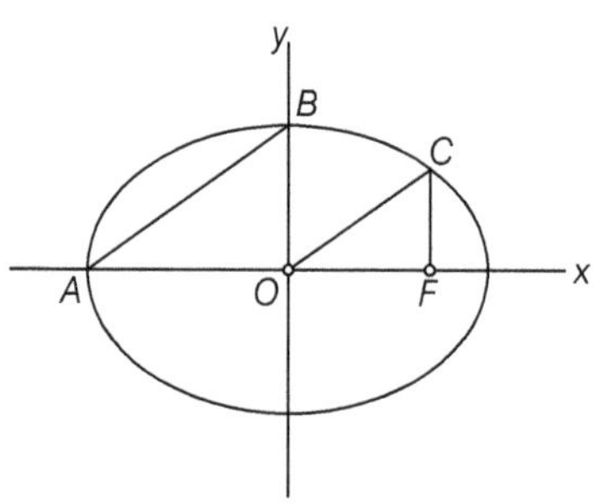

Fig. 15.

468. Find the equation of an ellipse with semi-axes a, b and centre $C(x_0, y_0)$, if the axes of symmetry of the ellipse are parallel to the coordinate axes.

469. An ellipse touches the x-axis at the point $A(3, 0)$, and the y-axis at the point $B(0, -4)$. Write the equation of the ellipse, given that its axes of symmetry are parallel to the coordinate axes.

470. The point $C(-3, 2)$ is the centre of an ellipse which touches both coordinate axes. Write the equation of the ellipse, given that its axes of symmetry are parallel to the coordinate axes.

471. Show that each of the following equations represents an ellipse, and find the coordinates of its centre C, the semi-axes, the eccentricity, and the equations of the directrices:

1) $5x^2 + 9y^2 - 30x + 18y + 9 = 0$;
2) $16x^2 + 25y^2 + 32x - 100y - 284 = 0$;
3) $4x^2 + 3y^2 - 8x + 12y - 32 = 0$.

472. Identify and plot the curves represented by the following equations:

1) $y = -7 + \frac{2}{5}\sqrt{16 + 6x - x^2}$; 2) $y = 1 - \frac{4}{3}\sqrt{-6x - x^2}$;

3) $x = -2\sqrt{-5 - 6y - y^2}$; 4) $x = -5 + \frac{2}{3}\sqrt{8 + 2y - y^2}$.

473. Write the equation of the ellipse satisfying the following conditions:

1) the major axis equals 26, and the foci are $F_1(-10, 0)$ and $F_2(14, 0)$;
2) the minor axis equals 2, and the foci are $F_1(-1, -1)$, $F_2(1, 1)$;
3) the foci are $F_1\left(-2, \frac{3}{2}\right)$, $F_2\left(2, -\frac{3}{2}\right)$, and the eccentricity $\varepsilon = \frac{\sqrt{2}}{2}$;
4) the foci are $F_1(1, 3)$, $F_2(3, 1)$, and the distance between the directrices is $12\sqrt{2}$.

474. Find the equation of an ellipse, given the eccentricity $\varepsilon = \frac{2}{3}$, one focus $F(2, 1)$, and the equation $x - 5 = 0$ of the directrix corresponding to this focus.

475. Find the equation of an ellipse, given the eccentricity $\varepsilon = \frac{1}{2}$, one focus $F(-4, 1)$, and the equation $y + 3 = 0$ of the directrix corresponding to this focus.

476. The point $A(-3, -5)$ lies on an ellipse which has a focus $F(-1, -4)$ and whose corresponding directrix is given by the equation $x - 2 = 0$. Write the equation of the ellipse.

477. Find the equation of an ellipse, given the eccentricity $\varepsilon = \frac{1}{2}$, one focus $F(3, 0)$, and the equation $x + y - 1 = 0$ of the directrix corresponding to this focus.

478. The point $M_1(2, -1)$ lies on an ellipse which has a focus $F(1, 0)$ and whose corresponding directrix is given by the equation

$$2x - y - 10 = 0.$$

Write the equation of the ellipse.

479. The point $M_1(3, -1)$ is an end point of the minor axis of an ellipse whose foci lie on the line $y + 6 = 0$. Write the equation of the ellipse, if its eccentricity $\varepsilon = \frac{\sqrt{2}}{2}$.

480. Find the points of intersection of the line $x + 2y - 7 = 0$ and the ellipse $x^2 + 4y^2 = 25$.

481. Find the points of intersection of the line $3x + 10y - 25 = 0$ and the ellipse $\frac{x^2}{25} + \frac{y^2}{4} = 1$.

482. Find the points of intersection of the line $3x - 4y - 40 = 0$ and the ellipse $\frac{x^2}{16} + \frac{y^2}{9} = 1$.

483. In each of the following, determine whether the given line cuts, touches, or fails to meet the given ellipse:

1) $2x - y - 3 = 0$, $\frac{x^2}{16} + \frac{y^2}{9} = 1$; 2) $2x + y - 10 = 0$, $\frac{x^2}{9} + \frac{y^2}{4} = 1$;

3) $3x + 2y - 20 = 0$, $\frac{x^2}{40} + \frac{y^2}{10} = 1$.

484. Determine the values of m for which the line $y = -x + m$:

1) cuts the ellipse $\frac{x^2}{20} + \frac{y^2}{5} = 1$;

2) touches the ellipse;

3) passes outside the ellipse.

485. Find the condition under which the line $y = kx + m$ touches the ellipse $\frac{x^2}{a^2} + \frac{y^2}{b^2} = 1$.

486. Write the equation of the tangent line to the ellipse $\frac{x^2}{a^2} + \frac{y^2}{b^2} = 1$ at the point $M_1(x_1, y_1)$.

487. Prove that the tangent lines to the ellipse $\frac{x^2}{a^2}+\frac{y^2}{b^2}=1$ at the end points of a diameter are parallel. (A diameter of an ellipse is defined as a chord passing through the centre.)

488. Write the equations of the tangent lines to the ellipse

$$\frac{x^2}{10}+\frac{2y^2}{5}=1$$

which are parallel to the line

$$3x+2y+7=0.$$

489. Write the equations of the tangent lines to the ellipse

$$x^2+4y^2=20$$

which are perpendicular to the line $2x-2y-13=0$.

490. Draw the tangent lines to the ellipse

$$\frac{x^2}{30}+\frac{y^2}{24}=1$$

which are parallel to the line

$$4x-2y+23=0,$$

and calculate the distance d between them.

491. On the ellipse

$$\frac{x^2}{18}+\frac{y^2}{8}=1,$$

find the point M_1 nearest to the line

$$2x-3y+25=0,$$

and calculate the distance d form M_1 to this line.

492. From the point $A\left(\frac{10}{3},\frac{5}{3}\right)$, tangent lines are drawn to the ellipse

$$\frac{x^2}{20}+\frac{y^2}{5}=1.$$

Write their equations.

493. From the point $C(10,-8)$, tangent lines are drawn to the ellipse

$$\frac{x^2}{25}+\frac{y^2}{16}=1.$$

Write the equation of the chord joining the points of contact.

494. From the point $P(-16,9)$, tangent lines are drawn to the ellipse

$$\frac{x^2}{4}+\frac{y^2}{3}=1.$$

Calculate the distance d from P to that chord of the ellipse which joins the points of contact.

495. An ellipse passes through the point $A(4,-1)$ and touches the line $x+4y-10=0$. Find the equation of this ellipse, given that its axes coincide with the coordinate axes.

496. Find the equation of the ellipse whose axes are coincident with the coordinate axes and which touches the two lines $3x-2y-20=0$, $x+6y-20=0$.

497. Prove that the product of the distances from the centre of an ellipse to the intersection point of any of its tangents with the focal axis, and to the foot of the perpendicular dropped from the point of contact to the focal axis,is a constant equal to the square of the length of the semi-major axis of the ellipse.

498. Prove that the product of the distances from the foci to any tangent to an ellipse is equal to the square of the length of the semi-minor axis.

499. The line $x-y-5=0$ touches an ellipse whose foci are at the points $F_1(-3,0)$ and $F_2(3,0)$. Write the equation of the ellipse.

500. Find the equation of an ellipse, if its foci are symmetrically situated on the x-axis with respect to the origin, $3x+10y-25=0$ is the equation of a tangent to the ellipse, and if its semi-minor axis $b=2$.

501. Prove that the tangent line to an ellipse at a point M makes equal angles with the focal radii F_1M, F_2M and passes externally to the angle F_1MF_2.

502. From the left-hand focus of the ellipse

$$\frac{x^2}{45}+\frac{y^2}{20}=1,$$

a ray of light is sent at an obtuse angle $\alpha\,(\tan\alpha=-2)$ to the axis Ox. Upon reaching the ellipse, the ray is reflected from it. Write the equation of the straight line along which the reflected ray travels.

503. Determine the points of intersection of the two ellipses

$$x^2+9y^2-45=0,\; x^2+9y^2-6x-27=0.$$

504. Verify that the two ellipses

$$n^2x^2+m^2y^2-m^2n^2=0,$$
$$m^2x^2+n^2y^2-m^2n^2=0\;(m\neq n)$$

intersect in four points which lie on a circle with centre at the origin, and determine the radius R of this circle.

505. Two planes, α and β, make an angle $\phi=30°$. Determine the semi-axes of the ellipse obtained by projecting a circle of radius $R=10$, lying in the plane α, on the plane β.

506. An ellipse whose semi-minor axis equals 6 is the projection of a circle of radius $R=12$. Find the angle ϕ between the planes in which the ellipse and the circle lie.

507. The directing curve of a circular cylinder is a circle of radius $R = 8$. Determine the semi-axes of the ellipse which is the section of this cylinder by a plane making an angle $\phi = 30°$ with the axis of the cylinder.

508. The directing curve of a circular cylinder is a circle of radius $R = \sqrt{3}$. Determine the angle which a cutting plane must make with the axis of the cylinder in order that the section should be an ellipse with semi-major axis $a = 2$.

509. A uniform compression (elongation) of the plane towards the x-axis is defined as the transformation of points in the plane (Fig. 16) which carries an arbitrary point $M(x, y)$ into a point $M'(x', y')$ such that

$$x' = x, y' = qy,$$

where $q > 0$ is a constant called the coefficient of compression.

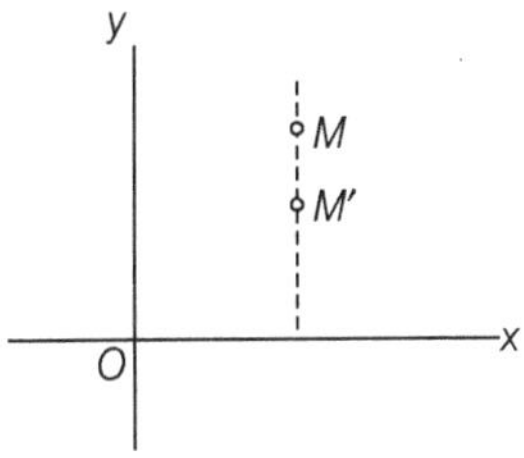

Fig. 16.

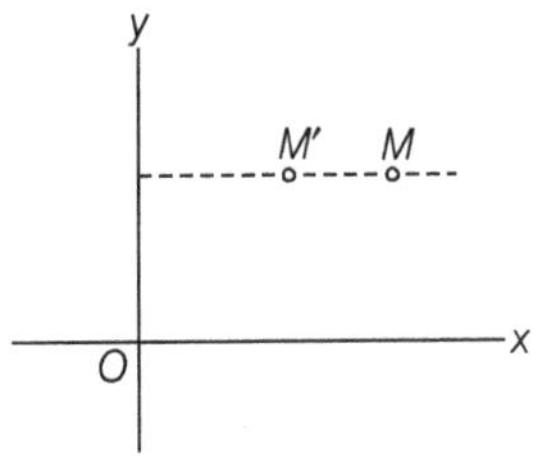

Fig. 17.

Similarly, a uniform compression of the plane towards the y-axis is determined (Fig. 17) by the equations

$$x' = qx, y' = y.$$

Find the curve into which the circle

$$x^2 + y^2 = 25$$

is transformed by a uniform compression of the plane towards the x-axis, if the coefficient of compression $q = \frac{4}{5}$.

510. Find the equation of the curve into which the ellipse $\frac{x^2}{16} + \frac{y^2}{9} = 1$ is transformed by a uniform compression of the plane towards the axis Oy, if the coefficient of compression is equal to $\frac{3}{4}$.

511. Find the equation of the curve into which the ellipse $\frac{x^2}{49} + \frac{y^2}{9} = 1$ is transformed by two consecutive uniform compressions of the plane (towards the x-axis and then towards the y-axis), if the respective coefficients of compression are $\frac{4}{3}$ and $\frac{6}{7}$.

512. Determine the coefficient q of the uniform compression of the plane towards the axis Ox, under which the ellipse $\frac{x^2}{36}+\frac{y^2}{9}=1$ is transformed into the ellipse $\frac{x^2}{36}+\frac{y^2}{16}=1$.

513. Determine the coefficient q of the uniform compression of the plane towards the axis Oy, under which the ellipse $\frac{x^2}{81}+\frac{y^2}{25}=1$ is transformed into the ellipse $\frac{x^2}{36}+\frac{y^2}{25}=1$.

514. Determine the coefficients q_1 and q_2 of the two consecutive uniform compressions of the plane towards the axes Ox and Oy, under which the ellipse $\frac{x^2}{25}+\frac{y^2}{9}=1$ is transformed into the circle $x^2+y^2=16$.

19. The Hyperbola

A hyperbola is the locus of points, the difference of whose distances from two fixed points (called the foci) in the plane is numerically a constant; this constant is usually denoted by $2a$. The foci of a hyperbola are designated as F_1, F_2, and the distance between them as $2c$. By the definition of the hyperbola, $2a < 2c$, or $a < c$.

Let there be given a hyperbola. If the axes of a rectangular cartesian coordinate system are chosen so that the foci of the given hyperbola are symmetrically situated on the x-axis with respect to the origin, then the equation of the hyperbola (referred to this coordinate system) has the form

$$\frac{x^2}{a^2}-\frac{y^2}{b^2}=1, \tag{1}$$

where $b=\sqrt{c^2-a^2}$. An equation of the form (1) is called the canonical equation of a hyperbola. When the coordinate system is chosen as indicated above, then the coordinate axes are the axes of symmetry of the hyperbola, and the origin is its centre of symmetry (Fig. 18). The axes of symmetry of a hyperbola are referred to simply as its axes, and the centre of symmetry as the centre of the hyperbola. A hyperbola intersects one of its axes; the points of intersection are called the vertices of the hyperbola. In Fig. 18. the vertices of the hyperbola are the points A' and A.

The rectangle with sides $2a$ and $2b$, which is symmetrical with respect to the axes of a hyperbola and tangent to it at the vertices, is called the fundamental rectangle of the hyperbola.

The term "axes of the hyperbola" is also applied to the line segments of lengths $2a$ and $2b$, which join the midpoints of the opposite

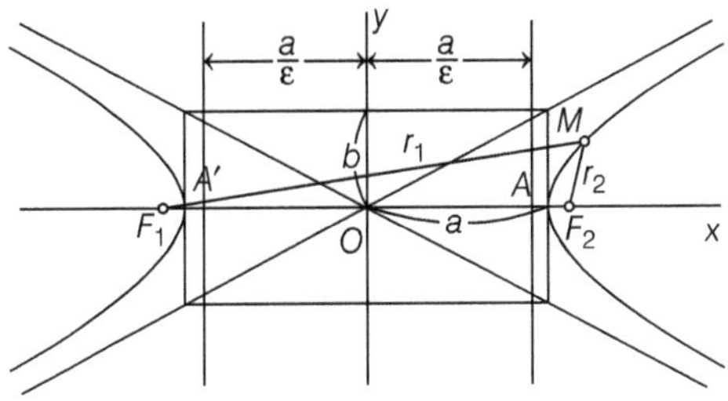

Fig. 18.

sides of the fundamental rectangle. The diagonals (extended indefinitely) of the fundamental rectangle are the asymptotes of the hyperbola; the equations of the asymptotes are

$$y = \frac{b}{a}x, \; x = -\frac{b}{a}x.$$

The equation
$$-\frac{x^2}{a^2} + \frac{y^2}{b^2} = 1 \qquad (2)$$

represents a hyperbola symmetrical with respect to the coordinate axes and having its foci on the y-axis, like equation (1), equation (2) is called the canonical equation of a hyperbola; in this case, the constant difference of the distances from an arbitrary point of a hyperbola to its foci is equal to $2b$.

Two hyperbolas represented, in the same coordinate system, by the equations

$$\frac{x^2}{a^2} - \frac{y^2}{b^2} = 1, \; -\frac{x^2}{a^2} + \frac{y^2}{b^2} = 1$$

are said to be conjugate.

A hyperbola with equal semi-axes ($a = b$) is called an equilateral hyperbola; its canonical equation is of the form

$x^2 - y^2 = a^2$ or $-x^2 + y^2 = a^2$.

The number

$$\varepsilon = \frac{c}{a},$$

where a is the distance from the centre of a hyperbola to its vertex, is called the eccentricity of the hyperbola. Obviously, $\varepsilon > 1$ for every hyperbola. Let $M(x, y)$ be an arbitrary point of a hyperbola; then the segments F_1M and F_2M (see Fig. 18) are called the focal radii of the point M. The focal radii of points of the right-hand branch of a hyperbola are calculated from the formulas

$$r_1 = \varepsilon x + a, \; r_2 = \varepsilon x - a;$$

the focal radii of points on the left-hand branch are calculated from the formulas

$$r_1 = -\varepsilon x - a,\ r_2 = -\varepsilon x + a.$$

In the case of a hyperbola represented by equation (1), the straight lines

$$x = -\frac{a}{\varepsilon},\ x = \frac{a}{\varepsilon}$$

are called the directrices of the hyperbola (see fig. 18). In the case of hyperbola represented by equation (2), the directrices are determined by the equations

$$y = -\frac{b}{\varepsilon},\ y = \frac{b}{\varepsilon}.$$

Each of the directrices possesses the following property: If r is the distance from an arbitrary point of a hyperbola to one of its foci, and d is the distance from the same point to the directrix associated with that focus, then the ratio $\frac{r}{d}$ is a constant equal to the eccentricity of the hyperbola: $\frac{r}{d} = \varepsilon$.

515. Write the equation of the hyperbola whose foci are symmetrically situated on the x-axis with respect to the origin, and which satisfies the following conditions:

1) the axes $2a = 10$ and $2b = 8$;
2) the distance between the foci $2c = 10$ and the axis $2b = 8$;
3) the distance between the foci $2c = 6$, and the eccentricity $\varepsilon = \frac{3}{2}$;
4) the axis $2a = 16$ and the eccentricity $\varepsilon = \frac{5}{4}$;
5) the equations of the asymptotes are $y = \pm\frac{4}{3}x$, and the distance between the foci $2c = 20$;
6) the distance between the directrices is equal to $22\frac{2}{13}$, and the distance between the foci $2c = 26$;
7) the distance between the directrices is $\frac{32}{5}$, and the axis $2b = 6$;
8) the distance between the directrices is $\frac{8}{3}$, and the eccentricity $\varepsilon = \frac{3}{2}$;
9) the equations of the asymptotes are $y = \pm\frac{3}{4}x$, and the distance between the directrices is $12\frac{4}{5}$.

516. Find the equation of the hyperbola whose foci are symmetrically situated on the y-axis with respect to the origin, and which satisfies the following conditions:

1) the semi-axes $a = 6$, $b = 18$ (the letter a denotes here the semi-axis of the hyperbola lying on the x-axis);
2) the distance between the foci $2c = 10$, and the eccentricity $\varepsilon = \frac{5}{3}$;
3) the equations of the asymptotes are $y = \pm\frac{12}{5}x$, and the distance between the vertices equals 48;
4) the distance between the directrices is $7\frac{1}{7}$, and the eccentricity $\varepsilon = \frac{7}{5}$;
5) the equations of the asymptotes are $y = \pm\frac{4}{3}x$, and the distance between the directrices is $6\frac{2}{5}$.

517. Determine the semi-axes a and b of each of the following hyperbolas:

1) $\frac{x^2}{9} - \frac{y^2}{4} = 1$;
2) $\frac{x^2}{16} - y^2 = 1$;
3) $x^2 - 4y^2 = 16$;
4) $x^2 - y^2 = 1$;
5) $4x^2 - 9y^2 = 25$;
6) $25x^2 - 16y^2 = 1$;
7) $9x^2 - 64y^2 = 1$.

518. Given the hyperbola $16x^2 - 9y^2 = 144$ Find:

1) its semi-axes a and b; 2) its foci; 3) the eccentricity; 4) the equations of the asymptotes; 5) the equations of the directrices.

519. Given the hyperbola $16x^2 - 9y^2 = -144$. Find:

1) the semi-axes a and b; 2) the foci; 3) the eccentricity; 4) the equations of the asymptotes; 5) the equations of the directrices.

520. Calculate the area of the triangle formed by the asymptotes of the hyperbola $\frac{x^2}{4} - \frac{y^2}{9} = 1$ and by the line $9x + 2y - 24 = 0$.

521. Identify and plot the curves represented by the following equations:

1) $y = +\frac{2}{3}\sqrt{x^2 - 9}$,
2) $y = -3\sqrt{x^2 + 1}$,
3) $x = -\frac{4}{3}\sqrt{y^2 + 9}$,
4) $y = +\frac{2}{5}\sqrt{x^2 + 25}$.

522. Given the point $M_1(10, -\sqrt{5})$ on the hyperbola $\frac{x^2}{80} - \frac{y^2}{20} = 1$. Find the equations of the straight lines along which the focal radii of M_1 lie.

523. Verify that the point $M_1\left(-5, \frac{9}{4}\right)$ lies on the hyperbola $\frac{x^2}{16} - \frac{y^2}{9} = 1$, and determine the focal radii of M_1.

524. The eccentricity ε of a hyperbola is 2, the distance (focal radius) of its point M from one focus is equal to 16. Calculate the distance of the point M from the directrix associated with this focus.

525. The eccentricity ε of a hyperbola is 3, the distance from a point M of the hyperbola to a directrix equals 4. Find the distance from the point M to the focus associated with this directrix.

526. The eccentricity ε of a hyperbola is 2, its centre lies at the origin, and one of the foci is $F(12, 0)$. Calculate the distance from a point M_1 of the hyperbola to the directrix associated with the given focus, if the abscissa of M_1 is 13.

527. the eccentricity ε of a hyperbola is $\frac{3}{2}$, its centre lies at the origin, and one of the directrices is represented by the equation $x = -8$. Calculate the distance from a point M_1 on the hyperbola to the focus associated with the given directrix, if the abscissa of M_1 is equal to 10.

528. Determine those points of the hyperbola $\frac{x^2}{64} - \frac{y^2}{36} = 1$ whose distance from the right-hand focus is 4.5.

529. Determine those points of the hyperbola $\frac{x^2}{9} - \frac{y^2}{16} = 1$ whose distance from the left-hand focus is 7.

530. Through the left-hand focus of the hyperbola $\frac{x^2}{144} - \frac{y^2}{25} = 1$, a perpendicular is drawn to the axis containing the vertices. Determine the distances from the foci to the points in which this perpendicular cuts the hyperbola.

531. Using a pair of compasses alone, construct the foci of the hyperbola $\frac{x^2}{16} - \frac{y^2}{25} = 1$ (assuming that the coordinate axes have been drawn and a unit segment chosen).

532. Write the equation of a hyperbola whose foci are symmetrically situated on the x-axis with respect to the origin, given:

1) the points $M_1(6, -1)$ and $M_2(-8, 2\sqrt{2})$ of the hyperbola;
2) the point $M_1(-5, 3)$ of the hyperbola and the eccentricity $\varepsilon = \sqrt{2}$;
3) the point $M_1\left(\frac{9}{2}, -1\right)$ of the hyperbola and the equations $y = \pm\frac{2}{3}x$ of the asymptotes;

4) the point $M_1\left(-3, \frac{5}{2}\right)$ of the hyperbola and the equations $x = \pm\frac{4}{3}$ of the directrices;

5) the equations $y = \pm\frac{3}{4}x$ of the asymptotes and the equations $x = \pm\frac{16}{5}$ of the directrices.

533. Determine the eccentricity of an equilateral hyperbola.

534. Determine the eccentricity of a hyperbola, if the line segment between its vertices subtends an angle of 60° at each focus of the conjugate hyperbola.

535. The foci of a hyperbola coincide with the foci of the ellipse $\frac{x^2}{25} + \frac{y^2}{9} = 1.$ Find the equation of the hyperbola if its eccentricity $\varepsilon = 2$.

536. Find the equation of the hyperbola whose foci lie at the vertices of the ellipse $\frac{x^2}{100} + \frac{y^2}{64} = 1$, and whose directrices pass through the foci of this ellipse.

537. Prove that the distance from a focus of the hyperbola $\frac{x^2}{a^2} - \frac{y^2}{b^2} = 1$ to its asymptote is equal to b.

538. Prove that the product of the distances of any point on the hyperbola $\frac{x^2}{a^2} - \frac{y^2}{b^2} = 1$ from its two asymptotes is a constant equal to $\frac{a^2b^2}{a^2 + b^2}$.

539. Prove that the area of a parallelogram formed by the asymptotes of the hyperbola $\frac{x^2}{a^2} - \frac{y^2}{b^2} = 1$

and by the straight lines drawn through any point of the hyperbola parallel to the asymptotes is a constant equal to $\frac{ab}{2}$.

540. Write the equation of a hyperbola, if its semi-axes are a and b, the centre is $C(x_0, y_0)$, and if the foci lie on a line:

1) parallel to the axis Ox; 2) parallel to the axis Oy.

541. Verify that each of the following equations represents a hyperbola, and find: the coordinates of its centre C, the semi-axes, the eccentricity, the equations of the asymptotes, and the equations of the directrices:

1) $16x^2 - 9y^2 - 64x - 54y - 161 = 0$;

2) $9x^2 - 16y^2 + 90x + 32y - 367 = 0$;

3) $16x^2 - 9y^2 - 64x - 18y + 199 = 0$.

542. Identify and plot the curves represented by the following equations:

1) $y = -1 + \frac{2}{3}\sqrt{x^2 - 4x - 5}$, 2) $y = 7 - \frac{3}{2}\sqrt{x^2 - 6x + 13}$,

3) $x = 9 - 2\sqrt{y^2 + 4y + 8}$, 4) $x = 5 - \frac{3}{4}\sqrt{y^2 + 4y - 12}$.

543. Write the equations of the hyperbola satisfying the following conditions:

1) the distance between its vertices is 24, and the foci are $F_1(-10, 2)$, $F_2(16, 2)$;
2) the foci are $F_1(3, 4)$, $F_2(-3, -4)$, and the distance between the directrices equals 3.6;
3) the angle between the asymptotes si 90°, and the foci are $F_1(4, -4)$, $F_2(-2, 2)$.

544. Find the equation of a hyperbola, given its eccentricity $\varepsilon = \frac{5}{4}$, one focus $F(5, 0)$, and the equation $5x - 16 = 0$ of the directrix associated with this focus.

545. Find the equation of a hyperbola, given its eccentricity $\varepsilon = \frac{13}{12}$, one focus $F(0, 13)$, and the equation $13y - 144 = 0$ of the directrix associated with this focus.

546. The point $A(-3, -5)$ lies on a hyperbola which has $F(-2, -3)$ as its focus and whose directrix corresponding to this focus is represented by the equation $x + 1 = 0$.

Find the equation of the hyperbola.

547. Find the equation of a hyperbola, given its eccentricity $\varepsilon = \sqrt{5}$, one focus $F(2, -3)$ and the equation

$$3x - y + 3 = 0$$

of the directrix corresponding to this focus.

548. The point $M_1(1, -2)$ lies on a hyperbola which has $F(-2, 2)$ as its focus and whose directrix corresponding to this focus is represented by the equation $2x - y - 1 = 0$.

Write the equation of the hyperbola.

549. Given the equation $x^2 - y^2 = a^2$ of an equilateral hyperbola. Find its equation in the new system when its asymptotes are taken as the coordinate axes.

550. In each of the following, show that the equation represents a hyperbola, find its centre, semi-axes, the equations of the asymptotes and draw the figure:

1) $xy = 18$; 2) $2xy - 9 = 0$; 3) $2xy + 25 = 0$.

551. Find the points of intersection of the line $2x - y - 10 = 0$ and the hyperbola $\frac{x^2}{20} - \frac{y^2}{5} = 1.$

552. Find the points of intersection of the line $4x - 3y - 16 = 0$ and the hyperbola $\frac{x^2}{25} - \frac{y^2}{16} = 1.$

553. Find the points of intersection of the line $2x - y + 1 = 0$ and the hyperbola $\frac{x^2}{9} - \frac{y^2}{4} = 0.$

554. In each of the following, determine whether the given line cuts, touches, or fails to meet the given hyperbola:

1) $x - y - 3 = 0, \quad \frac{x^2}{12} - \frac{y^2}{3} = 1;$

2) $x - 2y + 1 = 0, \quad \frac{x^2}{16} - \frac{y^2}{9} = 1;$

3) $7x - 5y = 0, \quad \frac{x^2}{25} - \frac{y^2}{16} = 1.$

555. Determine the values of m for which the line $y = \frac{5}{2}x + m$:

1) cuts the hyperbola $\frac{x^2}{9} - \frac{y^2}{36} = 1;$

2) touches this hyperbola;

3) passes outside the hyperbola.

556. Find the condition under which the line $y = kx + m$ touches the hyperbola $\frac{x^2}{a^2} - \frac{y^2}{b^2} = 1.$

557. Find the equation of the tangent line to the hyperbola $\frac{x^2}{a^2} - \frac{y^2}{b^2} = 1$ at its point $M_1(x_1, y_1)$.

558. Prove that the tangent lines to a hyperbola at the end points of a diameter of the hyperbola are parallel.

559. Write the equations of the tangent lines to the hyperbola $\frac{x^2}{20} - \frac{y^2}{5} = 1$ which are perpendicular to the line $4x + 3y - 7 = 0$.

560. Write the equations of the tangent lines to the hyperbola $\frac{x^2}{16} - \frac{y^2}{64} = 1$ which are parallel to the line $10x - 3y + 9 = 0$.

561. Draw the lines tangent to the hyperbola $\frac{x^2}{16} - \frac{y^2}{8} = -1$ and parallel to the line $2x + 4y - 5 = 0$; compute the distance d between them.

562. On the hyperbola $\frac{x^2}{24} - \frac{y^2}{18} = 1$, find the point M_1 nearest to the line $3x + 2y + 1 = 0$, and compute the distance d from M_1 to this line.

563. Find the equations of the tangent lines drawn from the point $A(-1, -7)$ to the hyperbola $x^2 - y^2 = 16$.

564. From the point $C(1, -10)$, tangent lines are drawn to the hyperbola $\frac{x^2}{8} - \frac{y^2}{32} = 1$. Write the equation of the chord joining the points of contact.

565. From the point $P(1, -5)$, tangent lines are drawn to the hyperbola $\frac{x^2}{3} - \frac{y^2}{5} = 1$. Calculate the distance d from P to that chord of the hyperbola which joins the points of contact.

566. A hyperbola passes through the point $A(\sqrt{6}, 3)$ and touches the line $9x + 2y - 15 = 0$. Write the equation of this hyperbola, given that its axes coincide with the coordinate axes.

567. Find the equation of the hyperbola whose axes are coincident with the coordinate axes and which touches the two lines $5x - 6y - 16 = 0$, $13x - 10y - 48 = 0$.

568. Show that the points of intersection of the ellipse $\frac{x^2}{20} + \frac{y^2}{5} = 1$ and the hyperbola $\frac{x^2}{12} - \frac{y^2}{3} = 1$ are the vertices of a rectangle and find the equations of the sides of this rectangle.

569. Given the hyperbola $\frac{x^2}{a^2} - \frac{y^2}{b^2} = 1$ and any tangent line to it; P is the point of intersection of this tangent line with the axis Ox, and Q is the projection of the point of contact on the axis Ox. Prove that $OP \cdot OQ = a^2$.

570. Prove that the foci of a hyperbola lie on opposite sides of any tangent line to the hyperbola.

571. Prove that the product of the distances from the foci to any tangent line to the hyperbola $\frac{x^2}{a^2} - \frac{y^2}{b^2} = 1$ is a constant equal to b^2.

572. The line

$$2x - y - 4 = 0$$

touches a hyperbola whose foci are at points $F_1(-3, 0)$ and $F_2(3, 0)$. Write the equation of the hyperbola.

573. Find the equation of a hyperbola, given that its foci are symmetrically situated on the x-axis with respect to the origin, that $15x + 16y - 36 = 0$ is the equation of a tangent line to this hyperbola and that the distance between its vertices $2a = 8$.

574. Prove that the tangent line to a hyperbola at a point M makes equal angles with the focal radii F_1M, F_2M and passes within the angle F_1MF_2.

575. From the right-hand focus of the hyperbola

$$\frac{x^2}{5} - \frac{y^2}{4} = 1,$$

a ray of light is sent at an angle $\alpha\left(\pi < \alpha < \frac{3}{2}\pi;\ \tan\alpha = 2\right)$ to the axis Ox. Upon reaching the hyperbola, the ray is reflected from it. Find the equation of the straight line along which the reflected ray travels.

576. Prove that an ellipse and a hyperbola which have the same foci intersect at right angles.

577. Find the equation of the curve into which the hyperbola $\frac{x^2}{16} - \frac{y^2}{9} = 1$ is transformed by a uniform compression of the plane towards the axis Ox, if the coefficient of compression is $\frac{4}{3}$.

Hint. See Problem 509.

578. Find the equation of the curve into which the hyperbola $\frac{x^2}{25} - \frac{y^2}{9} = 1$ is transformed by a uniform compression of the plane towards the axis Oy, if the coefficient of compression is $\frac{4}{5}$.

579. Find the equation of the curve into which the hyperbola $x^2 - y^2 = 9$ is transformed by two consecutive uniform compressions of the plane (towards the x-axis and then towards the y-axis), if the respective coefficients of compression are $\frac{2}{3}$ and $\frac{5}{3}$.

580. Determine the coefficient q of the uniform compression of the plane towards the axis Ox, under which the hyperbola $\frac{x^2}{25} - \frac{y^2}{36} = 1$ is transformed into the hyperbola $\frac{x^2}{25} - \frac{y^2}{16} = 1$.

581. Determine the coefficient q of the uniform compression of the plane towards the axis Oy, under which the hyperbola $\frac{x^2}{4} - \frac{y^2}{9} = 1$ is transformed into the hyperbola $\frac{x^2}{16} - \frac{y^2}{9} = 1$.

582. Determine the coefficients q_1 and q_2 of the two consecutive uniform compressions of the plane towards the axes Ox and Oy, under which the hyperbola $\frac{x^2}{49} - \frac{y^2}{16} = 1$ is transformed into the hyperbola $\frac{x^2}{25} - \frac{y^2}{64} = 1$.

20. The Parabola

A parabola is the locus of points whose distance from a fixed point (called the focus) in the plane is equal to their distance from a fixed straight line (called the directrix). The focus of a parabola is denoted by F, and the distance from the focus to the directrix by p. The number p is called the parameter of a parabola.

Let there be given a parabola; let the x-axis of the chosen rectangular certesian system of coordinates pass through the focus of

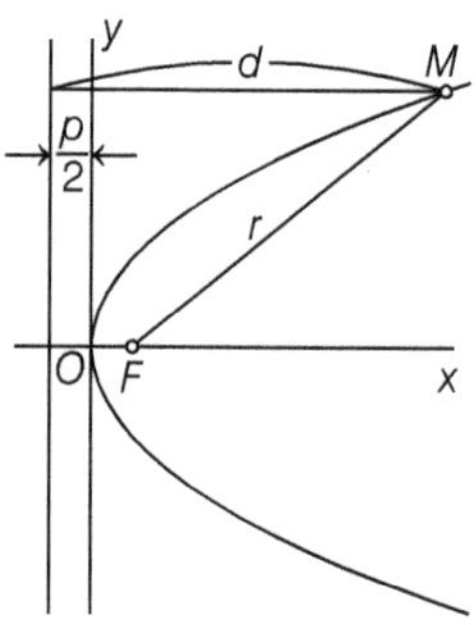

Fig. 19.

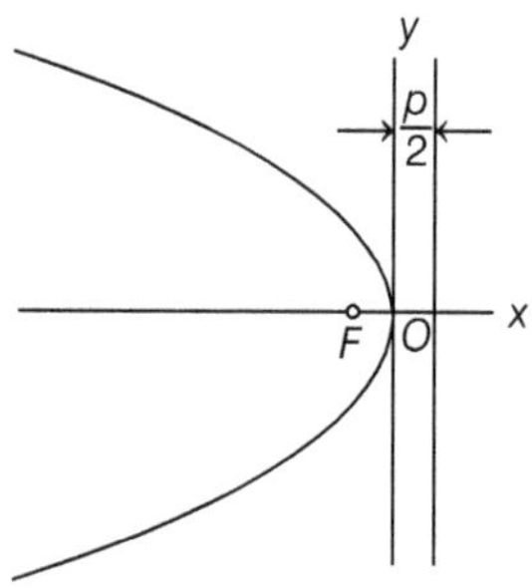

Fig. 20.

the given parabola perpendicular to the directrix and be directed from the directrix to the focus, and let the origin be placed midway between the focus and the directrix (Fig. 19). In this coordinate system, the given parabola will be represented by the equation

$$y^2 = 2px. \tag{1}$$

Equation (1) is called the canonical equation of the parabola. In the same coordinate system, the directrix of the given parabola has as its equation

$$x = -\frac{p}{2}.$$

The focal radius of an arbitrary point $M(x, y)$ of the parabola (that is, the length of the segment FM) may be calculated from the formula

$$r = x + \frac{p}{2}$$

A parabola has one axis of symmetry, which is called the axis of the parabola and which cuts the parabola in a single point; this point of intersection of the parabola and its axis is called the vertex of the parabola. When the coordinate system is chosen as indicated above, the axis of the parabola coincides with the x-axis, the vertex is at the origin, and the entire parabola lies in the right half-plane.

If the coordinate system is chosen so that the x-axis coincides with the axis of the parabola, the origin coincides with the vertex,

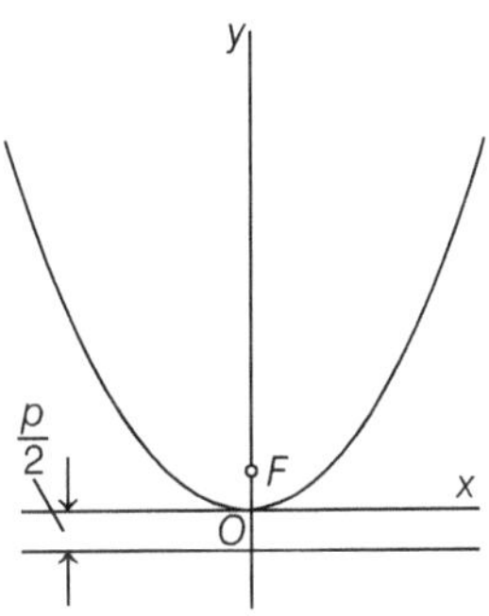

Fig. 21.

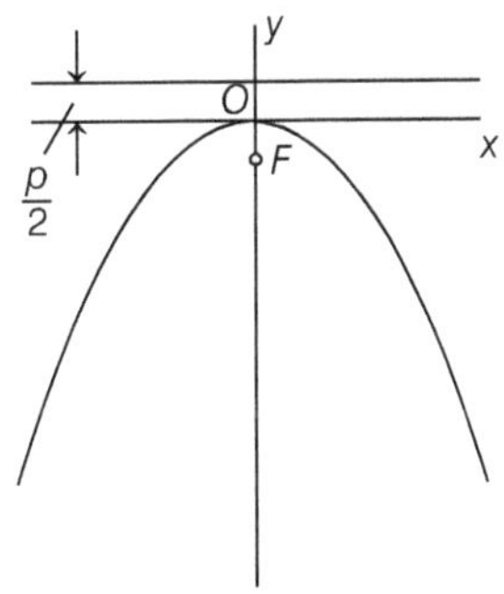

Fig. 22.

but the parabola lies in the left half-plane (Fig. 20), then the equation of the parabola is of the form

$$y^2 = -2px. \tag{2}$$

In the case when the vertex is at the origin and the axis of the parabola is coincident with the y-axis, the parabola will be represented by the equation

$$x^2 = 2py \tag{3}$$

if the parabola lies in the upper half-plane (Fig. 21), and by the equation

$$x^2 = -2py \tag{4}$$

if it lies in the lower half-plane (Fig. 22).

Each of equations. (2), (3), (4), as well as equation. (1), is referred to as the canonical equation of a parabola.

583. Find the equation of a parabola with vertex at the origin, if:

1) the parabola is symmetrically situated in the right half-plane with respect to the axis Ox, and its parameter $p=3$;
2) the parabola is symmetrically situated in the left half-plane with respect to the axis Ox, and its parameter $p=0.5$;
3) the parabola is symmetrically situated in the upper half-plane with respect to the axis Oy, and its parameter $p=\frac{1}{4}$;
4) the parabola is symmetrically situated in the lower half-plane with respect to the axis Oy, and its parameter $p=3$.

584. Determine the value of the parameter and the position (with respect to the coordinate axes) of the following parabolas:

1) $y^2=6x$; 2) $x^2=5y$; 3) $y^2=-4x$; 4) $x^2=-y$.

585. Find the equation of a parabola with vertex at the origin, if:

1) the parabola is symmetrically situated with respect to the axis Ox and passes through the point $A(9, 6)$;
2) the parabola is symmetrically situated with respect to the axis Ox and passes through the point $B(-1, 3)$;
3) the parabola is symmetrically situated with respect to the axis Oy and passes through the point $C(1, 1)$;
4) the parabola is symmetrical with respect to the axis Oy and passes through the point $D(4, -8)$.

586. A steel cable hangs from its two end supports; these supports are at the same level, and the distance between them is 20m. At a (horizontal) distance of 2 m from each support, the sag of the cable equals 14.4 cm. Determine the dip of the cable (i.e., the value of its sag midway between the supports) if the cable approximately forms an arc of a parabola.

587. Write the equation of the parabola which has the focus $F(0, -3)$ and passes through the origin, and whose axis coincides with the y-axis.

588. Identify and plot the curves represented by the following equations:

1) $y=+2\sqrt{x}$, 2) $y=+\sqrt{-x}$,
3) $y=-3\sqrt{-2x}$, 4) $y=-2\sqrt{x}$,
5) $x=+\sqrt{5y}$, 6) $x=-5\sqrt{-y}$,
7) $x=-\sqrt{3y}$, 8) $x=+4\sqrt{-y}$.

589. Find the focus F and the equation of the directrix of the parabola $y^2=24x$.

590. Calculate the focal radius of a point M of the parabola $y^2=20x$, if the abscissa of M is equal to 7.

591. Calculate the focal radius of a point M of the parabola $y^2 = 12x$, if the ordinate of M is equal to 6.

592. On the parabola $y^2 = 16x$, find the points whose focal radius is equal to 13.

593. Write the equation of a parabola, if its focus is $F(-7, 0)$ and the equation of the directrix is $x - 7 = 0$.

594. Write the equation of a parabola, if its vertex is at the point (α, β), its parameter equals p, its axis is parallel to the x-axis, and if the parabola opens:

1) in the positive direction of the x-axis;
2) in the negative direction of the x-axis.

595. Write the equation of a parabola, if its vertex is at the point (α, β), its parameter equals p, its axis is parallel to the y-axis, and if the parabola opens:

1) in the positive direction of the y-axis (that is upwards)
2) in the negative direction of the y-axis (that is downwards)

596. Show that each of the following equations represents a parabola, and find the coordinates of its vertex A, the value of the parameter p, and the equation of the directrix :

1) $y^2 = 4x - 8$, 2) $y^2 = 4 - 6x$,
3) $x^2 = 6y + 2$, 4) $x^2 = 2 - y$.

597. Show that each of the following equations represents a parabola, and find the coordinates of its vertex A and the value of the parameter p:

1) $y = \frac{1}{4}x^2 + x + 2$, 2. $y = 4x^2 - 8x + 7$,
3) $y = -\frac{1}{6}x^2 + 2x - 7$.

598. Show that each of the following equations represents a parabola, and find the coordinates of its vertex A and the value of the parameter p:

1) $x = 2y^2 - 12y + 14$, 2) $x = -\frac{1}{4}y^2 + y$,
3) $x = -y^2 + 2y - 1$.

599. Identify and plot the curves represented by the following equations:

1) $y = 3 - 4\sqrt{x - 1}$ 2) $x = -4 + 3\sqrt{y + 5}$
3) $x = 2 - \sqrt{6 - 2y}$ 4) $y = -5 + \sqrt{-3x - 21}$

600. Find the equation of the parabola with focus $F(7, 2)$ and directrix $x - 5 = 0$.

601. Find the equation of the parabola whose focus is $F(4, 3)$ and whose directrix is $y + 1 = 0$.

602. Find the equation of the parabola with focus $F(2, -1)$ and directrix $x - y - 1 = 0$.

603. Given the vertex $A(6, -3)$ of a parabola and the equation $3x - 5y + 1 = 0$ of its directrix. Find the focus F of the parabola.

604. Given the vertex $A(-2, -1)$ of a parabola and the equation $x + 2y - 1 = 0$ of its directrix. Write the equation of the parabola.

605. Determine the points of intersection of the line $x + y - 3 = 0$ and the parabola $x^2 = 4y$.

606. Determine the points of intersection of the line $3x + 4y - 12 = 0$ and the parabola $y^2 = -9x$.

607. Determine the points of intersection of the line $3x - 2y + 6 = 0$ and the parabola $y^2 = 6x$.

608. In each of the following, determine whether the given line cuts, touches, or fails to meet the given parabola:

1) $x - y + 2 = 0$, $y^2 = 8x$;
2) $8x + 3y - 15 = 0$, $x^2 = -3y$;
3) $5x - y - 15 = 0$, $y^2 = -5x$.

609. Determine the values of the slope k for which the line $y = kx + 2$:

1) cuts the parabola $y^2 = 4x$; 2) touches the parabola;
3) passes outside the parabola.

610. Find the condition for the line $y = kx + b$ to touch the parabola $y^2 = 2px$.

611. Prove that one, and only one, tangent line of a given slope $k \neq 0$ can be drawn to the parabola $y^2 = 2px$.

612. Write the equation of the tangent line to the parabola $y^2 = 2px$ at the point $M_1(x_1, y_1)$.

613. Write the equation of the line tangent to the parabola $y^2 = 8x$ and parallel to the line $2x + 2y - 3 = 0$.

614. Write the equation of the line tangent to the parabola $x^2 = 16y$ and perpendicular to the line $2x + 4y + 7 = 0$.

615. Draw the line tangent to the parabola $y^2 = 12x$ and parallel to the line $3x - 2y + 30 = 0$; calculate the distance d between this tangent and the given line.

616. On the parabola $y^2 = 64x$, find the point M_1 nearest to the line $4x + 3y - 14 = 0$ and calculate the distance d from M_1 to this line.

617. Write the equations of the tangent lines drawn to the parabola $y^2 = 36x$ from the point $A(2, 9)$.

618. A tangent line is drawn to the parabola $y^2 = 2px$. Prove that the vertex of the parabola lies midway between the point in which the tangent cuts the axis Ox and the projection of the point of contact on the axis Ox.

619. From the point $A(5, 9)$, tangent lines are drawn to the parabola $y^2 = 5x$. Find the equation of the chord joining the points of contact.

620. From the point $P(-3, 12)$, tangent lines are drawn to the parabola $y^2 = 10x$. Calculate the distance d from the point P to that chord of the parabola which joins the points of contact.

621. Determine the points of intersection of the ellipse $\frac{x^2}{100} + \frac{y^2}{225} = 1$ and the parabola $y^2 = 24x$.

622. Find the points of intersection of the hyperbola $\frac{x^2}{20} - \frac{y^2}{5} = -1$ and the parabola $y^2 = 3x$.

623. Find the points of intersection of the two parabolas

$y = x^2 - 2x + 1,\ x = y^2 - 6y + 7.$

624. Prove that the tangent line to a parabola at a point M makes equal angles with the focal radius of M and the ray drawn from M parallel to the axis of the parabola in the direction in which the parabola opens.

625. From the focus of the parable $y^2 = 12x$, a ray of light is sent at an acute angle $\alpha \left(\tan\alpha = \frac{3}{4}\right)$ to the axis Ox. Upon reaching the parabola, the ray is reflected from it. Find the equation of the straight line along which the reflected ray travels.

626. Prove that two parabolas which have a common axis and a common focus situated between their vertices intersect at right angles.

627. Prove that, if two parabolas with mutually perpendicular axes intersect in four points, these points of intersection lie on a circle.

21. The Polar Equation of the Ellipse, Hyperbola and Parabola

The polar equation of an ellipse, one branch of a hyperbola, and a parabola has the form (common to all the three curves):

$$\rho = \frac{p}{1 - \varepsilon\cos\theta}, \qquad \text{...(i)}$$

where ρ, θ are the polar coordinates of an arbitrary point of the curve, p is the focal parameter (half the focal chord perpendicular to the axis of the curve), and ε is the eccentricity (in the case of a parabola, $\varepsilon = 1$). The polar coordinate system is here chosen so that the pole coincides with the focus, and the polar axis is directed along the axis of the curve, away from the directrix associated with this focus.

628. Given the ellipse $\frac{x^2}{25} + \frac{y^2}{16} = 1$. Find its polar equation if the direction of the polar axis agrees with the positive direction of the x-axis and the pole is

1) at the left-hand focus of the ellipse;
2) at the right-hand focus of the ellipse.

629. Given the hyperbola $\frac{x^2}{16} - \frac{y^2}{9} = 1$. Find the polar equation of its right-hand branch, if the direction of the polar axis agrees with the positive direction of the x-axis and the pole is

1) at the right-hand focus of the hyperbola;
2) at the left-hand focus of the hyperbola.

630. Given the hyperbola $\frac{x^2}{25} - \frac{y^2}{144} = 1$. Find the polar equation of its left-hand branch, if the direction of the polar axis agrees with the positive direction of the x-axis and the pole is

1) at the left-hand focus of the hyperbola;
2) at the right-hand focus of the hyperbola.

631. Given the parabola $y^2 = 6x$. Find its polar equation if the direction of the polar axis agrees with the positive direction of the x-axis and the pole is at the focus of the parabola.

632. Identify the curves represented by the following polar equations

1) $\rho = \frac{5}{1 - \frac{1}{2}\cos\theta}$, 2) $\rho = \frac{6}{1 - \cos\theta}$, 3) $\rho = \frac{10}{1 - \frac{3}{2}\cos\theta}$,

4) $\rho = \frac{12}{2 - \cos\theta}$, 5) $\rho = \frac{5}{3 - 4\cos\theta}$, 6) $\rho = \frac{1}{3 - 3\cos\theta}$,

633. Show that the equation $\rho = \frac{144}{13 - 5\cos\theta}$ represents an ellipse, and find its semi-axes.

634. Show that the equation $\rho = \frac{18}{4 - 5\cos\theta}$ represents the right-hand branch of a hyperbola, and find its semi-axes.

635. Show that the equation $\rho = \frac{21}{5 - 2\cos\theta}$ represents an ellipse, and write the polar equations of its directrices.

636. Show that the equation $\rho = \dfrac{16}{3 - 5\cos\theta}$ represents the right-hand branch of a hyperbola, and write the polar equations of the directrices and asymptotes of this hyperbola.

637. On the ellipse $\rho = \dfrac{12}{3 - \sqrt{2}\cos\theta}$, find the points whose polar radius is 6.

638. On the hyperbola $\rho = \dfrac{15}{3 - 4\cos\theta}$, find the points whose polar radius is 3.

639. On the parabola $\rho = \dfrac{p}{1 - \cos\theta}$, find the points:

1) with the smallest polar radius;
2) with a polar radius equal to the parameter of the parabola.

640. Given the ellipse $\dfrac{x^2}{a^2} + \dfrac{y^2}{b^2} = 1$. Find its polar equation if the direction of the polar axis agrees with the positive direction of the x-axis and the pole is at the centre of the ellipse.

641. Given the hyperbola $\dfrac{x^2}{a^2} - \dfrac{y^2}{b^2} = 1$. Find its polar equation if the direction of the polar axis agrees with the positive direction of the x-axis and the pole is at the centre of the hyperbola.

642. Given the parabola $y^2 = 2px$. Find its polar equation if the direction of the polar axis agrees with the positive direction of the x-axis and the pole is at the vertex of the parabola.

22. Diameters of Curves of the Second Order

The midpoints of parallel chords of a second-order curve lie on a straight line (the proof will be found in a course in analytic geometry). This straight line is called a diameter of the second-order curve. The diameter bisecting a chord (and hence bisecting all chords parallel to this chord) is said to be conjugate to the chord (and to all chords parallel to this chord). All diameters of an ellipse or a hyperbola pass through the centre. If an ellipse is represented by the equation

$$\frac{x^2}{a^2} + \frac{y^2}{b^2} = 1, \tag{1}$$

then its diameter conjugate to chords of slope k is determined by the equation

$$y = -\frac{b^2}{a^2 k}x. \tag{2}$$

If a hyperbola is represented by the equation

$$\frac{x^2}{a^2} - \frac{y^2}{b^2} = 1, \tag{3}$$

then its diameter conjugate to chords of slope k is determined by the equation
$$y = \frac{b^2}{a^2k}x.$$

All diameters of a parabola are parallel to its axis. If a parabola is represented by the equation $y^2 = 2px$,

then its diameter conjugate to chords of slope k is determined by the equation $y = \frac{p}{k}$.

If one diameter of an ellipse or a hyperbola bisects the chords parallel to a second diameter, then this second diameter bisects the chords parallel to the first. Two such diameters are called conjugate diameters.

If k and k' are the slopes of two conjugate diameters of the ellipse, (1) then $kk' = -\frac{b^2}{a^2}$. (3)

If k and k' are the slopes of two conjugate diameters of the hyperbola (2), then $kk' = \frac{b^2}{a^2}$. (4)

Relations (3) and (4) are referred to as the conditions for conjugate diameters of an ellipse and a hyperbola, respectively.

That diameter of a second-order curve which is perpendicular to its conjugate chords is called a principal diameter of the curve.

643. Find the equation of that diameter of the ellipse $\frac{x^2}{25} + \frac{y^2}{16} = 1$ which bisects its chord lying on the line $2x - y - 3 = 0$.

644. Find the equation of that chord of the ellipse $\frac{x^2}{16} + \frac{y^2}{9} = 1$ which passes through the point $A(1, -2)$ and is bisected at this point.

645. Write the equations of two conjugate diameters of the ellipse $x^2 + 4y^2 = 1$, if one of them makes an angle of $45°$ with the axis Ox.

646. Write the equations of two conjugate diameters of the ellipse $4x^2 + 9y^2 = 1$, if one of them is parallel to the line $x + 2y - 5 = 0$.

647. Write the equations of two conjugate diameters of the ellipse $x^2 + 3y^2 = 1$, if one of them is perpendicualr to the line $3x + 2y - 7 = 0$.

648. An ellipse has been drawn; construct its centre by ruler and compass.

649. Prove that the axes of an ellipse form the only pair of its principal diameters.

650. By using the properties of conjugate diameters, prove that every diameter of a circle is its principal diameter.

651. (a) An isosceles triangle is inscribed in an ellipse, so that its vertex coincides with one of the vertices of the ellipse. Prove that the base of the triangle is parallel to one of the axes of the ellipse.

(b) Prove that the sides of a rectangle inscribed in an ellipse are parallel to the axes of the ellipse.

(c) An ellipse is drawn; construct its principal diameters by ruler and compass.

652. Prove that the chords of an ellipse which join its arbitrary point to the ends of any diameter of the ellipse are parallel to a pair of conjugate diameters of the ellipse.

653. (a) Prove that the sum of the squares of two conjugate semi-diameters of an ellipse is a constant (equal to the sum of the squares of its semi-axes).

(b) Prove that the area of a parallelogram constructed on two conjugate semi-diameters of an ellipse is a constant (equal to the area of the rectangle constructed on the semi-axes of the ellipse).

654. Write the equation of that diameter of the hyperbola $\frac{x^2}{5}-\frac{y^2}{4}=1$ which bisects its chord lying on the line $2x-y+3=0$

655. Given the hyperbola $\frac{x^2}{3}-\frac{y^2}{7}=1$. Find the equation of its chord passing through the point $A(3,-1)$ and bisected at that point.

656. Write the equations of two conjugate diameters of the hyperbola $x^2-4y^2=4$, if one of these diameters passes through the point $A(8,1)$.

657. Find the equations of two conjugate diameters of the hyperbola $\frac{x^2}{4}-\frac{y^2}{6}=1$ which form an angle of 45°.

658. A hyperbola is drawn; construct its centre by ruler and compass.

659. Prove that the axes of a hyperbola form the only pair of its principal diameters.

660. A hyperbola is drawn; construct its principal diameters by ruler and compass.

661. Find the equation of that diameter of the parabola $y^2=12x$ which bisects its chord lying on the line $3x+y-5=0$.

662. Given the parabola $y^2=20x$. Find the equation of its chord passing through the point $A(2,5)$ and bisected at that point.

663. Prove that the axis of a parabola is its only principal diameter.

664. A parabola is drawn; construct its principal diameter by ruler and compass.

5

SIMPLIFICATION OF THE GENERAL EQUATION OF A CURVE OF THE SECOND ORDER. THE EQUATIONS OF SOME CURVES ENCOUNTERED IN MATHEMATICS AND ITS APPLICATIONS.

23. The Centre of a Curve of the Second Order

A curve which is represented by an equation of the second degree in a cartesian coordinate system is called a curve of the second order. The general equation of the second degree (in two variables) is customarily written in the form

$$Ax^2 + 2Bxy + Cy^2 + 2Dx + 2Ey + F = 0 \qquad (1)$$

The centre of a curve is defined as that point in the plane with respect to which the points of the curve form symmetric pairs. Curves of the second order which have a single centre are called central curves.

A point $S(x_0, y_0)$ is the centre of a curve represented by equation (1) if, and only if, the coordinates of S satisfy the equations

$$\left.\begin{aligned} Ax_0 + By_0 + D &= 0, \\ Bx_0 + Cy_0 + E &= 0. \end{aligned}\right\} \qquad (2)$$

Denote by δ the determinant of the system:

$$\delta = \begin{vmatrix} A & B \\ B & C \end{vmatrix}.$$

The quantity δ is formed from the coefficients of the highest terms of equation (1) and is said to be the discriminant of the highest terms of this equation.

If $\delta \neq 0$, the system (2) is consistent and determinate, i.e., it has a unique solution. In this case, the coordinates of the centre can be determined from the formulas

$$x_0 = \frac{\begin{vmatrix} B & D \\ C & E \end{vmatrix}}{\begin{vmatrix} A & B \\ B & C \end{vmatrix}}, \qquad y_0 = \frac{\begin{vmatrix} D & A \\ E & B \end{vmatrix}}{\begin{vmatrix} A & B \\ B & C \end{vmatrix}}.$$

The inequality $\delta \neq 0$ is the condition for a central curve of the second order.

If $S(x_0, y_0)$ is the centre of second-order curve, then, after transforming the coordinates by the formulas $x = \tilde{x} + x_0, y = \tilde{y} + y_0$

(which corresponds to moving the origin to the centre of the curve), the equation of the curve will assume the form $A\tilde{x}^2 + 2B\tilde{x}\tilde{y} + C\tilde{y}^2 + \tilde{F} = 0$,

where A, B, C are the same as in the given equation (1), and $\tilde{F}$ is determined from the formula $\tilde{F} = Dx_0 + Ey_0 + F$.

In the case $\delta \neq 0$, we also have the following formula:

$$\tilde{F} = \frac{\Delta}{\delta}, \text{ where } \Delta = \begin{vmatrix} A & B & D \\ B & C & E \\ D & E & F \end{vmatrix}.$$

The determinant Δ is referred to as the discriminant of the left-hand member of the general equation of the second degree.

665. In each of the following, determine whether the given curve is a central curve (that is, has a single centre) has no centre, or has infinitely many centres:

1) $3x^2 - 4xy - 2y^2 + 3x - 12y - 7 = 0$;
2) $4x^2 + 5xy + 3y^2 - x + 9y - 12 = 0$;
3) $4x^2 - 4xy + y^2 - 6x + 8y + 13 = 0$;
4) $4x^2 - 4xy + y^2 - 12x + 6y - 11 = 0$;
5) $x^2 - 2xy + 4y^2 + 5x - 7y + 12 = 0$;
6) $x^2 - 2xy + y^2 - 6x + 6y - 3 = 0$;
7) $4x^2 - 20xy + 25y^2 - 14x + 2y - 15 = 0$;
8) $4x^2 - 6xy -`9y^2 + 3x - 7y + 12 = 0$.

666. Show that each of the following curves is a central curve, and find the coordinates of its centre:

1) $3x^2 + 5xy + y^2 - 8x - 11y - 7 = 0$;
2) $5x^2 + 4xy + 2y^2 + 20x + 20y - 18 = 0$;

3) $9x^2 - 4xy - 7y^2 - 12 = 0$;
4) $2x^2 - 6xy + 5y^2 + 22x - 36y + 11 = 0$.

667. Show that each of the following curves has an infinite number of centres, and find the equation of the locus of its centres:
1) $x^2 - 6xy + 9y^2 - 12x + 36y + 20 = 0$;
2) $4x^2 + 4xy + y^2 - 8x - 4y - 21 = 0$;
3) $25x^2 - 10xy + y^2 + 40x - 8y + 7 = 0$.

668. In each of the following, show that the given equation represents a central curve, and transform the equation by moving the origin to the centre of the curve:
1) $3x^2 - 6xy + 2y^2 - 4x + 2y + 1 = 0$;
2) $6x^2 + 4xy + y^2 + 4x - 2y + 2 = 0$;
3) $4x^2 + 6xy + y^2 - 10x - 10 = 0$;
4) $4x^2 + 2xy + 6y^2 + 6x - 10y + 9 = 0$.

669. Find the values of m and n for which the equation $mx^2 + 12xy + 9y^2 + 4x + ny - 13 = 0$ represents:
1) a central curve;
2) a curve having no centre;
3) a curve having an infinite number of centres.

670. Given the curve $4x^2 - 4xy + y^2 + 6x + 1 = 0$. Determine the values of the slope k for which the line $y = kx$
1) intersects the given curve in a single point;
2) touches the curve;
3) intersects the curve in two points;
4) has no points in common with the curve.

671. Write the equation of the second-order curve which has its centre at the origin, passes through the point $M(6, -2)$ and touches the line $x - 2 = 0$ at the point $N(2, 0)$.

672. The point $P(1, -2)$ is the centre of a second order curve which passes through the point $Q(0, -3)$ and touches the axis Ox at the origin. Find the equation of the curve.

24. Reducing the Equation of a Central Curve of the Second Order to Its Simplest Form

Let there be given the equation

$$Ax^2 + 2Bxy + Cy^2 + 2Dx + 2Ey + F = 0, \qquad (1)$$

which represents a central curve of the second order ($\delta = AC - B^2 \neq 0$). After moving the origin to the centre $S(x_0, y_0)$ of the curve and transforming equation (1) by the formulas

$$x = \tilde{x} + x_0, \quad y = \tilde{y} + y_0,$$

we obtain

$$A\tilde{x}^2 + 2B\tilde{x}\tilde{y} + C\tilde{y}^2 + \tilde{F} = 0. \tag{2}$$

To determine $\tilde{F}$, we can use the formula

$$\tilde{F} = Dx_0 + Ey_0 + F$$

or

$$\tilde{F} = \frac{\Delta}{\delta}.$$

A further simplification of equation (2) is achieved by the coordinate transformation

$$\left.\begin{aligned} \tilde{x} &= x'\cos\alpha - y'\sin\alpha, \\ \tilde{y} &= x'\sin\alpha + y'\cos\alpha, \end{aligned}\right\} \tag{3}$$

which corresponds to a rotation of the axes through an angle α. If the angle α is chosen so that

$$B\tan^2\alpha - (C - A)\tan\alpha - B = 0, \tag{4}$$

then the equation of the curve in the new coordinates will take the form

$$A'x'^2 + C'y'^2 + \tilde{F} = 0, \tag{5}$$

where $A' \neq 0, C' \neq 0$.

Note. Equation (4) permits us to find $\tan\alpha$, whereas formulas (3) contain $\sin\alpha$ and $\cos\alpha$. But $\sin\alpha$ and $\cos\alpha$ can be determined from $\tan\alpha$ by using the trigonometric formulas

$$\sin\alpha = \frac{\tan\alpha}{\pm\sqrt{1+\tan^2\alpha}}, \quad \cos\alpha = \frac{1}{\pm\sqrt{1+\tan^2\alpha}},$$

The coefficients of equations (1) and (5) are connected by the important relations $A'C' = AC - B^2$, $A' + C' = A + C$,

which enable us to determine the coefficients A' and C' without transforming the coordinates.

An equation of the second degree is said to be of the elliptic type if $\delta > 0$, of the hyperbolic type if $\delta < 0$, and of the parablic type if $\delta = 0$.

The equation of a central curve can only be of the elliptic or the hyperbolic type.

Every equation of the elliptic type represents either an ordinary ellipse, or a degenerate ellipse (that is, a single point), or else an imaginary ellipse (that is, no geometric object at all). Every equation of the hyperbolic type represents either an ordinary hyperbola or a degenerate hyperbola (that is, a pair of intersecting straight lines).

673. In each of the following, determine the type of the given equation*; reduce the equation to its simplest form by a translation of the coordinate axes; determine the geometric object represented by the equation and draw this object, showing both the old and the new coordinate axes.

1) $4x^2+9y^2-40x+36y+100=0$; 2) $9x^2-16y^2-54x-64y-127=0$;
3) $9x^2+4y^2+18x-8y+49=0$; 4) $4x^2-y^2+8x-2y+3=0$;
5) $2x^2+3y^2+8x-6y+11=0$.

674. In each of the following, reduce the given equation to its simplest form; determine the type of the equation; determine the geometric object represented by the equation and draw this object, showing both the old and the new coordinate axes.

1) $32x^2+52xy-7y^2+180=0$; 2) $5x^2-6xy+5y^2-32=0$;
3) $17x^2-12xy+8y^2=0$; 4) $5x^2+24xy-5y^2=0$;
5) $5x^2-6xy+5y^2+8=0$.

675. Determine the type of each of the following equations by calculating the discriminant of its highest terms:

1) $2x^2+10xy+12y^2-7x+18y-15=0$;
2) $3x^2-8xy+7y^2+8x-15y+20=0$;
3) $25x^2-20xy+4y^2-12x+20y-17=0$;
4) $5x^2+14xy+11y^2+12x-7y+19=0$;
5) $x^2-4xy+4y^2+7x-12=0$; 6) $3x^2-2xy-3y^2+12y-15=0$.

676. In each of the following, reduce the given equation to the canonical form; determine the type of the equation and the geometric object represented by the equation; draw this geometric object, showing the original, auxiliary and new coordinate axes.

1) $3x^2+10xy+3y^2-2x-14y-13=0$;
2) $25x^2-14xy+25y^2+64x-64y-224=0$;
3) $4xy+3y^2+16x+12y-36=0$;
4) $7x^2+6xy-y^2+28x+12y+28=0$;
5) $19x^2+6xy+11y^2+38x+6y+29=0$;
6) $5x^2-2xy+5y^2-4x+20y+20=0$.

* That is, determine whether the equation is of the elliptic, hyperbolic, or parabolic type.

677. The same as Problem 676 for the following:

1) $14x^2 + 24xy + 21y^2 - 4x + 18y - 139 = 0$;
2) $11x^2 - 20xy - 4y^2 - 20x - 8y + 1 = 0$;
3) $7x^2 + 60xy + 32y^2 - 14x - 60y + 7 = 0$;
4) $50x^2 - 8xy + 35y^2 + 100x - 8y + 67 = 0$;
5) $41x^2 + 24xy + 34y^2 + 34x - 112y + 129 = 0$;
6) $29x^2 - 24xy + 36y^2 + 82x - 96y - 91 = 0$;
7) $4x^2 + 24xy + 11y^2 + 64x + 42y + 51 = 0$;
8) $41x^2 + 24xy + 9y^2 + 24x + 18y - 36 = 0$.

678. Without transforming the coordinates, show that each of the following equations represents an ellipse, and find the values of its semi-axes:

1) $41x^2 + 24xy + 9y^2 + 24x + 18y - 36 = 0$;
2) $8x^2 + 4xy + 5y^2 + 16x + 4y - 28 = 0$;
3) $13x^2 + 18xy + 37y^2 - 26x - 18y + 3 = 0$;
4) $13x^2 + 10xy + 13y^2 + 46x + 62y + 13 = 0$.

679. Without transforming the coordinates, show that each of the following equations represents a single point (a degenerate ellipse) and find its coordinates:

1) $5x^2 - 6xy + 2y^2 - 2x + 2 = 0$;
2) $x^2 + 2xy + 2y^2 + 6y + 9 = 0$;
3) $5x^2 + 4xy + y^2 - 6x - 2y + 2 = 0$;
4) $x^2 - 6xy + 10y^2 + 10x - 32y + 26 = 0$.

680. Without transforming the coordinates, show that each of the following equations represents a hyperbola and find the values of its semi-axes:

1) $4x^2 + 24xy + 11y^2 + 64x + 42y + 51 = 0$;
2) $12x^2 + 26xy + 12y^2 - 52x - 48y + 73 = 0$;
3) $3x^2 + 4xy - 12x + 16 = 0$;
4) $x^2 - 6xy - 7y^2 + 10x - 30y + 23 = 0$.

681. Without transforming the coordinates, show that each of the following equations represents a pair of intersecting straight lines (a degenerate hyperbola) and find their equations:

1) $3x^2 + 4xy + y^2 - 2x - 1 = 0$;
2) $x^2 - 6xy + 8y^2 - 4y - 4 = 0$;
3) $x^2 - 4xy + 3y^2 = 0$;
4) $x^2 + 4xy + 3y^2 - 6x - 12y + 9 = 0$.

682. Without transforming the coordinates, determine the geometric objects represented by the following equations:

1) $8x^2 - 12xy + 17y^2 + 16x - 12y + 3 = 0$;
2) $17x^2 - 18xy - 7y^2 + 34x - 18y + 7 = 0$;
3) $2x^2 + 3xy - 2y^2 + 5x + 10y = 0$;
4) $6x^2 - 6xy + 9y^2 - 4x + 18y + 14 = 0$;
5) $5x^2 - 2xy + 5y^2 - 4x + 20y + 20 = 0$.

683. Prove that, for every elliptic equation, neither of the coefficients A and B can vanish, and that these coefficients agree in sign.

684. Prove that an elliptic ($\delta > 0$) equation of the second degree represents an ellipse if, and only if, A and Δ differ in sign.

685. Prove that an elliptic ($\delta > 0$) equation of the second degree represents an imaginary ellipse if, and only if, A and δ agree in sign.

686. Prove that an elliptic ($\delta > 0$) equation of the second degree represents a degenerate ellipse (a point) if, and only if, $\Delta = 0$.

687. Prove that a hyperbolic ($\delta < 0$) equation of the second degree represents a hyperbola if, and only if, $\Delta \neq 0$.

688. Prove that a hyperbolic ($\delta < 0$) equation of the second degree represents a degenerate hyperbola (a pair of intersecting lines) if, and only if, $\Delta = 0$.

25. Reducing a Parabolic Equation to Its Simplest Form

Let the equation

$$Ax^2 + 2Bxy + Cy^2 + 2Dx + 2Ey + F = 0 \tag{1}$$

be of the parabolic type, that is, let it satisfy the condition

$\delta = AC - B^2 = 0.$

In this case, the curve represented by equation (1) either has no centre, or else has an infinite number of centres. It is advisable to begin the simplification of a parabolic equation by rotating the coordinate axes, this means that equation (1) should first be transformed by using the formulas

$$\left.\begin{aligned} x &= x' \cos\alpha - y' \sin\alpha, \\ y &= x' \sin\alpha + y' \cos\alpha. \end{aligned}\right\} \tag{2}$$

The angle α is found from the equation

$$B\tan^2\alpha - (C - A)\tan\alpha - B = 0; \tag{3}$$

then, in the new coordinates, equation (1) reduces to the form

$$A' x'^2 + 2D' x' + 2E' y' + F = 0, \qquad (4)$$

where $A' \neq 0$, or to the form

$$C' y'^2 + 2D' x' + 2E' y' + F = 0, \qquad (5)$$

where $C' \neq 0$.

Equations (4) and (5) are further simplified by a translation of the (rotated) axes.

689. In each of the following, show that the given equation is of the parabolic type; reduce the equation to its simplest form; determine the geometric object represented by the equation; draw this geometric object, showing the original, auxiliary and new coordinate axes.

1) $9x^2 - 24xy + 16y^2 - 20x + 110y - 50 = 0$;
2) $9x^2 + 12xy + 4y^2 - 24x - 16y + 3 = 0$;
3) $16x^2 - 24xy + 9y^2 - 160x + 120y + 425 = 0$.

690. The same as Problem 689 for the following:

(1) $9x^2 + 24xy + 16y^2 - 18x + 226y + 209 = 0$;
(2) $x^2 - 2xy + y^2 - 12x + 12y - 14 = 0$;
(3) $4x^2 + 12xy + 9y^2 - 4x - 6y + 1 = 0$.

691. Prove that, for every parabolic equation, the coefficients A and C cannot differ in sign and are not both zero.

692. Prove that every parabolic equation can be written in the form

$(\alpha x + \beta y)^2 + 2Dx + 2Ey + F = 0$.

Prove also that elliptic and parabolic equations cannot be put in this form.

693. In each of the following, show that the given equation is of the parabolic type and write it in the form indicated in Problem 692.

1) $x^2 + 4xy + 4y^2 + 4x + y - 15 = 0$;
2) $9x^2 - 6xy + y^2 - x + 2y - 14 = 0$;
3) $25x^2 - 20xy + 4y^2 + 3x -`y + 11 = 0$;
4) $16x^2 + 16xy + 4y^2 - 5x + 7y = 0$;
5) $9x^2 -`42xy + 49y^2 + 3x - 2y - 24 = 0$.

694. Prove that, if an equation of the second degree is a parabolic equation written in the form $(\alpha x + \beta y)^2 + 2Dx + 2Ey + F = 0$, then the discriminant of its left-hand member is determined by the formula $\Delta = -(D\beta - E\alpha)^2$.

695. Prove that the parabolic equation $(\alpha x + \beta y)^2 + 2Dx + 2Ey + F = 0$ can be reduced, by means of the transformation

$$x = x' \cos\theta - y' \sin\theta,\ y = x' \sin\theta + y' \cos\theta, \qquad \tan\theta = -\frac{\alpha}{\beta},$$

to the form $C' y'^2 + 2D' x' + 2E' y' + F' = 0$,

where $C' = \alpha^2 + \beta^2, D' = \pm \sqrt{\frac{-\Delta}{\alpha^2 + \beta^2}}$,

Δ being the discriminant of the left-hand member of the given equation.

696. Prove that a parabolic equation represents a parabola if, and only if, $\Delta \neq 0$. Prove that, in this case, the parameter of the parabola is determined by the formula $P = \sqrt{\frac{-\Delta}{(A + C)^2}}$.

697. Without transforming the coordinates, show that each of the following equations represents a parabola, and find the parameter of the parabola:

1) $9x^2 + 24xy + 16y^2 - 120x + 90y = 0$;

2) $9x^2 - 24xy + 16y^2 - 54x - 178y + 181 = 0$;

3) $x^2 - 2xy + y^2 + 6x - 14y + 29 = 0$;

4) $9x^2 - 6xy + y^2 - 50x + 50y - 275 = 0$.

698. Prove that an equation of the second degree is the equation of a degenerate curve if, and only if, $\Delta = 0$.

699. Without transforming the coordinates, show that each of the following equations represents a pair of parallel straight lines, and find their equations

1) $4x^2 + 4xy + y^2 - 12x - 6y + 5 = 0$;

2) $4x^2 - 12xy + 9y^2 + 20x - 30y - 11 = 0$;

3) $25x^2 - 10xy + y^2 + 10x - 2y - 15 = 0$.

700. Without transforming the coordinates, show that each of the following equations represents a straight line (a pair of coinciding lines), and find the equation of the line

1) $x^2 - 6xy + 9y^2 + 4x - 12y + 4 = 0$

2) $9x^2 + 30xy + 25y^2 + 42x + 70y + 49 = 0$

3) $16x^2 - 16xy + 4y^2 - 72x + 36y + 81 = 0$

26. The Equations of Some Curves Encountered in Mathematics and Its Applications

701. Find the equation of the locus of points, the product of whose distances from the two given points

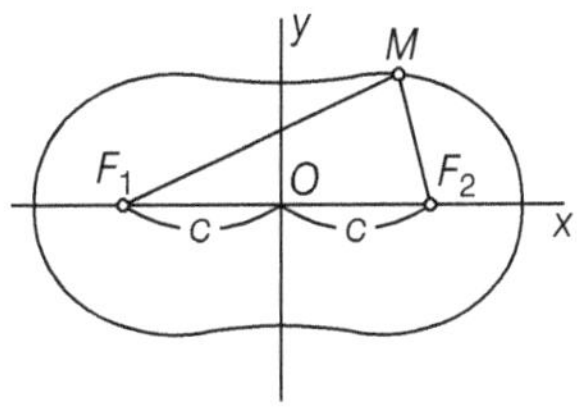

Fig. 23.

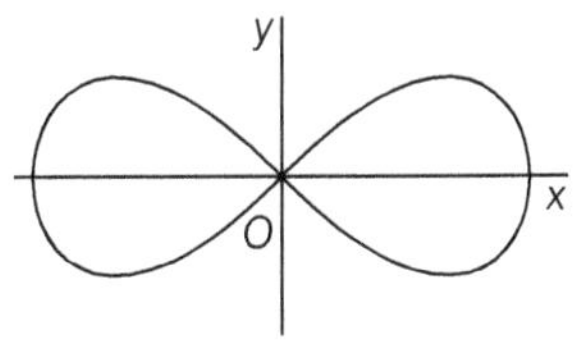

Fig.24.

$F_1(-c, 0)$ and $F_2(c, 0)$ is a constant equal to a^2. This locus is called the oval of Cassini (Fig. 23).

702. Find the equation of the locus of points, the product of whose distances from the two given points $F_1(-a, 0)$ and $F_2(a, 0)$ is a constant equal to a^2. This locus is called the *lemniscate* (Fig 24). (Find the equation of the lemniscate first directly, and then by considering it as a special form of the oval of Cassini.) Find also the polar equation of the lemniscate when the polar axis coincides with the positive x-axis and the pole is at the origin.

703. Find the equation of the locus of the feet of perpendiculars dropped from the origin to straight lines, each of which forms with the coordinate axes a triangle of area S.

Hint. Write first the equation in polar coordinates, placing the pole at the origin and making the polar axis coincide with the positive x-axis.

704. Prove that the locus of Problem 703 is a lemniscate (see Problem 702).

Hint. Rotate the coordinate axes through an angle of 45°.

705. A ray a, whose initial position coincides with the polar axis, revolves about the pole O with constant angular speed ω. In the given polar coordinate system, find the equation of the path traced by a point M which starts from O and moves along the ray a with uniform speed v. (The required locus is the spiral of Archimedes, Fig. 25.)

706. Given the line $x = 2r$ and the circle of radius r which passes through the origin O and touches this line. A ray drawn from O cuts the given circle at B and the given line at C (Fig. 26); a segment $OM - BC$ is laid off on this ray. As the ray revolves, the length of the segment OM varies and the point M describes a curve called the *cissoid*. Write the equation of the cissoid.

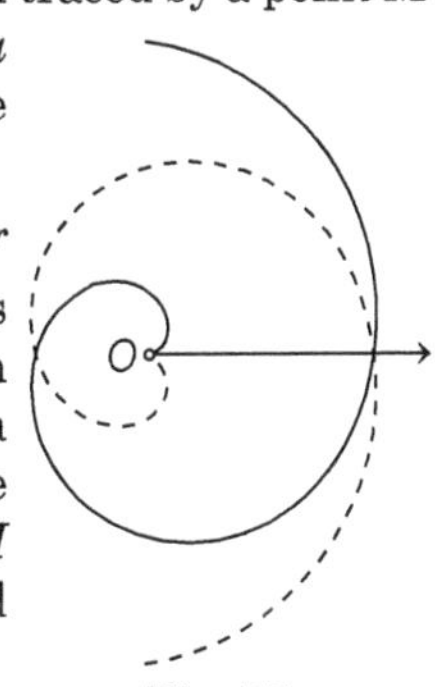

Fig.25.

707. Given the line $x = a$ $(a > 0)$ and the circle of diameter a passing through the origin O and touching this line. A ray drawn from O cuts the circle at A and the line at B (Fig. 27). From the points A and B, straight lines are drawn parallel to the axes O_y and O_x, respectively. As the ray revolves, the point of intersection M of these straight lines describes a curve called the *versiera.* Find its equation.

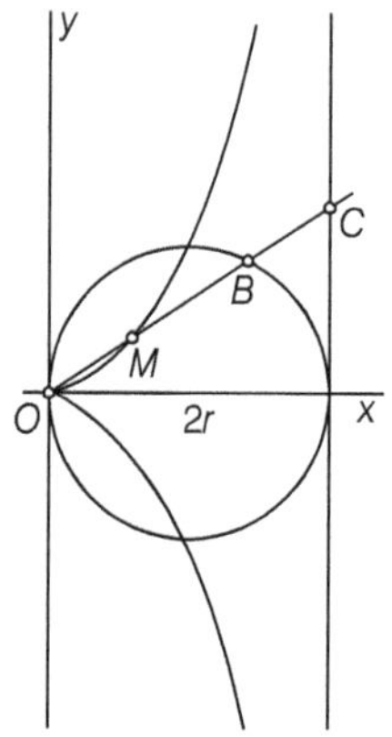

Fig. 26.

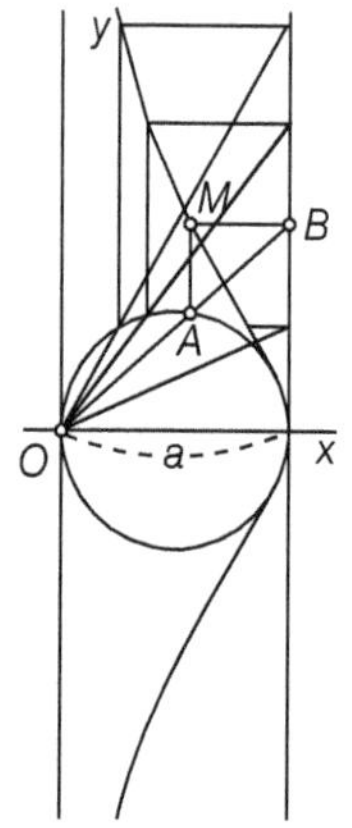

Fig. 27.

708. A ray AB (Fig. 28) is drawn from the point $A(-a, 0)$, where $a > 0$; from the point B, segments BM and BN of a length b $(b = \text{constant})$ are laid off in either direction along the ray. As the ray revolves, the points M and N describe a curve called the *conchoid.* Write its equation in polar coordinates, with the pole placed at A and the polar axis going in the positive direction of the axis Ox, and then transform the result to the given rectangular cartesian system of coordinates.

709. A ray AB (Fig. 29) is drawn from the point $A(-a, 0)$, where $a > 0$; from the point B, segments BM and BN, each equal to OB, are laid off in either direction along the ray. As the ray revolves, the points M and N describe a curve called the ***strophoid.*** Write its equation in polar coordinates, with the pole placed at A and the polar axis going in the positive direction of the axis Ox, and then transform the result to the given rectangular cartesian system of coordinates.

710. A ray is drawn from the origin to cut the given circle $x^2 + y^2 = 2ax$ $(a > 0)$ at a point B (Fig. 30); from the point B, equal segments BM and BN of a constant length b are laid off in either direction along the ray. As the ray revolves, the points M and N describe a curve called the *limacon of Pascal.* Write its equation in polar cordinates, placing the pole at the origin and letting the polar axis coincide with the positive x-axis, and then transform the result to the given rectangular cartesian system of coordinates.

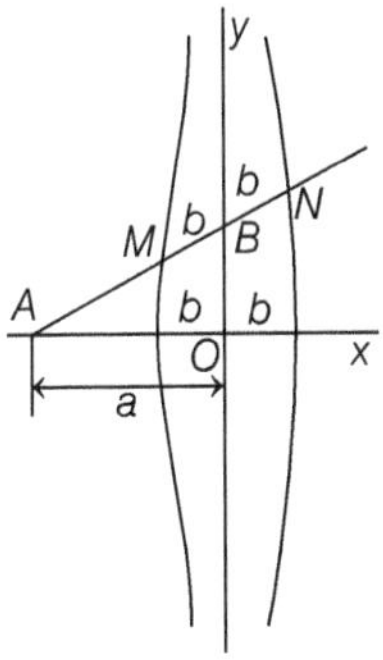

Fig. 28.

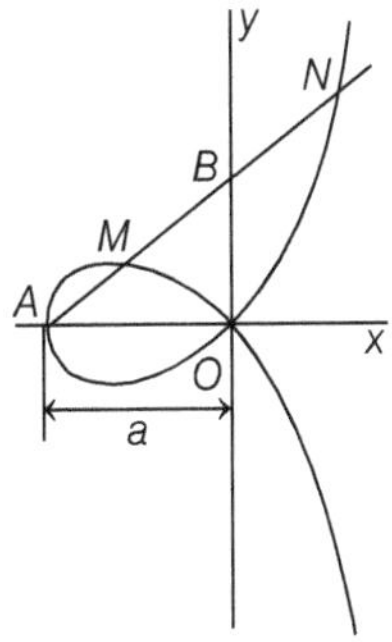

Fig.29.

711. A line segment of length $2a$ moves so that its end points always lie on the coordinate axes. Write the polar equation of the path traced (Fig. 31) by the foot M of the perpendicular dropped from the origin to the segment (placing the pole at the origin and letting the polar axis coincide with the positive x-axis), and transform the result to the given rectangular cartesian system of coordinates. The point M describes a curve called the *four-leaved rose.*

712. A line segment of length a moves so that its end points always lie on the coordinate axes (Fig 32.) Through the end points of the segment, straight lines are drawn parallel to the coordinate axes, these lines intersect at a point P. Find the euqation of the path traced by the foot M of the perpendicular dropped from P to the segment. This path is called the *astroid*.

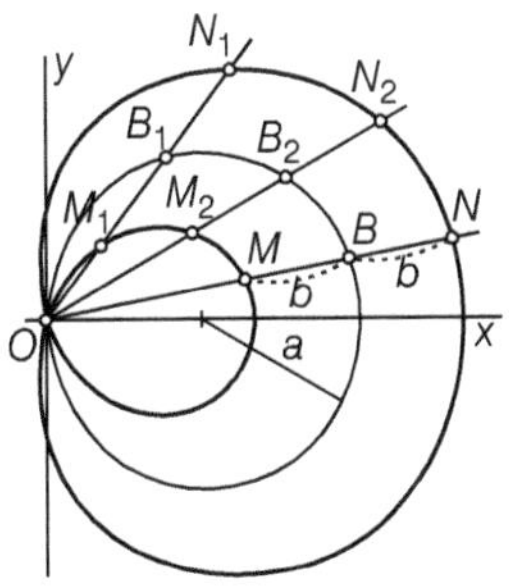

Fig. 30.

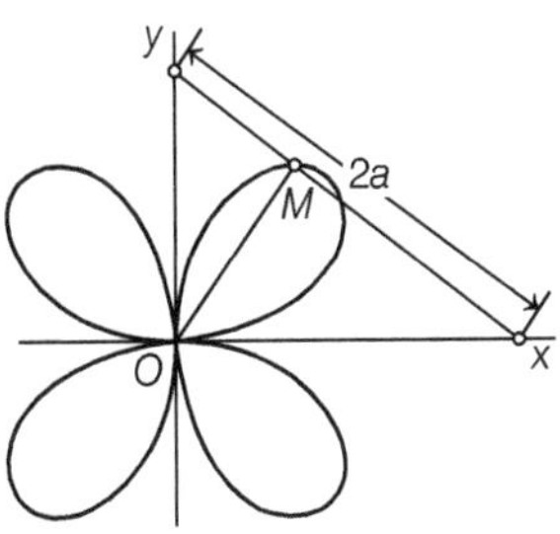

Fig. 31.

Hint. Write first the parametric equations of the astroid, choosing the parameter t as indicated in Fig. 32 (and then eliminate the parameter t).

713. A ray OB meets the circle $x^2 + y^2 = ax$ at a point B; from this point a perpendicular BC is dropped to the axis Ox. Next, a perpendicular CM is drawn from the point C to the ray OB. Derive the polar equation of the path traced by the point M (placing the pole at the origin O and letting the polar axis coincide with the positive x-axis), and then transform the result to the given rectangular cartesian system of coordinates.

714. A thread wound around the circle $x^2 + y^2 = a^2$ is unwound so as to be always tangent to the circle at the variable point B of contact (Fig. 33). Find the parametric equations of the curve described by the free end of the thread, if the initial position of this end is at the point $A(a, 0)$, where $a > 0$. This curve is called the *involute of the circle.*

715. A circle of radius a rolls, without slipping, on the axis Ox; the path traced by a point M on the circumference of the circle is called the *cycloid* (Fig. 34). Derive the parametric equations of the cycloid, using as parameter the angle t through which the rolling circle turns about

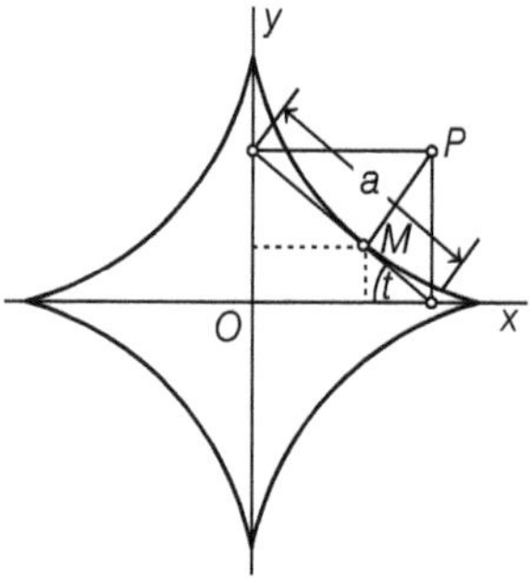

Fig. 32.

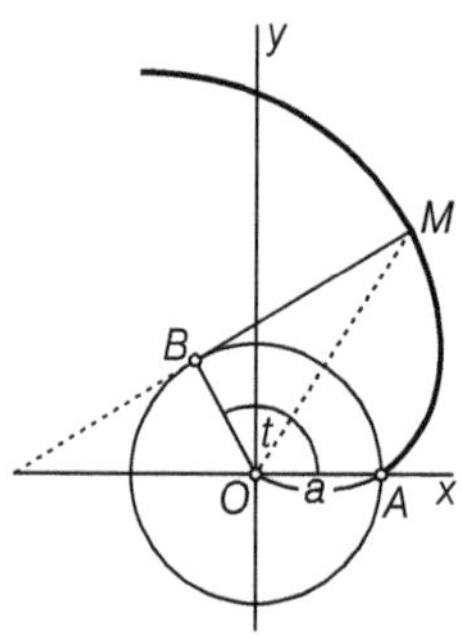

Fig.33.

its centre, and letting the point M coincide with the origin at the initial moment ($t = 0$). Eliminate the parameter t from the resulting equations.

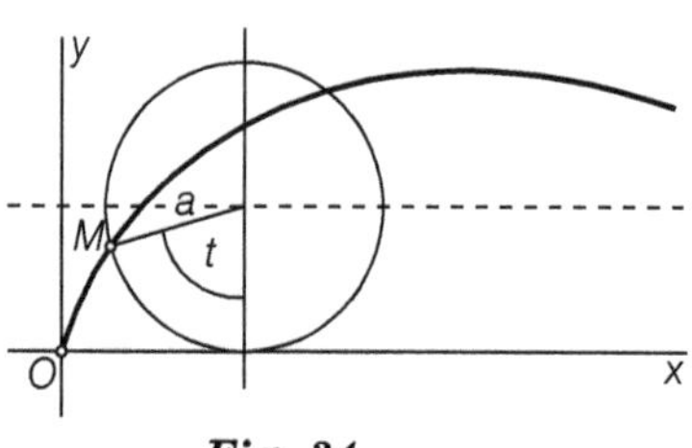

Fig. 34.

716. A circle of radius a rolls, without slipping, on the outside of the circle $x^2 + y^2 = a^2$, the path traced by a point M on the circumference of the rolling circle is called the *cardioid* (Fig. 35). Derive the parametric equations of the cardoid, using as parameter the angle t of inclination (with respect to the axis Ox) of that radius of the fixed circle which is drawn to the point of contact with the rolling circle. At the initial moment ($t = 0$), let the point M be on the x-axis, to the right of O. Transform the result to polar coordinates, placing the pole at A and letting the polar axis go in the positive direction of the x-axis. Prove that the cardioid is a special form of the limacon of Pascal (see Problem 710).

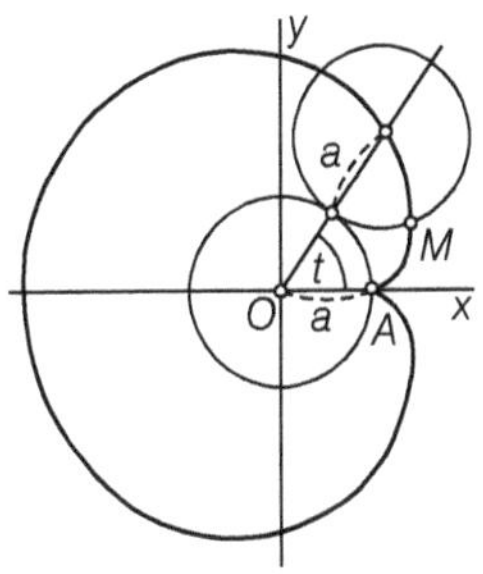

Fig. 35.

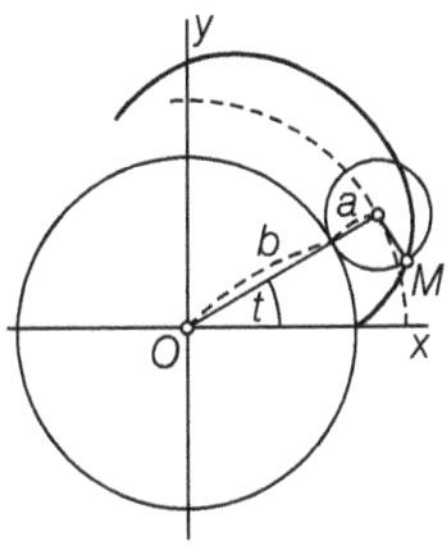

Fig. 36.

717. A circle of radius a rolls, without slipping, on the outside of the circle $x^2 + y^2 = b^2$; the path traced by a point M on the circumference of the rolling circle is called the *epicycloid* (Fig. 36) . Derive the parametric equations of the epicycloid, using as parameter the angle t of inclination (with respect to the axis Ox) of that radius of the fixed circle which is drawn to the point of contact of the two circles; at the initial moment ($t = 0$), let the point M be on the axis Ox, to the right of O. Prove that the cardioid (see Problem 716) is a special form of the epicycloid.

718. A circle of radius a rolls, without slipping, on the inside of the circle $x^2 + y^2 = b^2$; the path traced by a point M on the circumference of the rolling circle is called the *hypocycloid* (Fig. 37) Derive the parametric equations of the hypocyloid, using as parameter the angle t of inclination (with respect to the axis Ox) of that radius of the fixed circle which is drawn to the point of contact of the two circles; at the initial moment ($t = 0$), let the point M be on the axis Ox, to the right of O. Prove that the astroid (see Problem 712) is a special form of the hypocycloid.

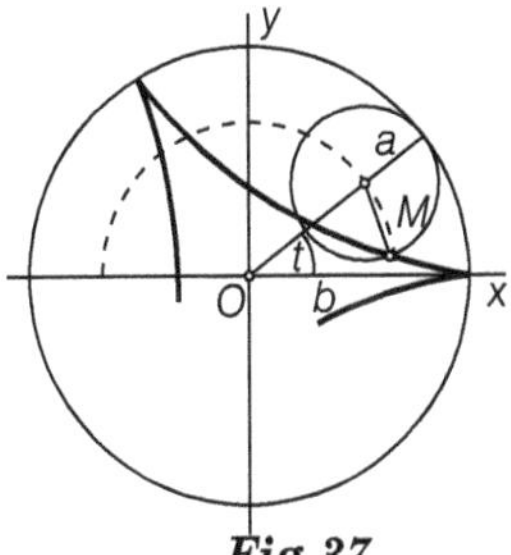

Fig.37.

PART TWO

SOLID ANALYTIC GEOMETRY

6

SOME ELEMENTARY PROBLEMS OF SOLID ANALYTIC GEOMETRY

27. Rectangular Cartesian Coordinates in Space

A rectangular cartesian coordinate system in space is determined by the choice of a linear unit (for measurement of lengths) and of three concurrent and mutually perpendicular axes, numbered in any order.

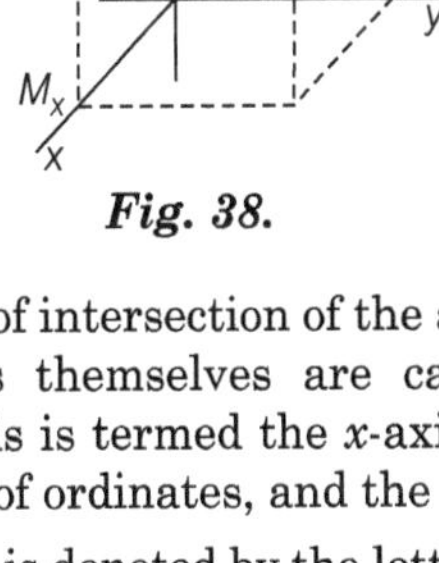

Fig. 38.

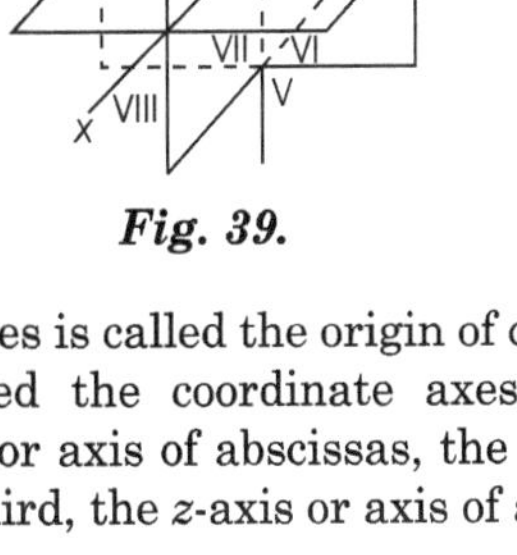

Fig. 39.

The point of intersection of the axes is called the origin of coordinates, and the axes themselves are called the coordinate axes. The first coordinate axis is termed the x-axis or axis of abscissas, the second, the y-axis or axis of ordinates, and the third, the z-axis or axis of applicates.

The origin is denoted by the letter O, and the coordinate axes by Ox, Oy and Oz, respectively.

Let, M be an arbitrary point in space, and let M_x, M_y and M_z be its projections on the coordinate axes (Fig. 38).

The coordinates of the point M in the given system are defined as the numbers

$$x = OM_x, y = OM_y, z = OM_z$$

(Fig. 38), where OM_x is the value of the segment $\overline{OM_x}$ of the x-axis, OM_y is the value of the segment $\overline{OM_y}$ of the y-axis, and OM_z is the value of the segment $\overline{OM_z}$ of the z-axis. The number x is called the abscissa, y the ordinate, and z the applicate of the point M. The notation $M(x, y, z)$ means that the point M has coordinates x, y, z.

The plane Oyz divides all space into two half-spaces; the half-space containing the positive half of the axis Ox is termed the near half-space, and the other half-space is termed the far half-space. Similarly, the plane Oxz divides space into two half-spaces, of which the one containing the positive half of the axis Oy is termed the right half-space, and the other, the left half-space. Finally, the plane Oxy also divides all space into two half-spaces, of which the one containing the positive half of the axis Oz is termed the upper half-space, and the other, the lower half-space.

The three planes Oxy, Oxz and Oyz jointly divide space into eight parts, called octants and numbered as shown in Fig. 39.

719. Construct (in axonometric projection) the following points from their cartesian coordinates: $A(3, 4, 6)$, $B(-5, 3, 1)$, $C(1, -3, -5)$, $D(0, -3, 5)$, $E(-3, -5, 0)$ and $F(-1, -5, -3)$.

720. Given the points $A(4, 3, 5)$, $B(-3, 2, 1)$, $C(2, -3, 0)$ and $D(0, 0, -3)$. Find the coordinates of their projections: 1) on the plane Oxy; 2) on the plane Oxz; 3) on the plane Oyz; 4) the x-ax is; 5) on the y-axis; 6) on the z-axis.

721. Find the coordinates of the points symmetric to the points $A(2,3,1)$, $B(5,-3,2)$, $C(-3, 2-1)$ and $D(a, b, c)$ with respect to: 1) the plane Oxy; 2) the plane Oxz; 3) the plane Oyz; 4) the $x-$ax is; 5) the $y-$axis; 6) the $z-$axis; (7) the origin.

722. Given the following four vertices of a cube: $A(-a, -a, -a)$, $B(a, -a, -a)$, $C(-a, a, -a)$ and $D(a, a, a)$. Determine its remaining vertices.

723. Which octants can contain the points whose coordinates satisfy one of the following conditions: 1) $x - y = 0$; 2) $x + y = 0$; 3) $x - z = 0$; 4) $x + z = 0$; 5) $y - z = 0$; 6) $y + z = 0$?

724. Name the octants that can contain the points for which : 1) $xy > 0$; 2) $xz < 0$; 3) $yz > 0$; 4) $xyz > 0$; 5) $xyz < 0$.

725. Find the centre of the sphere of radius $R = 3$ which touches all the three coordinate planes and is situated: 1) in the second octant; 2) in the fifth octant; 3) in the sixth octant; 4) in the seventh octant; 5) in the eighth octant.

28. The Distance Between Two Points. The Division of a Line Segment in a Given Ratio

The distance d between two points $M_1(x_1, y_1, z_1)$ and $M_2(x_2, y_2, z_2)$ is given by the formula

$$d = \sqrt{(x_2 - x_1)^2 + (y_2 - y_1)^2 + (z_2 - z_1)^2}.$$

The coordinates x, y, z of the point M which divides the segment $\overline{M_1M_2}$ bounded by the points $M_1(x_1, y_1, z_1)$ and $M_2(x_2, y_2, z_2)$ in a ratio λ are determined from the formulas

$$x = \frac{x_1 + \lambda x_2}{1 + \lambda}, \; y = \frac{y_1 + \lambda y_2}{1 + \lambda}, \; z = \frac{z_1 + \lambda z_2}{1 + \lambda}.$$

In particular, by setting $\lambda = 1$, we obtain the coordinates of the midpoint of a given segment:

$$x = \frac{x_1 + x_2}{2}, \; y = \frac{y_1 + y_2}{2}, \; z = \frac{z_1 + z_2}{2}.$$

726. Given the points $A(1, -2, -3)$, $B(2, -3, 0)$, $C(3, 1, -9)$, $D(-1, 1, -12)$. Calculate the distance between 1) A and C; 2) B and D; 3) C and D.

727. Calculate the distances from the origin O to the points: $A(4, -2, -4)$, $B(-4, 12, 6)$, $C(12, -4, 3)$, $D(12, 16, -15)$.

728. Prove that the triangle with vertices $A(3, -1, 2)$, $B(0, -4, 2)$ and $C(-3, 2, 1)$ is isosceles.

729. Prove that the triangle with vertices $A_1(3, -1, 6)$, $A_2(-1, 7, -2)$ and $A_3(1, -3, 2)$ is a right triangle.

730. Determine whether any one of the interior angles of the triangle $M_1(4, -1, 4)$, $M_2(0, 7, -4)$, $M_3(3, 1, -2)$ is obtuse.

731. Prove that the interior angles of the triangle $M(3, -2, 5)$, $N(-2, 1, -3)$, $P(5, 1, -1)$ are acute angles.

732. On the x-axis, find the points whose distance from the point $A(-3, 4, 8)$ is equal to 12.

733. On the y-axis, find the point equidistant from the points $A(1, -3, 7)$ and $B(5, 7, -5)$.

734. Find the centre C and the radius R of a sphere which passes through the point $P(4, -1, -1)$ and touches all the three coordinate planes.

735. Given the vertices $M_1(3, 2, -5)$, $M_2(1, -4, 3)$, $M_3(-3, 0, 1)$ of a triangle. Find the midpoints of its sides.

736. Given the vertices $A(2, -1, 4)$, $B(3, 2, -6)$, $C(-5, 0, 2)$ of a triangle. Calculate the length of the median drawn from the vertex A.

737. The centre of gravity of a uniform rod is at $C(1,-1,5)$, and one end of the rod is the point $A(-2,-1,7)$. Determine the coordinates of the other end of the rod.

738. Given two vertices $A(2,-3,-5)$, $B(-1,3,2)$ of a parallelogram $ABCD$ and the point $E(4,-1,7)$ of intersection of its diagonals. Determine the other two vertices of the parallelogram.

739. Three vertices of a parallelogram $ABCD$ are $A(3,-4,7)$, $B(-5,3,-2)$ and $C(1,2,-3)$. Find the fourth vertex D, which is opposite to B.

740. Three vertices of a paralleogram $ABCD$ are $A(3,-1,2)$, $B(1,2,-4)$ and $C(-1,1,2)$. Find the fourth vertex D.

741. Find the coordinates of the points C, D, E, F which divide into five equal parts the line segment bounded by the points $A(-1,8,3)$ and $B(9,-7,-2)$.

742. Determine the coordinates of the ends of the segment which is divided into three equal parts at the points $C(2,0,2)$ and $D(5,-2,0)$.

743. Given the vertices $A(1,2,-1)$, $B(2,-1,3)$, $C(-4,7,5)$ of a triangle. Calculate the length of the bisector of the interior angle at the vertex B.

744. Given the vertices $A(1,-1,-3)$, $B(2,1,-2)$, $C(-5,2,-6)$ of a triangle. Calculate the length of the bisector of the exterior angle at the vertex A.

745. Equal masses are concentrated at the vertices $A(x_1, y_1, z_1)$, $B(x_2, y_2, z_2)$, $C(x_3, y_3, z_3)$, $D(x_4, y_4, z_4)$ of a tetrahedron. Find the coordinates of the centre of gravity for the system of these masses.

746. Masses m_1, m_2, m_3 and m_4 are placed at the vertices $A_1(x_1, y_1, z_1)$, $A_2(x_2, y_2, z_2)$, $A_3(x_3, y_3, z_3)$, $A_4(x_4, y_4, z_4)$ of a tetrahedron. Find the coordinates of the centre of gravity for the system of these masses.

747. A straight line passes through the two points $M_1(-1,6,6)$ and $M_2(3,-6,-2)$. Find the points at which the line pierces the coordinate planes.

7

VECTOR ALGEBRA

29. The Concept of a Vector. The Projections of a Vector

Directed line segments are also called geometric vectors, or simply vectors. In as much as a vector is a directed line segment, it will as before, be designated in the text by two capital letters with a bar over them, the first letter denoting the initial point, and the second letter, the terminal point of the vector. Another way of indicating a vector will be by a single small letter in half-dark type; in diagrams this letter will be placed at the head of the arrow representing the vector (see Fig. 40 showing a vector **a** with initial point A and terminal point B). Also, the initial point of a vector will often be called its point of application.

Vectors are called equal if they have equal lengths, lie on the same straight line or on parallel straight lines, and are similarly directed.

Fig. 40.

The number equal to the length of a vector (in a given scale) is called the modulus of the vecter. The modulus of a vector $\boldsymbol{a}$ is designated as $|\mathbf{a}|$ or $\boldsymbol{a}$. If $|\boldsymbol{a}| = 1$, then $\boldsymbol{a}$ is called a unit vector.

The unit vector having the same direction as a given vector **a** is said to be co-directional with the vector **a** and is usually denoted by the symbol $\mathbf{a}^0$.

The projection of a vector $\overline{AB}$ on an axis u is defined as the number equal to the value of the segment $\overline{A_1B_1}$ of the axis u, where the point A_1 is the projection of the point A on the axis u, and B_1 is the projection of B on the axis u.

The projection of a vector $\overline{AB}$ on an axis u is designated as proj $_u\overline{AB}$. If the vector is denoted by $\boldsymbol{a}$, then its projection on the axis u is designated as $\text{proj}_u\boldsymbol{a}$.

The projection of a vector $\boldsymbol{a}$ on an axis u is expressed, in terms of its modulus and its angle of inclination φ with respect to the axis u, by the formula

$$\text{proj}_u \boldsymbol{a} = |\boldsymbol{a}| \cdot \cos\varphi \tag{1}$$

The projections of an arbitrary vector $\boldsymbol{a}$ on the axes of a given coordinate system will henceforth be denoted by the letters X, Y, Z. The equality $\boldsymbol{a} = \{X, Y, Z\}$

will mean that the numbers X, Y, Z are the projections of the vector on the coordinate axes.

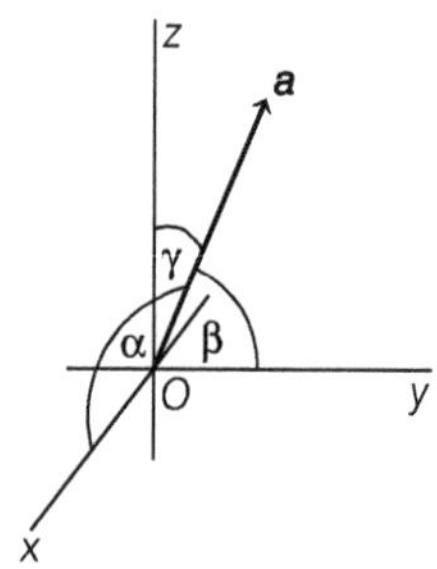

Fig. 41.

The projections of a vector on the coordinate axes are also called its (cartesian) coordinates. If two given points $M_1(x_1, y_1, z_1)$ and $M_2(x_2, y_2, z_2)$ are, respectively, the initial and the terminal point of a vector $\boldsymbol{a}$, then the coordinates X, Y, Z of $\boldsymbol{a}$ are determined from the formulas

$X = x_2 - x_1, Y = y_2 - y_1, Z = z_2 - z_1.$

The formula

$$|\boldsymbol{a}| = \sqrt{X^2 + Y^2 + Z^2} \tag{2}$$

enables us to determine the modulus of a vector from its coordinates.

If α, β, γ are the angles which a vector $\boldsymbol{a}$ makes with the coordinate axes (Fig. 41), then $\cos\alpha, \cos\beta, \cos\gamma$ are called the direction cosines of the vector $\boldsymbol{a}$. In consequence of formula (1),

$X = |\boldsymbol{a}|\cos\alpha, Y = |\boldsymbol{a}|\cos\beta, Z = |\boldsymbol{a}|\cos\gamma$. Hence, from formula (2), it follows that $\cos^2\alpha + \cos^2\beta + \cos^2\gamma = 1$.

This last relation permits us to determine any one of the angles α, β, γ when the other two angles are known.

748. Calculate the modulus of the vector $\boldsymbol{a} = \{6, 3, -2\}$.

749. Two coordinates of a vector $\boldsymbol{a}$ are $X = 4, Y = -12$. Determine its third coordinate Z if $|\boldsymbol{a}| = 13$.

750. Given the points $A(3, -1, 2)$ and $B(-1, 2, 1)$. Find the coordinates of the vector $\overline{AB}$ and $\overline{BA}$.

751. Find the point N which is the terminal point of the vector $\boldsymbol{a} = \{3, -1, 4\}$ whose intial point is at $M\,(1, 2, -3)$.

752. Find the initial point of the vector $\boldsymbol{a} = \{2, -3, -1\}$ whose terminal point is at $(1, -1, 2)$.

753. Given the modulus $|\,\boldsymbol{a}\,| = 2$ of a vector and the angles $\alpha = 45°, \beta = 60°$, $\gamma = 120°$. Calculate the projections of the vector $\boldsymbol{a}$ on the coordinate axes.

754. Calculate the direction cosines of the vector $\boldsymbol{a} = \{12, -15, -16\}$.

755. Calculate the direction cosines of the vector
$$\boldsymbol{a} = \left\{\frac{3}{13}, \frac{4}{13}, \frac{12}{13}\right\}.$$

756. Is it possible for a vector to make the following angles with the coordinate axes : 1) $\alpha = 45°, \beta = 60°, \gamma = 120°$; 2) $\alpha = 45°, \beta = 135°$, $\gamma = 60°$; 3) $\alpha = 90°, \beta = 150°, \gamma = 60°$?

757. Is it possible for a vector to make the following angles with two coordinate axes : 1) $\alpha = 30°, \beta = 45°$; (2) $\beta = 60°, \gamma = 60°$; 3) $\alpha = 150°$, $\gamma = 30°$?

758. A vector makes angles $\alpha = 120°$ and $\gamma = 45°$ with the axes Ox and Oz. What angle does it make with the axis Oy?

759. A vector $\boldsymbol{a}$ makes angles $\alpha = 60°, \beta = 120°$ with the coordinate axes Ox and Oy. Calculate its coordinates if $|\,\boldsymbol{a}\,| = 2$.

760. Determine the coordinates of a point M if the radius vector of M makes equal angles with the Cordinate axes and its modulus is 3.

30. Linear Operations on Vectors

The sum $\boldsymbol{a} + \boldsymbol{b}$ of two vectors $\boldsymbol{a}$ and $\boldsymbol{b}$ is defined as the vector extending from the initial point of the vector $\boldsymbol{a}$ to the terminal point of the vector $\boldsymbol{b}$, provided that the vector $\boldsymbol{b}$ has been drawn from the terminal point of the vector $\boldsymbol{a}$ (the triangle rule). The construction of the sum $\boldsymbol{a} + \boldsymbol{b}$ is shown in Fig. 42.

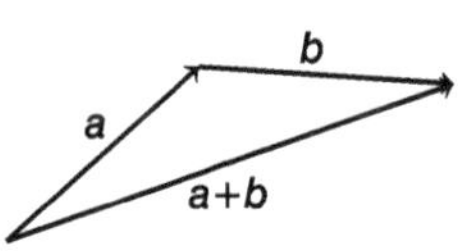

Fig. 42.

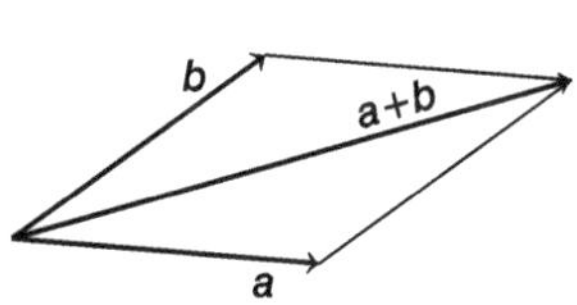

Fig. 43.

Another rule often used is the *parallelogram rule* (which is equivalent to the triangle rule): if vectors $\boldsymbol{a}$ and $\boldsymbol{b}$ drawn from a common initial point and a parallelogram is constructed on them, then the sum $\boldsymbol{a} + \boldsymbol{b}$ is the vector coincident with that diagonal of the parallelogram which extends from the common initial point of $\boldsymbol{a}$ and $\boldsymbol{b}$ (Fig. 43). It follows at once that $\boldsymbol{a} + \boldsymbol{b} = \boldsymbol{b} + \boldsymbol{a}$.

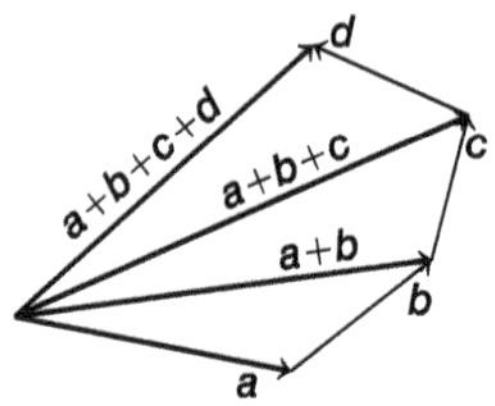

Fig. 44.

The addition of several vectors is carried out by successively applying the triangle rule (see Fig. 44 showing the construction of the sum of four vectors $\boldsymbol{a}, \boldsymbol{b}, \boldsymbol{c}, \boldsymbol{d}$).

The difference $\boldsymbol{a} - \boldsymbol{b}$ of two vectors $\boldsymbol{a}$ and $\boldsymbol{b}$ is defined as the vector which, added to the vector $\boldsymbol{b}$, gives the vector $\boldsymbol{a}$. If two vectors $\boldsymbol{a}$ and $\boldsymbol{b}$ are drawn from a common initial point, then their difference $\boldsymbol{a} - \boldsymbol{b}$ is the vector extending from the terminal point of $\boldsymbol{b}$ (the subtrahend) to the terminal point of $\boldsymbol{a}$ (the minuend). Two vectors of equal length which lie on the same straight line and have opposite directions are called the negatives of each other : if one of them is denoted by $\boldsymbol{a}$, then the other is denoted by $-\boldsymbol{a}$. It is easily seen that $\boldsymbol{a} - \boldsymbol{b} = \boldsymbol{a} + (-\boldsymbol{b})$. Hence, the construction of a vector difference is equivalent to the addition of the negative of the subtrahend to the minuend.

The product $\alpha\boldsymbol{a}$ (or $\boldsymbol{a}\,\alpha$) of a vector $\boldsymbol{a}$ by a number α is a vector defined as follows : its modulus is equal to the product of the modulus of $\mathbf{a}$ by the modulus of α; it is parallel to the vector $\boldsymbol{a}$ or lies on the same line as $\boldsymbol{a}$; It has the same direction as $\boldsymbol{a}$ if α is a positive number, or the opposite direction if α is a negative number.

Addtion of vectors and multiplication of vectors by numbers are called linear operation on vectors.

The two fundamental theorems on the projections of vectors are:

1. The projection of the sum of vectors on an axis is equal to the sum of their projections on this axis:
   $\text{proj}_u(\boldsymbol{a}_1 + \boldsymbol{a}_2 + \ldots + \boldsymbol{a}_n) = \text{proj}_u\boldsymbol{a}_1 + \text{proj}_u\boldsymbol{a}_2 + \ldots + \text{proj}_u\boldsymbol{a}_n$.
2. When a vector is multiplied by a number, the projection of the vector is multiplied by the same number:

   $\text{proj}_u(\alpha\mathbf{a}) = \alpha\,\text{proj}_u\mathbf{a}$.

   In particular, if

   $$\boldsymbol{a} = \{X_1, Y_1, Z_1\},\ \boldsymbol{a} = \{X_2, Y_2, Z_2\},$$

   then $\quad \boldsymbol{a} + \boldsymbol{b} = \{X_1 + X_2, Y_1 + Y_2, Z_1 + Z_2\}$

   and $\quad \boldsymbol{a} - \boldsymbol{b} = \{X_1 - X_2, Y_1 - Y_2, Z_1 - Z_2\}$.

If $\quad \boldsymbol{a} = \{X, Y, Z\}$, then

$$\alpha\boldsymbol{a} = \{\alpha X, \alpha Y, \alpha Z\}$$

for any number α.

Vectors lying on the same straight line or on parallel lines are said to be collinear. The condition for the collinearity of two vectors, $\boldsymbol{a} = \{X_1, Y_1, Z_1\}$, $\boldsymbol{b} = \{X_2, Y_2, Z_2\}$, is that their coordinates should be proportional

$$\frac{X_2}{X_1} = \frac{Y_2}{Y_1} = \frac{Z_2}{Z_1}.$$

The triad of vectors $\boldsymbol{i}$, $\boldsymbol{j}$, $\boldsymbol{k}$ is referred to as the coordinate basis if these vectors satisfy the following conditions

1) the vector $\boldsymbol{i}$ lies on the axis Ox, the vector $\mathbf{j}$ lies on the axis Oy, and the vector $\boldsymbol{k}$ lies on the axis Oz;
2) each of the vectors $\boldsymbol{i}$, $\boldsymbol{j}$, $\boldsymbol{k}$ points in the positive direction of the axis on which it lies;
3) $\boldsymbol{i}, \boldsymbol{j}, \boldsymbol{k}$ are unit vectors, that is, $|\boldsymbol{i}| = 1, |\boldsymbol{j}| = 1, |\boldsymbol{k}| = 1$.

Any vector $\mathbf{a}$ can always be resolved into components with respect to the basis $\boldsymbol{i}, \boldsymbol{j}, \boldsymbol{k}$ that is, can always be represented in the form

$$\boldsymbol{a} = X\boldsymbol{i} + Y\boldsymbol{j} + Z\boldsymbol{k}$$

the cofficients of this resolution are the coordinates of the vector $\boldsymbol{a}$ (that is, X, Y, Z are the projections of the vector $\boldsymbol{a}$ on the coordinate axes).

761. If $\boldsymbol{a}$ and $\boldsymbol{b}$ are two given vectors, construct each of the following vectors : 1) $\boldsymbol{a} + \boldsymbol{b}$; 2) $\boldsymbol{a} - \boldsymbol{b}$; 3) $\boldsymbol{b} - \boldsymbol{a}$; 4) $-\boldsymbol{a} - \boldsymbol{b}$.

762. Given that $|\boldsymbol{a}| = 13, |\boldsymbol{b}| = 19$ and $|\boldsymbol{a} + \boldsymbol{b}| = 24$; calculate $|\boldsymbol{a} - \boldsymbol{b}|$.

763. Given that $|\boldsymbol{a}| = 11, |\boldsymbol{b}| = 23$ and $|\boldsymbol{a} - \boldsymbol{b}| = 30$; find $|\boldsymbol{a} + \boldsymbol{b}|$.

764. Vectors $\mathbf{a}$ and $\mathbf{b}$ are perpendicular to each other; $|\boldsymbol{a}| = 5$ and $|\boldsymbol{b}| = 12$. Evaluate $|\boldsymbol{a} + \boldsymbol{b}|$ and $|\boldsymbol{a} - \boldsymbol{b}|$.

765. Vectors $\boldsymbol{a}$ and $\boldsymbol{b}$ make an angle $\varphi = 60°$; $|\boldsymbol{a}| = 5$ and $|\boldsymbol{b}| = 8$. Evaluate $|\boldsymbol{a} + \boldsymbol{b}|$ and $|\boldsymbol{a} - \boldsymbol{b}|$.

766. Vectors $\boldsymbol{a}$ and $\boldsymbol{b}$ make an angle $\varphi = 120°$, $|\boldsymbol{a}| = 3$ and $|\boldsymbol{b}| = 5$. Evaluate $|\boldsymbol{a} + \boldsymbol{b}|$ and $|\boldsymbol{a} - \boldsymbol{b}|$.

767. What conditions must be satisfied by vectors $\mathbf{a}$ and $\mathbf{b}$ in order that the following relations should hold : 1) $|\boldsymbol{a} + \boldsymbol{b}| = |\boldsymbol{a} - \boldsymbol{b}|$; 2) $|\boldsymbol{a} + \boldsymbol{b}| > |\boldsymbol{a} - \boldsymbol{b}|$; 3) $|\boldsymbol{a} + \boldsymbol{b}| < |\boldsymbol{a} - \boldsymbol{b}|$?

768. What condition must be satisfied by vectors $\boldsymbol{a}$ and $\boldsymbol{b}$ in order that the vector $\boldsymbol{a} + \boldsymbol{b}$ should bisect the angle between the vectors $\boldsymbol{a}$ and $\boldsymbol{b}$?

769. Given two vectors $\boldsymbol{a}$ and $\boldsymbol{b}$; construct each of the following vectors : 1) $3\boldsymbol{a}$; 2) $-\frac{1}{2}\boldsymbol{b}$; 3) $2\boldsymbol{a} + \frac{1}{3}\boldsymbol{b}$; 4) $\frac{1}{2}\boldsymbol{a} - 3\boldsymbol{b}$.

770. In a triangle ABC, the vector $\overline{AB} = \boldsymbol{m}$ and the vector $\overline{AC} = \boldsymbol{n}$. Construct each of the following vectors : 1) $\frac{\boldsymbol{m}+\boldsymbol{n}}{2}$; 2) $\frac{\boldsymbol{m}-\boldsymbol{n}}{2}$; 3) $\frac{\boldsymbol{n}-\boldsymbol{m}}{2}$; 4) $-\frac{\boldsymbol{m}+\boldsymbol{n}}{2}$. Using $\frac{1}{2}|\boldsymbol{n}|$ as a unit segment, construct also the vectors : 5) $|\boldsymbol{n}|\boldsymbol{m} + |\boldsymbol{m}|\boldsymbol{n}$, 6) $|\boldsymbol{n}|\boldsymbol{m} - |\boldsymbol{m}|\boldsymbol{n}$.

771. A point O is the centre of gravity of a triangle ABC. Prove that $\overline{OA} + \overline{OB} + \overline{OC} = 0$.

772. Given the following vectors, coincident with the sides of a regular pentagon $ABCDE$: $\overline{AB} = \boldsymbol{m}$, $\overline{BC} = \boldsymbol{n}$, $\overline{CD} = \boldsymbol{p}$, $DE = \boldsymbol{q}$ and $\overline{EA} = \boldsymbol{r}$. Construct the vectors

1) $\boldsymbol{m} - \boldsymbol{n} + \boldsymbol{p} - \boldsymbol{q} + \boldsymbol{r}$; 2) $\boldsymbol{m} + 2\boldsymbol{p} + \frac{1}{2}\boldsymbol{r}$; 3) $2\boldsymbol{m} + \frac{1}{2}\boldsymbol{n} - 3\boldsymbol{p} - \boldsymbol{q} + 2\boldsymbol{r}$.

773. Given the following vectors, coincident with the edges of a parallelepiped $ABCDA'B'C'D'$ (Fig. 45):

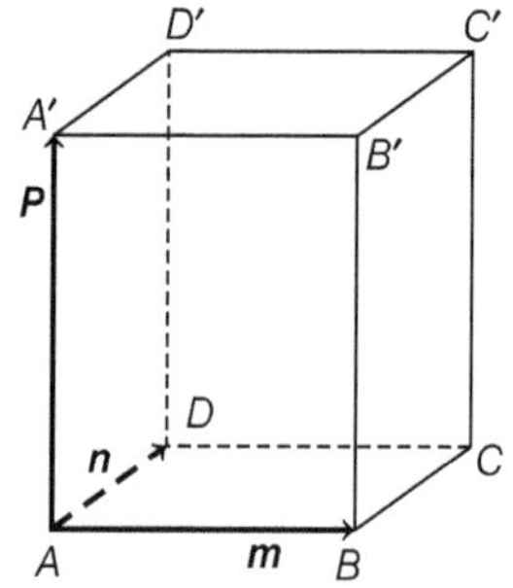

Fig. 45.

$\overline{AB} = \boldsymbol{m}$, $\overline{AD} = \boldsymbol{n}$ and $\overline{AA'} = \boldsymbol{p}$. Construct each of the following vectors :
1) $\boldsymbol{m} + \boldsymbol{n} + \boldsymbol{p}$; 2) $\boldsymbol{m} + \boldsymbol{n} + \frac{1}{2}\boldsymbol{p}$; 3) $\frac{1}{2}\boldsymbol{m} + \frac{1}{2}\boldsymbol{n} + \boldsymbol{p}$, 4) $\boldsymbol{m} + \boldsymbol{n} - \boldsymbol{p}$;
5) $-\boldsymbol{m} - \boldsymbol{n} + \frac{1}{2}\boldsymbol{p}$.

774. Three forces $\boldsymbol{M}, \boldsymbol{N}$ and $\boldsymbol{P}$ are applied at the same point and have mutually perpendicular directions. Find the value of their resultant $\boldsymbol{R}$, if $|\boldsymbol{M}| = 2$ kg, $|\boldsymbol{N}| = 10$ kg and $|\boldsymbol{P}| = 11$ kg weight.

775. Given the two vectors $\boldsymbol{a} = \{3, -2, 6\}$ and $\boldsymbol{b} = \{-2, 1, 0\}$. Determine the projections (on the coordinate axes) of the following vectors :
1) $\boldsymbol{a} + \boldsymbol{b}$; 2) $\boldsymbol{a} - \boldsymbol{b}$; 3) $2\boldsymbol{a}$; 4) $-\frac{1}{2}\boldsymbol{b}$; 5) $2\boldsymbol{a} + 3\boldsymbol{b}$, 6) $\frac{1}{3}\boldsymbol{a} - \boldsymbol{b}$.

776. Verify that $\boldsymbol{a} = \{2, -1, 3\}$ and $\boldsymbol{b} = \{-6, 3, -9\}$ are collinear vectors. Determine which of them has the greater length and find how many times it is longer than the other; determine whether the vectors are similarly or oppositely directed.

777. Determine the values of α, β for which the vectors $\boldsymbol{a} = -2\boldsymbol{i} + 3\boldsymbol{j} + \beta\boldsymbol{k}$ and $\boldsymbol{b} = \alpha\boldsymbol{i} + 6\boldsymbol{j} + 2\boldsymbol{k}$ are collinear.

778. Verify that the four points $A(3,-1,2)$, $B(1,2,-1)$, $C(-1,1,-3)$, $D(3,-5,3)$ are the vertices of a trapezoid.

779. Given the points $A(-1,5,-10)$, $B(5,-7,8)$, $C(2,2,-7)$ and $D(5,-4,2)$. Verify that the vectors $\overline{AB}$ and $\overline{CD}$ are collinear; determine which of them has the greater length and find how many times it is longer than the other, determine whether they are similarly or oppositely dicrected.

780. Find the unit vector co-directional with the vector $\boldsymbol{a} = \{6.-2,-3\}$.

781. Find the unit vector co-directional with the vector $\boldsymbol{a} = \{3,4,-12\}$.

782. Determine the moduli of the sum and the difference of the vectors $\boldsymbol{a} = \{3,-5,8\}$ and $\boldsymbol{b} = \{-1,1,-4\}$.

783. The resolution of a vector $\boldsymbol{c}$ with respect to the basis $\boldsymbol{i}, \boldsymbol{j}, \boldsymbol{k}$ is $\boldsymbol{c} = 16\boldsymbol{i} - 15\boldsymbol{j} + 12\boldsymbol{k}$. Determine the resolution, with respect to the same basis, of a vector $\boldsymbol{d}$ which is parallel and opposite in direction to the vector $\boldsymbol{c}$, if $|\boldsymbol{d}| = 75$.

784. The two vectors $\boldsymbol{a} = \{2,-3,6\}$ and $\boldsymbol{b} = \{-1,2,-2\}$ are drawn from the same point. Determine the coordinates of a vector $\boldsymbol{c}$ directed along the bisector of the angle between the vectors $\boldsymbol{a}$ and $\boldsymbol{b}$, if $|\boldsymbol{c}| = 3\sqrt{42}$.

785. The vectors $\overline{AB} = \{2,6,-4\}$ and $\overline{AC} = \{4,2,-2\}$ from two sides of a triangle ABC. Determine the coordinates of the vectors drawn from the vertices of the triangle and coincident with its medians AM, BN, CP.

786. * Prove that, if $\boldsymbol{p}$ and $\boldsymbol{q}$ any non-collinear vectors, then every vector lying in the plane of $\boldsymbol{p}$ and $\boldsymbol{q}$ can be expressed in the form

$$\boldsymbol{a} = \alpha\,\boldsymbol{p} + \beta\boldsymbol{q}.$$

Prove that the numbers α and β are uniquely determined by the vectors $\boldsymbol{a}, \boldsymbol{p}$ and $\boldsymbol{q}$. (The representation of a vector $\boldsymbol{a}$ in the form $\boldsymbol{a} = \alpha\boldsymbol{p} + \beta\boldsymbol{q}$ is called the resolution of $\boldsymbol{a}$ into components with respect to the basis $\boldsymbol{p}, \boldsymbol{q}$, the numbers α and β are called the coefficients of this resolution).

Proof Let us draw the vectors $\boldsymbol{a}, \boldsymbol{p}$ and $\boldsymbol{q}$ from a common initial point. Denote this initial point by O (Fig. 46), and the terminal point of the vector $\boldsymbol{a}$ by A. Draw a straight line through A parallel to the vector $\boldsymbol{q}$. Let A_p denote the point of intersection of this straight line with the line of action of the vector $\boldsymbol{p}$ Similary, by drawing a straight line through A parallel to the vector $\boldsymbol{p}$, we shall obtain A_q as the point of intersection of this straight line with the line of action of the vector $\boldsymbol{q}$.

---

* Problems 786 and 792 are essential for a proper understanding of the other problems. We therefore give a detailed solution of Problem 786.

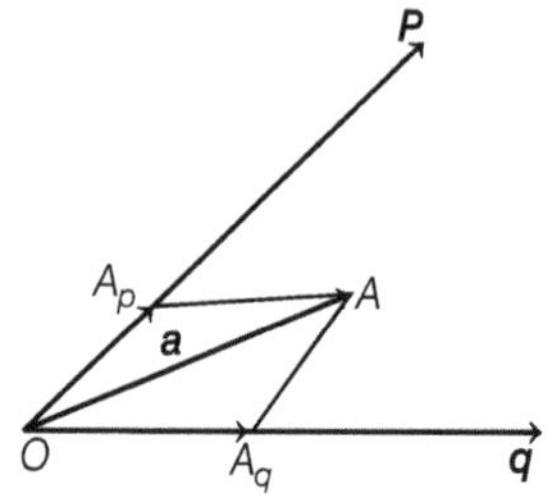

Fig. 46

According to the parallelogram rule,

$$a = \overline{OA} = \overline{OA_p} + \overline{OA_q}. \quad (1)$$

Since, the vectors $\overline{OA_p}$ and $\boldsymbol{p}$ lie on the same straight line, it follows that the vector $\overline{OA_p}$ can be obtained by multiplying the vector $\boldsymbol{p}$ by some number $\boldsymbol{a}$

$$\overline{OA_p} = \alpha \boldsymbol{p}. \quad ...(2)$$

Similarly,

$$\overline{OA_q} = \beta \boldsymbol{q}. \quad ...(3)$$

From relations (1), (2) and (3), we obtained: $\boldsymbol{a} = \alpha \boldsymbol{p} + \beta \boldsymbol{q}$. Thus, we have proved that the required resolution is possible. It remains to prove that the coefficients α and β of this resolution are determined uniquely:

Suppose that the vector $\boldsymbol{a}$ has two resolutions

$$\boldsymbol{a} = \alpha \boldsymbol{p} + \beta \boldsymbol{q}, \ \boldsymbol{a} = \alpha' \boldsymbol{p} + \beta' \boldsymbol{q},$$

and, say, $\alpha' \neq \alpha$. A member-for-member subtraction of the first relation from the second gives

$$(\alpha' - \alpha)\boldsymbol{p} + (\beta' - \beta)\boldsymbol{q} = 0, \quad \text{or} \quad \boldsymbol{p} = \frac{\beta - \beta'}{\alpha' - \alpha}\boldsymbol{q}$$

But this equality means the collinearity of the vectors $\boldsymbol{p}$ and $\boldsymbol{q}$ which are non-collinear by hypothesis. Hence, the inequality $\alpha' \neq \alpha$ is impossible. In similar fashion, we can prove that the inequality $\beta' \neq \beta$ is impossible. Thus, $\alpha' = \alpha, \beta' = \beta$, which means that a vector cannot have two different resolutions.

787. If $\boldsymbol{p} = \{2, -3\}$, $\boldsymbol{q} = \{1, 2\}$ are two given vectors in the plane, find the resolution of the vector $\boldsymbol{a} = \{9, 4\}$ with respect to the basis $\boldsymbol{p}, \boldsymbol{q}$.

788. If $\boldsymbol{a} = \{3, -2\}$, $\boldsymbol{b} = \{-2, 1\}$ and $\boldsymbol{c} = \{7, -4\}$ are three given vectors in the plane, determine the resolution of each of them with respect to the basis formed by the other two vectors.

789. Given the three vectors $\boldsymbol{a} = \{3, -1\}$, $\boldsymbol{b} = \{1, -2\}$, $\boldsymbol{c} = \{-1, 7\}$. Determine the resolution of the vector $\boldsymbol{p} = \boldsymbol{a} + \boldsymbol{b} + \boldsymbol{c}$ with respect to the basis $\boldsymbol{a}, \boldsymbol{b}$.

790. Determine the resolution of the vectors drawn from the vertices of a triangle ABC and coincident with its medians, when the vectors $\overline{AB} = \boldsymbol{b}$ and $\overline{AC} = \boldsymbol{c}$ (coincident with the sides of the triangle) are taken as the basis of the resolution.

791. If $A(1,-2), B(2,1), C(3,2), D(-2,3)$ are four given points in the plane, determine the resolution of the vectors $\overline{AD}, \overline{BD}, \overline{CD}$ and $\overline{AD} + \overline{BD} + \overline{CD}$, with respect to the basis formed by the vectors $\overline{AB}$ and $\overline{AC}$.

792. Prove that, if $\boldsymbol{p}$, $\boldsymbol{q}$ and $\boldsymbol{r}$ are any non-coplanar* vectors, then every vector $\boldsymbol{a}$ in space can be expressed in the form

$$\boldsymbol{a} = \alpha\boldsymbol{p} + \beta\boldsymbol{q} + \gamma\boldsymbol{r}.$$

Prove that the numbers α, β, γ are uniquely determined by the vectors $\boldsymbol{a}, \boldsymbol{p}, \boldsymbol{q}$ and $\boldsymbol{r}$. (The representation of a vector $\boldsymbol{a}$ in the form $\boldsymbol{a} = \alpha\boldsymbol{p} + \beta\boldsymbol{q} + \gamma\boldsymbol{r}$ is called the resolution of $\boldsymbol{a}$ into components with respect to the basis $\boldsymbol{p}, \boldsymbol{q}, \boldsymbol{r}$. The numbers α, β and γ are called the coefficients of this resolution.)

793. Given the three vectors $\boldsymbol{p} = \{3,-2,1\}$, $\boldsymbol{q} = \{-1,1,-2\}$, $\boldsymbol{r} = \{2,1,-3\}$. Find the resolution of the vector $\mathbf{c} = \{11,-6,5\}$ with respect to the basis $\boldsymbol{p}, \boldsymbol{q}, \boldsymbol{r}$.

794. Given the four vectors $\boldsymbol{q} = \{2,1,0\}$, $\boldsymbol{b} = \{1,-1,2\}$, $\mathbf{c} = \{2,2,-1\}$ and $\boldsymbol{d} = \{3,7,-7\}$. Find the resolution of each of these vectors with respect to the basis formed by the other three vectors.

31. The Scalar Product of Vectors

The scalar product of two vectors is defined as the number equal to the product of the moduli of these vectors by the cosine of their included angle.

The scalar product of vectors $\boldsymbol{a}, \boldsymbol{b}$ is denoted by the symbol $\boldsymbol{ab}$ (the order in which the factors are written is immaterial, that is, $\boldsymbol{ab} = \boldsymbol{ba}$).

Designating the angle between vectors $\boldsymbol{a}, \boldsymbol{b}$ as φ, we may express their scalar product by the formula

$$\boldsymbol{ab} = |\boldsymbol{a}| \cdot |\boldsymbol{b}| \cdot \cos\varphi. \qquad \ldots(1)$$

The scalar product of vectors $\mathbf{a}, \mathbf{b}$ may also be expressed by the formula

$$\boldsymbol{ab} = |\boldsymbol{a}| \cdot \text{proj}_{\mathbf{a}}\boldsymbol{b} \text{ or } \boldsymbol{ab} = |\boldsymbol{b}| \cdot \text{proj}_{\mathbf{b}}\boldsymbol{a}.$$

From, formula (1) it follows that $\boldsymbol{ab} > 0$ if φ is an acute angle; $\boldsymbol{ab} < 0$ if φ is an obtuse angle; $\boldsymbol{ab} = 0$ if, and only if, the vectors $\boldsymbol{a}$ and $\boldsymbol{b}$ are mutually perpendicular. (In particular, $\boldsymbol{ab} = 0$ if $\boldsymbol{a} = 0$ or $\boldsymbol{b} = 0$.)

---

* Three vectors are said to be non-coplanar if, drawn from a common initial point, they do not lie in the same plane.

The scalar product $\boldsymbol{aa}$ is called the scalar square of the vector **a** and is denoted by the symbol $\boldsymbol{a}^2$. From (1) it follows that the scalar square of a vector is equal to the square of its modulus:

$$\boldsymbol{a}^2 = |\boldsymbol{a}|^2.$$

Given the coordinates of vectors $\boldsymbol{a}$ and $\boldsymbol{b}$:

$$\boldsymbol{a} = \{X_1, Y_1, Z_1\}, \boldsymbol{b} = \{X_2, Y_2, Z_2\},$$

their scalar product can be calculated from the formula

$$\boldsymbol{ab} = X_1X_2 + Y_1Y_2 + Z_1Z_2.$$

Hence, a necessary and sufficient condition for the perpendicularity of the vectors is

$$X_1X_2 + Y_1Y_2 + Z_1Z_2 = 0.$$

The angle φ between vectors,

$$\boldsymbol{a} = \{X_1, Y_1, Z_1\} \text{ and } \boldsymbol{b} = \{X_2, Y_2, Z_2\}$$

is given by the formula $\cos\varphi = \dfrac{\boldsymbol{ab}}{|\boldsymbol{a}|\cdot|\boldsymbol{b}|}$ or, in terms of coordinates,

$$\cos\varphi = \frac{X_1X_2 + Y_1Y_2 + Z_1Z_2}{\sqrt{X_1^2 + Y_1^2 + Z_1^2}\sqrt{X_2^2 + Y_2^2 + Z_2^2}}.$$

The projection of an arbitrary vector $\boldsymbol{S} = \{X, Y, Z\}$ on an axis u is determined by the formula.

$$\text{proj}_u \boldsymbol{S} = \boldsymbol{S}\boldsymbol{e}$$

where $\boldsymbol{e}$ is the unit vector whose direction is that of the axis u. If the angles α, β, γ which the axis u makes with the coordinate axes are given, then $\mathbf{e} = \{\cos\alpha, \cos\beta, \cos\gamma\}$, and the projection of the vector $\boldsymbol{S}$ may be calculated from the formula.

$$\text{proj}_u \boldsymbol{S} = X\cos\alpha + Y\cos\beta + Z\cos\gamma.$$

795. Vectors $\boldsymbol{a}$ and $\boldsymbol{b}$ make an angle $\varphi = \frac{2}{3}\pi$; if $|\boldsymbol{a}| = 3, |\boldsymbol{b}| = 4$, calculate

1) $\boldsymbol{ab}$; 2) $\boldsymbol{a}^2$; 3) $\boldsymbol{b}^2$ (4) $(\boldsymbol{a} + \boldsymbol{b})^2$;
5) $(3\boldsymbol{a} - 2\boldsymbol{b})(\boldsymbol{a} + 2\boldsymbol{b})$; 6) $(\boldsymbol{a} - \boldsymbol{b})^2$; 7) $(3\boldsymbol{a} + 2\boldsymbol{b})^2$.

796. Vectors $\boldsymbol{a}$ and $\boldsymbol{b}$ are mutually perpendicular, and each of them makes an angle equal to $\frac{\pi}{3}$ with a third vector c, if $|\boldsymbol{a}| = 3, |\boldsymbol{b}| = 5, |\boldsymbol{c}| = 8$, calculate : 1) $(3\boldsymbol{a} - 2\boldsymbol{b}) \times (\boldsymbol{b} + 3\boldsymbol{c})$, 2) $(\boldsymbol{a} + \boldsymbol{b} + \boldsymbol{c})^2$; (3) $(\boldsymbol{a} + 2\boldsymbol{b} - 3\boldsymbol{c})^2$.

797. Prove the validity of the identity $(\boldsymbol{a} + \boldsymbol{b})^2 + (\boldsymbol{a} - \boldsymbol{b})^2 = 2(\boldsymbol{a}^2 + \boldsymbol{b}^2)$ and find its geometric meaning.

798. Prove that $-ab \le ab \le ab$. When does the equals sign hold?

799. If vectors $\boldsymbol{a}, \boldsymbol{b}, \boldsymbol{c}$ are each of them different from zero, determine their relative position such that the condition $(\boldsymbol{ab})\,\boldsymbol{c} = \boldsymbol{a}\,(\boldsymbol{bc})$ is satisfied.

800. Given the unit vectors $\boldsymbol{a}, \boldsymbol{b}$ and $\boldsymbol{c}$ satisfying the condition $\boldsymbol{a}+\boldsymbol{b}+\boldsymbol{c}=0$. Evaluate $\boldsymbol{ab}+\boldsymbol{bc}+\boldsymbol{ca}$.

801. Given three vectors $\boldsymbol{a}, \boldsymbol{b}, \boldsymbol{c}$ which satisfy the condition $\boldsymbol{a}+\boldsymbol{b}+\boldsymbol{c}=0$. If $|\boldsymbol{a}|=3, |\boldsymbol{b}|=1$ and $|\boldsymbol{c}|=4$, evaluate $\boldsymbol{ab}+\boldsymbol{bc}+\boldsymbol{ca}$.

802. Each of the angles between vectors $\boldsymbol{a}, \boldsymbol{b}, \boldsymbol{c}$ is equal to 60°. If $|\boldsymbol{a}|=4$, $|\boldsymbol{b}|=2$ and $|\boldsymbol{c}|=6$, find the modulus of the vector $\boldsymbol{p}=\boldsymbol{a}+\boldsymbol{b}+\boldsymbol{c}$.

803. Given that $|\boldsymbol{a}|=3, |\boldsymbol{b}|=5$. Determine the value of α for which the vectors $\boldsymbol{a}+\alpha\boldsymbol{b}, \boldsymbol{a}-\alpha\boldsymbol{b}$ will be mutually perpendicular.

804. What condition must be satisfied by vectors $\boldsymbol{a}$ and $\boldsymbol{b}$ in order that the vector $\boldsymbol{a}+\boldsymbol{b}$ should be perpendicular to the vector $\boldsymbol{a}-\boldsymbol{b}$?

805. Prove that the vector $\boldsymbol{p}=\boldsymbol{b}(\boldsymbol{ac})-\boldsymbol{c}(\boldsymbol{ab})$ is perpendicular to the vector $\boldsymbol{a}$.

806. Prove that the vector $\boldsymbol{p}=\boldsymbol{b}-\dfrac{\boldsymbol{a}(\boldsymbol{ab})}{\boldsymbol{a}^2}$ is perpendicular to the vector $\boldsymbol{a}$.

807. Given the vectors $\overline{AB}=\boldsymbol{a}$ and $\overline{AC}=\boldsymbol{c}$, coincident with two sides of a triangle ABC. Find the resolution (with respect to the basis $\boldsymbol{b}, \boldsymbol{c}$) of the vector drawn from the vertex B of the triangle and coinciding with the altitude BD.

808. Vectors $\boldsymbol{a}$ and $\boldsymbol{b}$ make an angle $\varphi=\dfrac{\pi}{6}$; if $|\boldsymbol{a}|=\sqrt{3}, |\boldsymbol{b}|=1$, calculate the angle α between the vectors $\boldsymbol{p}=\boldsymbol{a}+\boldsymbol{b}$ and $\boldsymbol{q}=\boldsymbol{a}-\boldsymbol{b}$.

809. Find the obtuse angle which is formed by the medians drawn from the vertices of the acute angles of a right-angled isosceles triangle.

810. Determine the locus of the terminal points of variable vector $\boldsymbol{x}$ drawn from a given origin A, if the vector $\boldsymbol{x}$ satisfies the condition $\boldsymbol{xa}=\alpha$, where $\boldsymbol{a}$ is a given vector and α is a given number.

811. Find the locus of the terminal points of a variable vector $\boldsymbol{x}$ drawn from a given origin A, if the vector $\boldsymbol{x}$ satisfies the conditions

$$\boldsymbol{xa}=\alpha, \ \boldsymbol{xb}=\beta,$$

where $\boldsymbol{a}, \boldsymbol{b}$ are given non-collinear vectors and α, β are given numbers.

812. Given the vectors $\boldsymbol{a}=\{4,-2,-4\}$, $\boldsymbol{b}=\{6,-3,2\}$. Evaluate : 1) $\boldsymbol{ab}$; 2) $\sqrt{\boldsymbol{a}^2}$; 3) $\sqrt{\boldsymbol{b}^2}$; 4) $(2\boldsymbol{a}-3\boldsymbol{b})(\boldsymbol{a}+2\boldsymbol{b})$; 5) $(\boldsymbol{a}+\boldsymbol{b})^2$; 6) $(\boldsymbol{a}-\boldsymbol{b})^2$.

813. Calculate the work done by the force $\boldsymbol{f}=\{3,-5,2\}$ whose point of application is displaced from the initial to the terminal point of the vector $\boldsymbol{S}(2,-5,-7)$.*

814. Given the points $A(-1,3,-7)$, $B(2,-1,5)$ and $C(0,1,-5)$. Evaluate: 1) $(2\overline{AB}-\overline{CB})(2\overline{BC}+\overline{BA})$; 2) $\sqrt{\overline{AB}^2}$; 3) $\sqrt{\overline{AC}^2}$; 4) find the coordinates of the vectors $(\overline{AB}\,\overline{AC})\,\overline{BC}$ and $\overline{AB}\,(\overline{AC}\,\overline{BC})$.

815. Calculate the work done by the force $\boldsymbol{f} = \{3, -2, -5\}$ whose point of application is given a rectilinear displacement from $A(2, -3, 5)$ to $B(3, -2, -1)$.

816. Given the three forces $\boldsymbol{M} = \{3, -4, 2\}, \boldsymbol{N} = \{2, 3, -5\}$ and $\boldsymbol{P} = \{-3, -2. 4\}$ applied at the same point. Calculate the work done by the resultant of these forces when its point of application experiences a rectilinear displacement from $M_1(5, 3, -7)$ to $M_2(4, -1, -4)$.

817. Given the vertices $A(1, -2, 2), B(1, 4, 0), C(-4, 1, 1), D(-5, -5, 3)$ of a quadrilateral. Prove that its diagonals AC and BD are perpendicular to each other.

818. Determine the value of α for which the vectors $\boldsymbol{a} = \alpha \boldsymbol{i} - 3\boldsymbol{j} + 2\boldsymbol{k}$ and $\boldsymbol{b} = \boldsymbol{i} + 2\boldsymbol{j} - \alpha \boldsymbol{k}$ are mutually perpendicular.

819. Calculate the cosine of the angle formed by the vectors $\boldsymbol{a} = \{2, -4, 4\}$ and $\boldsymbol{a} = \{-3, 2, 6\}$.

820. Given the vertices $A(-1, -2, 4), B(-4, -2, 0), C(3, -2, 1)$ of a triangle. Determine the interior angle at the vertex B.

821. Given the vertices $A(3, 2, -3), B(5, 1, -1), C(1, -2, 1)$ of a triangle. Determine the exterior angle at the vertex A.

$$w = \boldsymbol{fs}.$$

822. By calculating the interior angles of the triangle $A(1, 2, 1), B(3, -1, 7), C(7, 4, -2)$, verify that the triangle is isosceles.

823. A vector $\boldsymbol{x}$ is collinear with the vector $\boldsymbol{a} = \{6, -8, -7.5\}$ and makes an acute angle with the axis Oz. Find the coordinates of $\boldsymbol{x}$ if $|\boldsymbol{x}| = 50$.

824. Find the vector $\boldsymbol{x}$ collinear with the vector $\boldsymbol{a} = \{2, 1, -1\}$ and satisfying the condition $\boldsymbol{xa} = 3$.

825. A vector $\boldsymbol{x}$ is perpendicular to the vectors $\boldsymbol{a} = 3\boldsymbol{i} + 2\boldsymbol{j} + 2\boldsymbol{k}$ and $\boldsymbol{b} = 18\boldsymbol{i} - 22\boldsymbol{j} - 5\boldsymbol{k}$ and makes an obtuse angle with the axis Oy. Find the coordinates of $\boldsymbol{x}$ if $|\boldsymbol{x}| = 14$.

826. Find the vector $\boldsymbol{x}$ perpendicular to the vectors $\boldsymbol{a} = \{2, 3, -1\}$ and $\boldsymbol{b} = \{1, -2, 3\}$ and satisfying the condition $\boldsymbol{x}(2\boldsymbol{i} - \boldsymbol{j} + \boldsymbol{k}) = -6$.

827. Given the two vectors $\boldsymbol{a} = \{3, -1, 5\}$ and $\boldsymbol{b} = \{1, 2, -3\}$. Find the vector $\boldsymbol{x}$ perpendicular to the axis Oz and satisfying the conditions $\boldsymbol{xa} = 9$, $\boldsymbol{xb} = -4$.

---

* If a vector $\boldsymbol{f}$ represents a force whose point of application is given a displacement from the initial to the terminal point of a vector $\boldsymbol{S}$, then the work w done by this force is determined by the relation

828. Given the three vectors $\boldsymbol{a} = 2\boldsymbol{i} - \boldsymbol{j} + 3\boldsymbol{k}$, $\boldsymbol{b} = \boldsymbol{i} - 3\boldsymbol{j} + 2\boldsymbol{k}$ and $\boldsymbol{c} = 3\boldsymbol{i} + 2\boldsymbol{j} - 4\boldsymbol{k}$. Find the vector $\boldsymbol{x}$ satisfying the conditions $\boldsymbol{xa} = -5$, $\boldsymbol{xb} = -11$, $\boldsymbol{xc} = 20$.

829. Find the projection of the vector $\boldsymbol{S} = \{4, -3, 2\}$ on the axis making equal acute angles with the coordinate axes.

830. Find the projection of the vector $\boldsymbol{S} = \{\sqrt{2}, -3, -5\}$ on the axis which makes angles $\alpha = 45°$, $\gamma = 60°$ with the coordinate axes Ox, Oz and an acute angle β with the axis Oy.

831. Given the two points $A(3, -4, -2)$, $B(2, 5, -2)$. Find the projection of the vector $\overline{AB}$ on the axis which makes angles $\alpha = 60°$, $\beta = 120°$ with the coordinate axes Ox, Oy and an obtuse angle γ with the axis Oz.

832. Calculate the projection of the vector $\boldsymbol{a} = \{5, 2, 5\}$ on the axis of the vector $\boldsymbol{b} = \{2, -1, 2\}$.

833. Given the three vectors $\boldsymbol{a} = 3\boldsymbol{i} - 6\boldsymbol{j} - \boldsymbol{k}$, $\boldsymbol{b} = \boldsymbol{i} + 4\boldsymbol{j} - 5\boldsymbol{k}$ and $\boldsymbol{c} = 3\boldsymbol{i} - 4\boldsymbol{j} + 12\boldsymbol{k}$. Calculate $\text{proj}_c(\boldsymbol{a} + \boldsymbol{b})$.

834. Given the three vectors

$$\boldsymbol{a} = \{1, -3, 4\}, \boldsymbol{b} = \{3, -4, 2\} \text{ and } \boldsymbol{c} = \{-1, 1, 4\}.$$

Calculate $\text{proj}_{\mathbf{b}+\mathbf{c}}\, \boldsymbol{a}$.

835. Given the three vectors $\boldsymbol{a} = -2\boldsymbol{i} + \boldsymbol{j} + \boldsymbol{k}$, $\boldsymbol{b} = \boldsymbol{i} + 5\boldsymbol{j}$ and $\boldsymbol{c} = 4\boldsymbol{i} + 4\boldsymbol{j} - 2\boldsymbol{k}$.
Calculate $\text{proj}_c(3\boldsymbol{a} - 2\boldsymbol{b})$.

836. The force determined by the vector $\boldsymbol{R} = \{1, -8, -7\}$ is resolved along three mutually perpendicular directions, one of which is the direction of the vector $\boldsymbol{a} = 2\boldsymbol{i} + 2\boldsymbol{j} + \boldsymbol{k}$. Find the component of the force R in the direction of the vector $\boldsymbol{a}$.

837. Given the two points $M(-5, 7, -6)$ and $N(7, -9, 9)$. Calculate the projection of the vector $\boldsymbol{a} = \{1, -3, 1\}$ on the axis of the vector $\overline{MN}$.

838. Given the points $A(-2, 3, -4)$, $B(3, 2, 5)$, $C(1, -1, 2)$, $D(3, 2, -4)$. Calculate $\text{proj}_{\overline{CD}} \overline{AB}$.

32. The Vector Product of Vectors

The vector product of a vector $\boldsymbol{a}$ by a vector $\boldsymbol{b}$ is defined as the vector denoted by the symbol $[\boldsymbol{ab}]$ and determined by the following three conditions

1) the modulus of the vector $[\boldsymbol{ab}]$ is equal to $|\boldsymbol{a}| \cdot |\boldsymbol{b}| \sin\varphi$, where φ is the angle between the vectors $\boldsymbol{a}$ and $\boldsymbol{b}$
2) the vector $[\boldsymbol{ab}]$ is perpendicular to each of the vectors $\boldsymbol{a}$ and $\boldsymbol{b}$;

3) the direction of the vector $[ab]$ is determined by the "right-hand" rule : if the vectors a, b and $[ab]$ are drawn from the same initial point, then the vector $[ab]$ must be directed analogous to the middle finger of the right hand whose thumb extends in the direction of the first factor (i.e. the vector a), while its forefinger extends in the direction of the seconed factor (i.e. the vector b).

The vector product depends on the order of its factors, namely,

$[ab] = -[ba]$.

The modulus of the vector product $[ba]$ is equal to the area S of the parallelogram constructed on the vectors a and b : $|[ab]| = S$.

The vector product itself can be expressed by the formula $[ab] = Se$,

where e is the unit vector co-directional with the vector product.

Vectors a and b have their vector product $[ab]$ zero if, and only if, they are collinear. In particular, $[aa] = 0$.

Given the coordinates of vectors **a** and **b** (in a right-handed coordinate system) : $a = \{X_1, Y_1, Z_1\}$, $b = \{X_2, Y_2, Z_2\}$,

the vector product of the vector a by the vector b is determined from the formula

$$[ab] = \left\{\begin{vmatrix} Y_1 & Z_1 \\ Y_2 & Z_2 \end{vmatrix}, -\begin{vmatrix} X_1 & Z_1 \\ X_2 & Z_2 \end{vmatrix}, \begin{vmatrix} X_1 & Y_1 \\ X_2 & Y_2 \end{vmatrix}\right\}$$

$$[ab] = \begin{vmatrix} i & j & k \\ X_1 & Y_1 & Z_1 \\ X_2 & Y_2 & Z_2 \end{vmatrix}.$$

839. Vectors a and b make an angle $\varphi = \frac{\pi}{6}$. If $|a| = 6$, $|b| = 5$, evaluate $|[ab]|$.

840. Given that $|a| = 10$, $|b| = 2$ and $ab = 12$. Evaluate $|[ab]|$.

841. Given that $|a| = 3$, $|b| = 26$ and $[ab] = 72$. Find ab.

842. Vectors a and b are mutually perpendicular. If $|a| = 3$, $|b| = 4$, evaluate:

1) $|[(a + b)(a - b)]|$; 2) $|[(3a - b)(a - 2b)]|$.

843. Vectors a and b make an angle $\varphi = \frac{2}{3}\pi$. If $|a| = 1$, $|b| = 2$, evaluate:

1) $[ab]^2$; 2) $[(2a + b)(a + 2b)]^2$; 3) $[(a + 3b)(3a - b)]^2$.

844. What condition must be satisfied by vectors a, b in order that the vectors $a + b$ and $a - b$ should be collinear?

845. Prove the identity $[ab]^2 + (ab)^2 = a^2b^2$.

846. Prove that $[ab]^2 \le a^2b^2$. When does the equals sign hold?

847. Given four arbitrary vectors $\boldsymbol{p}$, $\boldsymbol{q}$, $\boldsymbol{r}$, $\boldsymbol{n}$. Prove that the vectors

$\boldsymbol{a} = [\boldsymbol{pn}], \boldsymbol{b} = [\boldsymbol{qn}], \boldsymbol{c} = [\boldsymbol{rn}]$

are coplanar (i.e., will lie in the same plane when they are drawn from a common initial point).

849. Vectors $\boldsymbol{a}$, $\boldsymbol{b}$ and $\mathbf{c}$ satisfy the condition

$$\boldsymbol{a} + \boldsymbol{b} + \mathbf{c} = 0.$$

Prove that $[\boldsymbol{ab}] = [\boldsymbol{bc}] = [\boldsymbol{ca}]$.

849. Vectors $\boldsymbol{a}$, $\boldsymbol{b}$, $\boldsymbol{c}$ and $\boldsymbol{d}$ are connected by the relations

$$[\boldsymbol{ab}] = [\boldsymbol{cd}], [\boldsymbol{ac}] = [\boldsymbol{bd}].$$

Prove that the vectors $\boldsymbol{a} - \boldsymbol{d}$ and $\boldsymbol{b} - \boldsymbol{c}$ are collinear.

850. Given the vectors $\boldsymbol{a} = \{3, -1, -2\}$ and $\boldsymbol{b} = \{1, 2, -1\}$.

Find the coordinates of the vector products:

1) $[\boldsymbol{ab}]$; 2) $[(2\boldsymbol{a} + \boldsymbol{b})\, \boldsymbol{b}]$; 3) $[(2\boldsymbol{a} - \boldsymbol{b})(2\boldsymbol{a} + \boldsymbol{b})]$.

851. Given the pooints $A(2, -1, 2)$, $B(1, 2, -1)$ and $C(3, 2, 1)$. Find the coordinates of the vector products :

1) $[\overline{AB}\, \overline{BC}]$; 2) $[(\overline{BC} - 2\overline{CA})\, \overline{CB}]$.

852. The force $\boldsymbol{f} = \{3, 2, -4\}$ is applied at the point $A(2, -1, 1)$. Determine the moment of this force about the origin*.

853. The force $\boldsymbol{P} = \{2, -4, 5\}$ is applied at the point $M_0(4, -2, 3)$. Determine the moment of this force about the point $A(3, 2, -1)$.

854. The force $\boldsymbol{Q} = \{3, 4, -2\}$ is applied at the point $C(2, -1, -2)$. Determine the value and the direction cosines of the moment of this force about the origin.

855. The force $\boldsymbol{P} = \{2, 2, 9\}$ is applied at the point $A(4, 2, -3)$. Determine the value and the direction cosines of the moment of this force about the point $C(2, 4, 0)$.

856. Given the three forces $\boldsymbol{M} = \{2, -1, -3,\}$ $\boldsymbol{N} = \{3, 2, -1\}$ and $\boldsymbol{P} = \{-4, 1, 3\}$, which are applied at the point $C(-1, 4, -2)$. Find the value and the direction cosines of the moment of the resultant of these forces about the point $A(2, 3, -1)$.

857. Given the points $A(1, 2, 0)$, $B(3, 0, -3)$, $C(5, 2, 6)$. Compute the area of the triangle ABC.

858. Given the vertices $A(1, -1, 2)$, $B(5, -6, 2)$, $C(1, 3, -1)$ of a triangle. Find the length of the altitude from the vertex B to the side AC.

* If $\boldsymbol{f}$ is the vector representing a force applied at a point M, and $\boldsymbol{a}$ is the vector extending from a point O to the point M, then the vector $[\boldsymbol{af}]$ represents the moment of the force $\boldsymbol{f}$ about the point O.

859. Calculate the sine of the angle formed by the vectors $\boldsymbol{a} = \{2, -2, 1\}$ and $\boldsymbol{b} = \{2, 3, 6\}$.

860. A vector $\boldsymbol{x}$ is perpendicular to the vectors $\boldsymbol{a} = \{4, -2, -3\}$ and $\boldsymbol{b} = \{0, 1, 3\}$, and makes an obtuse angle with the axis *Oy*. Find the coordinates of $\boldsymbol{x}$ if $|\boldsymbol{x}| = 26$.

861. A vector $\boldsymbol{m}$ is perpendicular to the axis *Oz* and to the vector $\boldsymbol{a} = \{8, -15, 3\}$, and makes an acute angle with the axis *Ox*. Find the coordinates of $\boldsymbol{m}$ if $|\boldsymbol{m}| = 51$.

862. Find the vector $\boldsymbol{x}$ perpendicular to the vectors $\boldsymbol{a} = \{2, -3, 1\}$, $\boldsymbol{b} = \{1, -2, 3\}$ and satisfying the condition $\boldsymbol{a} = (\boldsymbol{i} + 2\boldsymbol{j} - 7\boldsymbol{k}) = 10$.

863. Prove the identity $(l_1^2 + m_1^2 + n_1^2)(l_2^2 + m_2^2 + n_2^2)$

$$- (l_1 l_2 + m_1 m_2 + n_1 n_2)^2 = (m_1 n_2 - m_2 n_1)^2 + (l_2 n_1 - l_1 n_2)^2 + (l_1 m_2 - l_2 m_1)^2.$$

Hint. Use the identity of Problem 845.

864. Given the vectors $\boldsymbol{a} = \{2, -3, 1\}$, $\boldsymbol{b} = \{-3, 1, 2\}$ and $\boldsymbol{c} = \{1, 2, 3\}$.

Evaluate $[[\boldsymbol{ab}]\,\boldsymbol{c}]$ and $[\boldsymbol{a}\,[\boldsymbol{bc}]]$.

33. The Triple Scalar Product

Three vectors designated as the first, second and third vecctor are called a triad of vectors. The vectors of a triad are written in their order; for example, when we write $\boldsymbol{a}, \boldsymbol{b}, \boldsymbol{c}$, this means that $\boldsymbol{a}$ is regarded as the first, $\boldsymbol{b}$ as the second, and $\boldsymbol{c}$ as the third vector of the triad.

A triad of non-coplanar vectors $\boldsymbol{a}$, $\boldsymbol{b}$, $\boldsymbol{c}$, is called right-handed if its vectors, when drawn from a common initial point and taken in their order, are directed analogous to the thumb, forefinger and middle finger of the right hand, if the vectors $\boldsymbol{a}, \boldsymbol{b}, \boldsymbol{c}$ are directed analogous to the thumb, forefingher and middle finger of the left hand, the triad is called left-handed.

The triple scalar product of three vectors $\boldsymbol{a}, \boldsymbol{b}, \boldsymbol{c}$ is defined as the number obtained by the scalar multiplication of the vector product $[ab]$ and the vector $\boldsymbol{c}$; that is, $[\boldsymbol{ab}]\,\boldsymbol{c}$.

In virtue of the identity $[\boldsymbol{ab}]\,\boldsymbol{c} = \boldsymbol{a}\,[\boldsymbol{bc}]$, the triple scalar product $[\boldsymbol{ab}]\,\boldsymbol{c}$ is denoted by a simpler symbol : $\boldsymbol{abc}$. Thus, $\boldsymbol{abc} = [\boldsymbol{ab}]\,\boldsymbol{c}$, $\boldsymbol{abc} = \boldsymbol{a}\,[\boldsymbol{bc}]$.

The triple scalar product $\boldsymbol{abc}$ is equal to the volume of the parallelepiped constructed on the vectors $\boldsymbol{a}, \boldsymbol{b}, \boldsymbol{c}$; the sign of this volume is positive or negative according as the triad $\boldsymbol{a}, \boldsymbol{b}, \boldsymbol{c}$ is right-handed or left-handed. If (and only if) the vectors $\boldsymbol{a}, \boldsymbol{b}, \boldsymbol{c}$ are coplanar the triple scalar product $\boldsymbol{abc}$ is zero; in other words, the relation

$$\boldsymbol{abc} = 0$$

constitutes a necessary and sufficient condition for the coplanarity of the vectors $\boldsymbol{a}, \boldsymbol{b}, \boldsymbol{c}$.

Given the coordinates of vectors $\boldsymbol{a}, \boldsymbol{b}, \boldsymbol{c}$:

$\boldsymbol{a} = \{X_1, Y_1, Z_1\}, \boldsymbol{b} = \{X_2, Y_2, Z_2\}, \boldsymbol{c} = \{X_3, Y_3, Z_3\}$,

the triple scalar product $\boldsymbol{abc}$ is determined from the formula

$$\boldsymbol{abc} = \begin{vmatrix} X_1 & Y_1 & Z_1 \\ X_2 & Y_2 & Z_2 \\ X_3 & Y_3 & Z_3 \end{vmatrix}$$

It should be noted that the system of coordinate axes (as well as the vector triad $\boldsymbol{i}, \boldsymbol{j}, \boldsymbol{k}$) is assumed here to be right-handed.

865. In each of the following, determine whether the triad $\boldsymbol{a}, \boldsymbol{b}, \boldsymbol{c}$ is right-handed or left-handed :

1) $\boldsymbol{a} = \boldsymbol{k}, \boldsymbol{b} = \boldsymbol{i}, \boldsymbol{c} + \boldsymbol{j}$; 2) $\boldsymbol{a} = \boldsymbol{i}, \boldsymbol{b} = \boldsymbol{k}, \boldsymbol{c} = \boldsymbol{j}$,

3) $\boldsymbol{a} = \boldsymbol{j}, \boldsymbol{b} = \boldsymbol{i}, \boldsymbol{c} = \boldsymbol{k}$, 4) $\boldsymbol{a} = \boldsymbol{i} + \boldsymbol{j}, \boldsymbol{b} = \boldsymbol{j}, \boldsymbol{c} = \boldsymbol{k}$;

5) $\boldsymbol{a} = \boldsymbol{i} + \boldsymbol{j}, \boldsymbol{b} = \boldsymbol{i} - \boldsymbol{j}, \ \boldsymbol{c} = \boldsymbol{j}$,

6) $\boldsymbol{a} = \boldsymbol{i} + \boldsymbol{j}, \boldsymbol{b} = \boldsymbol{i} - \boldsymbol{j}, \boldsymbol{c} = \boldsymbol{k}$.

866. Vectors $\boldsymbol{a}, \boldsymbol{b}, \boldsymbol{c}$ are mutually perpendicular and form a right-handed triad. Evaluate $\boldsymbol{abc}$ if $|\boldsymbol{a}| = 4, |\boldsymbol{b}| = 2, |\boldsymbol{c}| = 3$.

867. A vector $\boldsymbol{c}$ is perpendicular to vectors $\boldsymbol{a}$ and $\boldsymbol{b}$; the angle between $\boldsymbol{a}$ and $\boldsymbol{b}$ is equal to 30°. Evaluate $\boldsymbol{abc}$ if $|\boldsymbol{a}| = 6, |\boldsymbol{b}| = 3, |\boldsymbol{c}| = 3$.

868. Prove that

$$|\boldsymbol{abc}| \le |\boldsymbol{a}||\boldsymbol{b}||\boldsymbol{c}|.$$

When does the equals sign hold?

869. Prove the identity $(\boldsymbol{a} + \boldsymbol{b})(\boldsymbol{b} + \boldsymbol{c})(\boldsymbol{c} + \boldsymbol{a}) = 2\boldsymbol{abc}$.

870. Prove the identity

$$\boldsymbol{ab}(\boldsymbol{c} + \lambda\boldsymbol{a} + \mu\boldsymbol{b}) = \boldsymbol{abc},$$

where λ and μ are any numbers.

871. Prove that vectors $\boldsymbol{a}, \boldsymbol{b}, \boldsymbol{c}$ which satisfy the condition

$$[\boldsymbol{ab}] + [\boldsymbol{bc}] + [\boldsymbol{ac}] = 0,$$

are coplanar.

872. Prove that a necessary and sufficient condition for the coplanarity of vectors $\boldsymbol{a}, \boldsymbol{b}, \boldsymbol{c}$ is given by the relation

$\alpha\boldsymbol{a} + \beta\boldsymbol{b} + \gamma\boldsymbol{c} = 0$,

where at least one of the numbers α, β, γ is different from zero.

873. Given the three vectors

$\boldsymbol{a} = \{1, -1, 3\}, \boldsymbol{b} = \{-2, 2, 1\}, \ \boldsymbol{c} = \{3, -2, 5\}$.

Evaluate $\boldsymbol{abc}$.

874. In each of the following, determine whether the vectors $\boldsymbol{a}, \boldsymbol{b}, \boldsymbol{c}$ are coplanar

1) $\boldsymbol{a} = \{2, 3, -1\}; \boldsymbol{b} = \{1, -1, 3\}; \boldsymbol{c} = \{1, 9, -11\}$;

2) $\boldsymbol{a} = \{3, -2, 1\}; \boldsymbol{b} = \{2, 1, 2\}; \boldsymbol{c} = \{3, -1, -2\}$;

3) $\boldsymbol{a} = \{2, -1, 2\}; \boldsymbol{b} = \{1, 2, -3\}; \boldsymbol{c} = \{3, -4, 7\}$.

875. Prove that the four points

$A(1, 2, -1), B(0, 1, 5), C(-1, 2, 1), D(2, 1, 3)$ lie in the same plane.

876. Calculate the volume of the tetrahedron whose vertices are at the points $A(2, -1, 1), B(5, 5, 4), C(3, 2, -1)$ and $D(4, 1, 3)$.

877. Given the vertices $A(2, 3, 1), B(4, 1, -2), C(6, 3, 7), D(-5, -4, 8)$ of a tetrahedron. Find the length of the altitude drawn from the vertex D.

878. A tetrahedron of volume $v = 5$ has three of its vertices at the points $A(2, 1, -2), B(3, 0, 1), C(2, -1, 3)$; the fourth vertex D lies on the axis Oy. Find the coordinates of D.

34. The Triple Vector Product

Suppose that the vector multiplication of two vectors $\boldsymbol{a}$ and $\boldsymbol{b}$ is followed by the vector multipliation of the resulting vector $[\boldsymbol{ab}]$ and a third vector $\boldsymbol{c}$. This gives the so-called triple vector product $[[\boldsymbol{ab}]\boldsymbol{c}]$ (clearly, $[[\boldsymbol{abc}]\,\boldsymbol{c}]$ is a vector). The vector multiplication of the vectors $\boldsymbol{a}$ and $[\boldsymbol{bc}]$ gives the triple vector product $[\boldsymbol{a}\,[\boldsymbol{bc}]]$.

In general,

$$[[\boldsymbol{ab}]\,\boldsymbol{c}] \neq [\boldsymbol{a}\,[\boldsymbol{bc}]].$$

Let us prove the identity

$$[[\boldsymbol{ab}]\,\boldsymbol{c}] = \boldsymbol{b}\,(\boldsymbol{ac}) - \boldsymbol{a}\,(\boldsymbol{bc}).$$

Proof : For convenience in calculations, let us place the axes of the (rectangular cartesian) coordinate system as follows : let the axis Ox be directed along the vector $\boldsymbol{a}$, and let the axis Oy lie in the plane of the vectors $\boldsymbol{a}$ and $\boldsymbol{b}$ (drawn from a common initial point). We then have:

$$\boldsymbol{a} = \{X_1, 0, 0\}, \boldsymbol{b} = \{X_2, Y_2, 0\}, \boldsymbol{c} = \{X_3, Y_3, Z_3\}.$$

We next find

$$[\boldsymbol{ab}] = \{0, , X_1Y_2\},$$

$$[[\boldsymbol{abc}]\,\boldsymbol{c}] = \{-X_1Y_2Y_3, X_1Y_2X_3, 0\}. \qquad (1)$$

On the other hand,

$$\boldsymbol{ac} = X_1X_3,\ \boldsymbol{b}\,(\boldsymbol{ac}) = \{X_1X_2X_3, X_1Y_2X_3, 0\},$$

$$\boldsymbol{bc} = X_2X_3 + Y_2Y_3,\ \boldsymbol{a}\,(\boldsymbol{bc}) = \{X_1X_2X_3 + X_1Y_2Y_3, 0, 0\}.$$

Hence,

$$\boldsymbol{b}(\boldsymbol{ac}) - \boldsymbol{a}(\boldsymbol{bc}) = \{-X_1Y_2Y_3, X_1Y_2X_3, 0\}. \quad ...(2)$$

Comparng the right-handed members of (1) and (2), we have

$$[[\boldsymbol{ab}]\,\boldsymbol{c}] = b(\boldsymbol{ac}) - a(\boldsymbol{ab}).$$

as was to be shown.

879. Prove the identity

$$[\boldsymbol{a}\,[\boldsymbol{bc}]] = \boldsymbol{b}(\boldsymbol{ac}) - \boldsymbol{c}(\boldsymbol{ab}).$$

880. Solve Problem 864 by using the identities given at the beginning of this section and the identity of Problem 879.

881. Given the vertices $A(2, -1, -3), B(1, 2, -4), C(3, -1, -2)$ of a triangle. Calculate the coordinates of a vector $\boldsymbol{h}$ collinear with the altitude drawn from the vertex A to the opposite side, if the vector $\boldsymbol{h}$ makes an obtuse angle with the axis Oy and if the modulus of $\boldsymbol{h}$ is equal to $2\sqrt{34}$.

882. If vectors $\boldsymbol{a}, \boldsymbol{b}, \boldsymbol{c}$ are each of them different from zero, determine their relative position for which the condition

$$[\boldsymbol{a}\,[\boldsymbol{bc}]] = [[\boldsymbol{ab}]\boldsymbol{c}]$$

is satisfied.

883. Prove the identities :

1) $[\boldsymbol{a}\,[\boldsymbol{bc}]] + [\boldsymbol{b}[\boldsymbol{ca}]] + [\boldsymbol{c}[\boldsymbol{ab}] = 0;$
2) $[\boldsymbol{ab}]\,[\boldsymbol{cd}] = (\boldsymbol{ac})(\boldsymbol{bd}) - (\boldsymbol{ad})(\boldsymbol{bc});$
3) $[\boldsymbol{ab}]\,[\boldsymbol{cd}] + [\boldsymbol{ac}]\,[\boldsymbol{db}] + [\boldsymbol{ad}]\,[\boldsymbol{bc}] = 0;$
4) $[[\boldsymbol{ab}]\,[\boldsymbol{cd}] = \boldsymbol{c}(\boldsymbol{abd}) - \boldsymbol{d}(\boldsymbol{abc}),$
5) $[\boldsymbol{ab}]\,[\boldsymbol{bc}]\,[\boldsymbol{ca}] = (\boldsymbol{abc})^2;$
6) $[\boldsymbol{a}\,[\boldsymbol{a}\,[\boldsymbol{a}\,[\boldsymbol{ab}]]]] = \boldsymbol{a}^4\boldsymbol{b}$, if the vectors $\boldsymbol{a}$ and $\boldsymbol{b}$ are mutually perpendicular;
7) $[\boldsymbol{a}\,[\boldsymbol{b}[\boldsymbol{cd}]]] = [\boldsymbol{ac}](\boldsymbol{bd}) - [\boldsymbol{ad}]\,(\boldsymbol{bc});$
8) $[\boldsymbol{a}\,[\boldsymbol{b}[\boldsymbol{acd}]]] = (\boldsymbol{acd})\boldsymbol{b} - (\boldsymbol{ab})[\boldsymbol{cd}];$
9) $[\boldsymbol{ab}]^2[\boldsymbol{ac}]^2 - ([\boldsymbol{ab}]\,[\boldsymbol{ac}])^2 = \boldsymbol{a}^2(\boldsymbol{abc})^2;$
10) $[\boldsymbol{ab}]\,[\boldsymbol{bc}]]\,[[\boldsymbol{bc}]\,[\boldsymbol{ca}]\,[\boldsymbol{ab}]] = (\boldsymbol{abc})^4;$
11) $(\boldsymbol{ab})\,[\boldsymbol{cd}] + (\boldsymbol{ac})\,[\boldsymbol{db}] + (\boldsymbol{ad})\,[\boldsymbol{bc}]\;\; \boldsymbol{a} = (\boldsymbol{bcd});$
12) $(\boldsymbol{abc})\,(\boldsymbol{ade}) = \begin{vmatrix} \boldsymbol{abd} & \boldsymbol{abe} \\ \boldsymbol{acd} & \boldsymbol{ace} \end{vmatrix}.$

884. Three non-coplanar vectors $\boldsymbol{a}, \boldsymbol{b}$ and $\boldsymbol{c}$ are drawn from a common intitial point. Prove that the plane passing through the terminal points of these vectors is perpendicular to the vector

$$[\boldsymbol{ab}] + [\boldsymbol{bc}] + [\boldsymbol{ca}].$$

8

THE EQUATION OF A SURFACE AND THE EQUATIONS OF A CURVE

35. The Equation of a Surface

The equation of a given surface (in a chosen coordinate system) is defined as the equation in three variables, $F(x, y, z) = 0$,

which is satisfied by the coordinates of all points lying on the surface and by the coordinates of no other point.

885. Given the points $M_1(2, -3, 6)$, $M_2(0, 7, 0)$, $M_3(3, 2, -4)$, $M_4(2\sqrt{2}, 4, -5)$, $M_5(1, -4, -5)$, $M_6(2, 6, -\sqrt{5})$. Determine which of them lie on the surface represented by the equation $x^2 + y^2 + z^2 = 49$. Identify the surface.

886. On the surface $x^2 + y^2 + z^2 = 9$, find the points:

1) with abscissa 1 and ordinate 2; 2) with abscissa 2 and ordinate 5; 3) with abscissa 2 and applicate 2; 4) with ordinate 2 and applicate 4.

887. Identify the geometric objects represented by the following equations in rectangular cartesian coordinates in space :

1) $x = 0$; 2) $y = 0$; 3) $z = 0$; 4) $x - 2 = 0$;
5) $y + 2 = 0$; 6) $z + 5 = 0$; 7) $x^2 + y^2 + z^2 = 25$;
8) $(x-2)^2 + (y+3)^2 + (z-5)^2 = 49$;
9) $x^2 + 2y^2 + 3z^2 = 0$; 10) $x^2 + 2y^2 + 3z^2 + 5 = 0$;
11) $x - y = 0$; 12) $x + z = 0$;
13) $y - z = 0$; 14) $xy = 0$;
15) $xz = 0$; 16) $yz = 0$;
17) $xyz = 0$; 18) $x^2 - 4x = 0$;
19) $xy - y^2 = 0$; 20) $yz + z^2 = 0$.

888. Given the two points $F_1(-c, 0, 0)$ and $F_2(c, 0, 0)$. Derive the equation of the locus of points, the sum of whose distances from the two given points is a constant equal to $2a$, provided that $a > 0, c > 0$; $a > c$.

Solution. Let M denote an arbitrary point in space, and let x, y, z be its coordinates. Since the point M may occupy any position, it follows that x, y and z are variables; they are called the current coordinates.

The point M lies on the given surface if, and only if,

$$MF_1 + MF_2 = 2a. \tag{1}$$

This is the definition of the surface, expressed in symbolic language.

Let us express MF_1 and MF_2 in terms of the current coordinates of the point M:

$$MF_1 = \sqrt{(x+c)^2 + y^2 + z^2},$$

$$MF_2 = \sqrt{(x-c)^2 + y^2 + z^2}.$$

Inserting these expressions in (1), we obtain the equation

$$\sqrt{(x+c)^2 + y^2 + z^2} + \sqrt{(x-c)^2 + y^2 + z^2} = 2a, \tag{2}$$

which connects the current coordinates x, y, z, and which is the equation of the given surface.

For, the condition (1) is fulfilled for every point M lying on the given surface, and hence the coordinates of such a point satisfy equation (2); on the other hand, the condition (1) is not fulfilled for any point not lying on the surface, and hence the coordinates of such a point do not satisfy equation (2). The problem is thus solved; the purpose of the remaining operations is to reduce the equation of the surface to a simpler form.

Transpose the second radical in (2) to the right-hand side:

$$\sqrt{(x+c)^2 + y^2 + z^2} = 2a - \sqrt{(x-c)^2 + y^2 + z^2};$$

squaring both sides of this equation and removing the parentheses, we obtain

$$x^2 + 2cx + c^2 + y^2 + z^2$$
$$= 4a^2 - 4a\sqrt{(x-c)^2 + y^2 + z^2} + x^2 - 2cx + c^2 + y^2 + z^2,$$

or $a\sqrt{(x-c)^2 + y^2 + z^2} = a^2 - cx.$

Clearing of radicals once again, we find

$$a^2x^2 - 2a^2cx + a^2c^2 + a^2y^2 + a^2z^2 = a^4 - 2a^2cx + c^2x^2$$

or $(a^2 - c^2)x^2 + a^2y^2 + a^2z^2 = a^2(a^2 - c^2).$ (3)

Since $a > c$, it follows that $a^2 - c^2 > 0$, let us denote the positive number $a^2 - c^2$ by b^2. Equation (3) will then assume the form
$b^2x^2 + a^2y^2 + a^2z^2 = a^2b^2$,

or $$\frac{x^2}{a^2} + \frac{y^2}{b^2} + \frac{z^2}{b^2} = 1. \qquad (4)$$

This surface is called an ellipsoid of revolution. Equation (4) is referred to as the canonical equation of this ellipsoid.

889. Derive the equation of a sphere with centre at the origin and radius r.

890. Derive the equation of a sphere with centre at $C(\alpha, \beta, \gamma)$ and radius r.

891. From the point $P(2, 6, -5)$, all possible rays are drawn to intersect the plane Oxz. Find the equation of the locus of their mid points.

892. From the point $A(3, -5, 7)$, all possible rays are drawn to intersect the plane Oxy. Find the equation of the locus of their mid points.

893. From the point $C(-3, -5, 9)$, all possible rays are drawn to intersect the plane Oyz. Find the equation of the locus of their midpoints.

894. Derive the equation of the locus of points, the difference of the squares of whose distances from the points $F_1(2, 3, -5)$ and $F_2(2, -7, -5)$ is a constant equal to 13.

895. Derive the equation of the locus of points, the sum of the squares of whose distances from two points $F_1(-a, 0, 0)$ and $F_2(a, 0, 0)$ is a constant equal to $4a^2$.

896. The points $A(-a, -a, -a)$, $B(a, -a, -a)$, $C(-a, a, -a)$ and $D(a, a, a)$ are the vertices of a cube. Write the equation of the locus of points, the sum of the squares of whose distances from the faces of this cube is a constant equal to $8a^2$.

897. Find the equation of the locus of points equidistant from the two points $M_1(1, 2, -3)$ and $M_2(3, 2, 1)$.

898. Derive the equation of the locus of points, the sum of whoe distances from the two given points $F_1(0, 0, -4)$ and $F_2(0, 0, 4)$ is a constant equal to 10.

899. Derive the equation of the locus of points, the difference of whose distances from the two given points $F_1(0, -5, 0)$ and $F_2(0, 5, 0)$ is a constant equal to 6.

36. The Equations of a Curve. The Problem of the Intersection of Three Surfaces

A space curve is represented by two simultaneous equations,

$$\begin{cases} F(x, y, z) = 0 \\ \Phi(x, y, z) = 0, \end{cases}$$

as the intersection of the two surfaces $F(x, y, z) = 0$ and $\Phi(x, y, z) = 0$.

If $F(x, y, z) = 0$, $\Phi(x, y, z) = 0$, $\Psi(x, y, z) = 0$ are the equations of three surfaces, their points of intersection are found by solving simultaneously the system

$$\begin{cases} F(x, y, z) = 0, \\ \Phi(x, y, z) = 0, \\ \Psi(x, y, z) = 0. \end{cases}$$

Each solution of this system for x, y, z gives the coordinates of one of the intersection points of the given surfaces.

900. Given the points $M_1(3, 4, -4)$, $M_2(-3, 2, 4)$, $M_3(-1, -4, 4)$ and $M_4(2, 3, -3)$. Determine which of them lie on the curve

$$\begin{cases} (x-1)^2 + y^2 + z^2 = 36, \\ y + z = 0. \end{cases}$$

901. In each of the following, determine whether the given curve passes through the origin:

1) $\begin{cases} x^2 + y^2 + z^2 - 2z = 0, \\ y = 0; \end{cases}$

2) $\begin{cases} (x-3)^2 + (y+1)^2 + (z-2)^2 = 25, \\ x + y = 0; \end{cases}$

3) $\begin{cases} (x-1)^2 + (y+2)^2 + (z+2)^2 = 9, \\ x - z = 0; \end{cases}$

902. On the curve

$$\begin{cases} x^2 + y^2 + z^2 = 49, \\ x^2 + y^2 + z^2 - 4z - 25 = 0, \end{cases}$$

find the points : 1) with abscissa 3; 2) with ordinate 2; 3) with applicate 8.

903. Identify the curves represented by the following equations:

1) $\begin{cases} x = 0, \\ y = 0; \end{cases}$ (2) $\begin{cases} x = 0, \\ z = 0; \end{cases}$ (3) $\begin{cases} y = 0, \\ z = 0; \end{cases}$ (4) $\begin{cases} x - 2 = 0, \\ y = 0; \end{cases}$

5) $\begin{cases} x+2=0, \\ y-3=0; \end{cases}$ 6) $\begin{cases} x-5=0, \\ z+2=0; \end{cases}$

7) $\begin{cases} y+2=0, \\ z-5=0; \end{cases}$ 8) $\begin{cases} x^2+y^2+z^2=9, \\ z=0; \end{cases}$

9) $\begin{cases} x^2+y^2+z^2=49, \\ y=0; \end{cases}$ 10) $\begin{cases} x^2+y^2+z^2=25, \\ x=0; \end{cases}$

11) $\begin{cases} x^2+y^2+z^2=20, \\ z-2=0. \end{cases}$

904. Find the equations of the curve of intersection of the plane Oxz and the sphere with centre at the origin and radius 3.

905. Find the equations of the curve of intersection of the sphere with centre at the origin and radius 5 and the plane parallel to the plane Oxz and situated in the left half-space at a distance of two units from Oxz.

906. Find the equations of the curve of intersection of the plane Oyz and the sphere with centre at $C(5, -2, 1)$ and radius 13.

907. Write the equations of the curve of intersection of two spheres, one of which is of radius 6 and with centre at the origin, and the other of radius 5 and with centre at $C(1, -2, 2)$.

908. Find the points of intersetion of the three surfaces $x^2+y^2+z^2=49$, $y-3=0, z+6=0$.

909. Find the points of intersection of the three surfaces $x^2+y^2+z^2=9$, $x^2+y^2+(z-2)^2=5, y-2=0$.

37. The Equation of a Cylindrical Surface with Elements Parallel to a Coordinate Axis

An equation in two variables of the form $F(x, y)=0$

represents, in a space coordinate system, a cylinderical surface with elements parallel to the axis Oz. In a plane coordinate system (with axes Ox, Oy) the equation $F(x, y)=0$ represents a curve, namely, the directing curve of the cylinder under consideration. In a space coordinate system, however, the same curve must be represented by two equations:

$$\begin{cases} F(x, y)=0, \\ z=0. \end{cases}$$

In like manner, the equation $F(x, z)=0$ represents (in space) a cylindrical surface with elements parallel to the axis Oy; the equation $F(y, z)=0$ represents a cylindrical surface with elements parallel to the axis Ox.

910. Identify the geometric objects represented, in a space coordinate system, by the following equations :

1) $x^2 + z^2 = 25;$ 2) $\dfrac{y^2}{25} + \dfrac{z^2}{16} = 1;$

3) $\dfrac{x^2}{16} - \dfrac{y^2}{9} = 1;$ 4) $x^2 = 6z;$

5) $x^2 - xy = 0;$ 6) $x^2 - z^2 = 0;$

7) $y^2 + z^2 = 0;$ 8) $x^2 + 4y^2 + 4 = 0;$

9) $x^2 + z^2 = 2z;$ 10) $y^2 + z^2 = -z.$

911. Find the equation of the cylinder which projects the circle

$$\begin{cases} x^2 + (y+2)^2 + (z-1)^2 = 25, \\ x^2 + y^2 + z^2 = 16 \end{cases}$$

on the plane : 1) Oxy; 2) Oxz; 3) Oyz.

912. Find the equations of the projection of the circle

$$\begin{cases} (x+1)^2 + (y+2)^2 + (z-2)^2 = 36, \\ x^2 + (y+2)^2 + (z-1)^2 = 25 \end{cases}$$

on the plane : 1) Oxy; 2) Oxz; 3) Oyz.

9

THE EQUATION OF A PLANE. THE EQUATIONS OF A STRAIGHT LINE. THE EQUATIONS OF QUADRIC SURFACES

38. The General Equation of a Plane. The Equation of the Plane Passing Through a Given Point and Having a Given Normal Vector

In cartesian coordinates, every plane is represented by an equation of the first degree and every equation of the first degree represents a plane.

Every (non-zero) vector perpendicular to a given plane is called its normal vector. The equation

$$A(x - x_0) + B(y - y_0) + C(z - z_0) = 0 \qquad (1)$$

represents the plane passing through the point $M_0(x_0, y_0, z_0)$ and having $\boldsymbol{n} = \{A, B, C\}$ as its normal vector.

By removing the parentheses and denoting the number $-Ax_0 - By_0 - Cz_0$ by the letter D, we can put equation (1) in the form

$$Ax + By + Cz + D = 0.$$

An equation of this form is called the general equation of plane.

913. Write the equation of the plane passing through the point $M_1(2, 1, -1)$ and having $\boldsymbol{n} = \{1, -2, 3\}$ as its normal vector.

914. Write the equation of the plane passing through tho origin and having $\boldsymbol{n} = \{5, 0, -3\}$ as its normal vector.

915. The point $P(2, -1, -1)$ is the foot of a perpendicular dropped from the origin to a plane. Find the equation of this plane.

916. Given the two points $M_1(3, -1, 2)$ and $M_2(4, -2, -1)$. Find the equation of the plane through the point M_1 and perpendicular to the vector $\overline{M_1M_2}$.

917. Write the equation of the plane through the point $M_1(3, 4, -5)$ and parallel to the two vectors $\boldsymbol{a}_1 = \{3, 1, -1\}$ and $\boldsymbol{a}_2 = \{1, -2, 1\}$.

918. Prove that the equation of the plane passing through the point $M_0(x_0, y_0, z_0)$ and parallel to the two vectors
$\boldsymbol{a}_1 = \{l_1, m_1, n_1\}$ and $\boldsymbol{a}_2 = \{l_2, m_2, n_2\}$
can be written in the form

$$\begin{vmatrix} x - x_0 & y - y_0 & z - z_0 \\ l_1 & m_1 & n_1 \\ l_2 & m_2 & n_2 \end{vmatrix} = 0.$$

919. Find the equation of the plane passing through the points $M_1(2, -1, 3)$ and $M_2(3, 1, 2)$ and parallel to the vector $\boldsymbol{a} = \{3, -1, -4\}$.

920. Prove that the equation of the plane passing through the points $M_1(x_1, y_1, z_1)$ and $M_2(x_2, y_2, z_2)$ and parallel to the vector

$$\boldsymbol{a} = \{l, m, n\}$$

can be written in the form

$$\begin{vmatrix} x - x_1 & y - y_1 & z - z_1 \\ x_2 - x_1 & y_2 - y_1 & z_2 - z_1 \\ l & m & n \end{vmatrix} = 0.$$

921. Find the equation of the plane passing through the three points $M_1(3, -1, 2)$, $M_2(4, -1, -1)$ and $M_3(2, 0, 2)$.

922. Prove that the equation of the plane passing through the three points $M_1(x_1, y_1, z_1)$, $M_2(x_2, y_2, z_2)$ and $M_3(x_3, y_3, z_3)$
can be written in the form

$$\begin{vmatrix} x - x_1 & y - y_1 & z - z_1 \\ x_2 - x_1 & y_2 - y_1 & z_2 - z_1 \\ x_3 - x_1 & y_3 - y_1 & z_3 - z_1 \end{vmatrix} = 0.$$

923. In each of the following, find the coordinates of a vector normal to the given plane and write the general expression for the coordinates of its arbitrary normal vector:

(1) $2x - y - 2z + 5 = 0$; (2) $x + 5y - z = 0$;

(3) $3x - 2y - 7 = 0$; (4) $5y - 3z = 0$;

(5) $x + 2 = 0$; (6) $y - 3 = 0$.

924. Determine which of the following pairs of equations represent parallel planes:

1) $2x - 3y + 5z - 7 = 0, 2x - 3y + 5z + 3 = 0$;

2) $4x + 2y - 4z + 5 = 0, 2x + y + 2z - 1 = 0$;

3) $x - 3z + 2 = 0, 2x - 6z - 7 = 0$.

925. Determine which of the following pairs of equations represent perpendicular planes:

1) $3x - y - 2z - 5 = 0, x + 9y - 3z + 2 = 0$;

2) $2x + 3y - z - 3 = 0, x - y - z + 5 = 0$;

3) $2x - 5y + z = 0, x + 2z - 3 = 0$.

926. Determine the values of l and m for which the following pairs of equations represent parallel planes:

1) $2x + ly + 3z - 5 = 0, mx - 6y - 6z + 2 = 0$;

2) $3x - y + lz - 9 = 0, 2x + my + 2z - 3 = 0$;

3) $mx + 3y - 2z - 1 = 0, 2x - 5y - lz = 0$.

927. In each of the following, determine the value of l for which the given pair of equations represents perpendicular planes:

1) $3x - 5y + lz - 3 = 0, x + 3y + 2z + 5 = 0$;

2) $5x + y - 3z - 2 = 0, 2x + ly - 3z + 1 = 0$;

3) $7x - 2y - z = 0, lx + y - 3z - 1 = 0$.

928. In each of the following, determine the dihedral angles formed by the two intersecting planes:

1) $x - y\sqrt{2} + z - 1 = 0, x + y\sqrt{2} - z + 3 = 0$;

2) $3y - z = 0, 2y + z = 0$;

3) $6x + 3y - 2z = 0, x + 2y + 6z - 12 = 0$;

4) $x + 2y + 2z - 3 = 0, 16x + 12y - 15z - 1 = 0$.

929. Write the equation of the plane passing through the origin and parallel to the plane $5x - 3y + 2z - 3 = 0$.

930. Write the equation of the plane passing through the point $M_1(3, -2, -7)$ and parallel to the plane $2x - 3z + 5 = 0$.

931. Find the equation of the plane passing through the origin and perpendicular to the two planes

$2x - y + 3z - 1 = 0, x + 2y + z = 0$.

932. Find the equation of the plane which passes through the point $M_1(2, -1, 1)$ and is perpendicular to the two planes

$$2x - z + 1 = 0, y = 0.$$

933. Prove that the equation of the plane which passes through the point $M_0(x_0, y_0, z_0)$ and is perpendicular to the planes

$A_1x + B_1y + C_1z + D_1 = 0, A_2x + B_2y + C_2z + D_2 = 0$

can be written in the form

$$\begin{vmatrix} x - x_0 & y - y_0 & z - z_0 \\ A_1 & B_1 & C_1 \\ A_2 & B_2 & C_2 \end{vmatrix} = 0.$$

934. Find the equation of the plane which passes through the two points $M_1(1,-1,-2)$ and $M_2(3,1,1)$ and is perpendicular to the plane $x-2y+3z-5=0$.

935. Prove that the equation of the plane which passes through two points $M_1(x_1, y_1, z_1), M_2(x_2, y_2, z_2)$ and is perpendicular to the plane

$$Ax+By+Cz+D=0$$

can be written in the form

$$\begin{vmatrix} x-x_1 & y-y_1 & z-z_1 \\ x_2-x_1 & y_2-y_1 & z_2-z_1 \\ A & B & C \end{vmatrix} = 0.$$

936. Show that the three planes $x-2y+z-7=0$, $2x+y-z+2=0$, $x-3y+2z-11=0$ have a point in common, and calculate the coordinates of this point.

937. Prove that the three planes $7x+4y+7z+1=0$, $2x-y-z+2=0$, $x+2y+3z-1=0$ pass through the same straight line.

938. Prove that the three planes $2x-y+3z-5=0$, $3x+y+2z-1=0$, $4x+3y+z+2=0$ intersect in three distinct and parallel lines.

939. Determine the values of a and b for which the planes $2x-y+3z-1=0, x+2y-z+b=0, x+ay-6z+10=0$:

(1) have a common point;

(2) pass through the same straight line;

(3) intersect in three distinct and parallel lines.

39. Incomplete Equations of Planes. The Intercept Equation of a Plane

Every first-degree equation

$$Ax+By+Cz+D=0$$

(in cartesian coordinates) represents a plane. If this equation lacks the constant term ($D=0$), then the plane passes through the origin. If the equation lacks a term containing one of the current coordinates (that is, if any one of the coefficients A, B, C is zero), then the plane is parallel to a coordinate axis, namely, to the axis of the lacking coordinate; if the constant term is also lacking, the plane passes through this axis. If the equation lacks two terms containing the current coordinates (that is, if any two of the coefficients A, B, C are zero), then the plane is parallel to a coordinate plane, namely, to the coordinate plane that passes through the axes of the lacking coordinates; if the constant term is also lacking, the plane coincides with this coordinate plane.

If the equation of a plane

$$Ax + By + Cz + D = 0,$$

has all its coefficients A, B, C, D different from zero, it can be reduced to the form

$$\frac{x}{a} + \frac{y}{b} + \frac{z}{c} = 1, \qquad (1)$$

where

$$a = -\frac{D}{A}, b = -\frac{D}{B}, c = -\frac{D}{C}$$

are the respective intercepts cut off by the plane on the xy and z-axes. Equation (1) is called the intercept equation of a plane.

940. Find the equation of the plane:

1) through the point $M_1(2, -3, 3)$ and parallel to the plane Oxy;
2) through the point $M_2(1, -2, 4)$ and parallel to the plane Oxz;
3) through the point $M_3(-5, 2, -1)$ and parallel to the plane Oyz.

941. Find the equation of the plane which passes:

1) through the axis Ox and the point $M_1(4, -1, 2)$;
2) through the axis Oy and the point $M_2(1, 4, -3)$;
3) through the axis Oz and the point $M_3(3, -4, 7)$.

942. Find the equation of the plane:

1) through the points $M_1(7, 2, -3)$ and M_2, $(5, 6, -4)$, and parallel to the axis Ox;
2) through the points $P_1(2, -1, 1)$ and $P_2(3, 1, 2)$, and parallel to the axis Oy;
3) through the points $Q_1(3, -2, 5)$ and $Q_2(2, 3, 1)$, and parallel to the axis Oz.

943. Find the points of intersection of the plane $2x - 3y - 4z - 24 = 0$ with the coordinate axes.

944. Given the equation $x + 2y - 3z - 6 = 0$ of a plane. Write its intercept equation.

945. Find the intercepts cutoff by the plane $3x - 4y - 24z + 12 = 0$ on the coordinate axes.

946. Calculate the area of the triangle cut by the plane

$$5x - 6y + 3z + 120 = 0$$

from the quadrant Oxy.

947. Calculate the volume of the pyramid formed by the plane $2x - 3y + 6z - 12 = 0$ and the coordinate planes.

948. A plane passes through the point $M_1(6,-10,1)$ and has the intercepts $a=-3$ and $c=2$ on the x-and z-axes, respectively. Write the intercept equation of the plane.

949. A plane passes through the points $M_1(1,2,-1), M_2(-3,2,1)$ and has the intercept $b=3$ on the y-axis. Write the intercept equation of the plane.

950. Write the equation of the plane which passes through the point $M_1(2,-3,-4)$ and makes equal (non-zero) intercepts on the coordinate axes.

951. Write the equation of a plane which passes through the points M_1 $(-1,4,-1), M_2(-13,2,-10)$ and whose (non-zero) intercepts on the x- and z-axes are equal in absolute value.

952. Write the equations of the planes which pass through the point $M_1(4,3,2)$ and whose (non-zero) intercepts on the x-y-and z-axes are all equal in absolute value.

953. Find the equation of the plane which has the intercept $c=-5$ on the z-axis and is perpendicular to the vector $\boldsymbol{n}=\{-2,1,3\}$.

954. Find the equation of the plane which is parallel to the vector $l=\{2,1,-1\}$ and has the intercepts $a=3, b=-2$ on the x-and y-axes, respectively.

955. Find the equation of the plane which is perpendicular to the plane $2x-2y+4z-5=0$ and has the intercepts $a=-2, b=\frac{2}{3}$ on the x-and y-axes, respectively.

40. The Normal Equation of a Plane. The Distance of a Point from a Plane.

The normal equation of a plane is its equation written in the form

$$x\cos\alpha + y\cos\beta + z\cos\gamma - p = 0, \qquad (1)$$

where $\cos\alpha, \cos\beta, \cos\gamma$ are the direction cosines of the normal to the plane, and p is the distance of the plane from the origin. When calculating the direction cosines of the normal, we assume it to be directed from the origin to the plane. (If the plane passes through the origin, the positive direction of the normal may be chosen at will.)

Let M^* be any point in space, and let d denote the distance of M^* from a given plane. The departure δ of the point M^* from the given plane is defined as the number $+d$ if the point M^* and the origin are on opposite sides of the plane, and as the number $-d$ if M^* and the origin are on the same side of the plane. (if M^* lies in the plane itself, the departure is equal to zero.)

For a point M^* having coordinates x^*, y^*, z^* and a plane represented by the normal equation

$$x\cos\alpha + y\cos\beta + z\cos\gamma - p = 0,$$

the departure of the point M^* from the plane is given by the formula

$$\delta = x^*\cos\alpha + y^*\cos\beta + z^*\cos\gamma - p.$$

Clearly, $d = |\delta|$.

The general equation of a plane,

$$Ax + By + Cz + D = 0,$$

is reduced to the normal form (1) by multiplication by a normalizing factor found from the formula

$$\mu = \pm\frac{1}{\sqrt{A^2 + B^2 + C^2}};$$

the normalizing factor must be taken with its sign opposite to that of the constant term of the equation to be normalized.

956. Determine which of the following equations of planes are in the normal form:

1) $\frac{1}{3}x - \frac{2}{3}y - \frac{2}{3}z - 5 = 0$; 2) $\frac{2}{3}x + \frac{1}{3}y - \frac{1}{3}z - 3 = 0$;

3) $\frac{6}{7}x - \frac{3}{7}y + \frac{2}{7}z + 5 = 0$; 4) $-\frac{6}{7}x + \frac{3}{7}y - \frac{2}{7}z - 5 = 0$;

5) $\frac{3}{5}x - \frac{4}{5}z - 3 = 0$; 6) $-\frac{5}{13}y + \frac{12}{13}z + 1 = 0$;

7) $\frac{5}{13}y - \frac{12}{13}z - 1 = 0$; 8) $\frac{4}{5}x - \frac{3}{5}y + 3 = 0$;

9) $x - 1 = 0$; 10) $y + 2 = 0$;

11) $-y - 2 = 0$; 12) $z - 5 = 0$.

957. In each of the following, reduce the given equation of a plane to the normal form:

1) $2x - 2y + z - 18 = 0$; 2) $\frac{3}{7}x - \frac{6}{7}y + \frac{2}{7}z + 3 = 0$;

3) $4x - 6y - 12z - 11 = 0$; 4) $-4x - 4y + 2z + 1 = 0$;

5) $5y - 12z + 26 = 0$; 6) $3x - 4y - 1 = 0$;

7) $y + 2 = 0$; 8) $-x + 5 = 0$;

9) $-z + 3 = 0$; 10) $2z - 1 = 0$.

958. For each of the following planes, calculate the angles α, β, γ which its normal makes with the coordinate axes, and the distance p from the origin:

1) $x + y\sqrt{2} + z - 10 = 0$; 2) $x - y - z\sqrt{2} + 16 = 0$;

3) $x + z - 6 = 0$; 4) $y - z + 2 = 0$;

5) $x\sqrt{3} + y + 10 = 0$; 6) $z - 2 = 0$;
7) $2x + 1 = 0$; 8) $2y + 1 = 0$;
9) $x - 2y + 2z - 6 = 0$; 10) $2x + 3y - 6z + 4 = 0$.

959. In each of the following, calculate the departure δ and the distance d of the given point from the given plane:
1) $M_1(-2, -4, 3)$, $2x - y + 2z + 3 = 0$;
2) $M_2(2, -1, -1)$, $16x - 12y + 15z - 4 = 0$;
3) $M_3(1, 2, -3)$, $5x - 3y + z + 4 = 0$;
4) $M_4(3, -6, 7)$, $4x - 3z - 1 = 0$;
5) $M_5(9, 2, -2)$, $12y - 5z + 5 = 0$.

960. Calculate the distance d of the point $P(-1, 1, -2)$ from the plane passing through the three points $M_1(1, -1, 1)$, $M_2(-2, 1, 3)$ and $M_3(4, -5, -2)$.

961. In each of the following, determine whether the point $Q(2, -1, 1)$ and the origin lie on the same side or on opposite sides of the given plane:
1) $5x - 3y + z - 18 = 0$; 2) $2x + 7y + 3z + 1 = 0$;
3) $x + 5y + 12z - 1 = 0$; 4) $2x - y + z + 11 = 0$;
5) $2x + 3y - 6z + 2 = 0$; 6) $3x - 2y + 2z - 7 = 0$.

962. Prove that the plane $3x - 4y - 2z + 5 = 0$ cuts the line segment bounded by the points $M_1(3, -2, 1)$ and $M_2(-2, 5, 2)$.

963. Prove that the plane $5x - 2y + z - 1 = 0$ does not cut the line segment bounded by the points $M_1(1, 4, -3)$ and $M_2(2, 5, 0)$.

964. In each of the following, calculate the distance between the parallel planes:
1) $x - 2y - 2z - 12 = 0$, $x - 2y - 2z - 6 = 0$;
2) $2x - 3y + 6z - 14 = 0$, $4x - 6y + 12z + 21 = 0$;
3) $2x - y + 2z + 9 = 0$, $4x - 2y + 4z - 21 = 0$;
4) $16x + 12y - 15z + 50 = 0$, $16x + 12y - 15z + 25 = 0$;
5) $30x - 32y + 24z - 75 = 0$, $15x - 16y + 12z - 25 = 0$;
6) $6x - 18y - 9z - 28 = 0$, $4x - 12y - 6z - 7 = 0$.

965. Two faces of a cube lie in the planes
$2x - 2y + z - 1 = 0$, $2x - 2y + z + 5 = 0$.
Find the volume of the cube.

966. On the axis Oy, find a point situated at a distance $d = 4$ from the plane $x + 2y - 2z - 2 = 0$.

967. On the axis Oz, find a point equidistant from the point $M(1, -2, 0)$ and the plane $3x - 2y + 6z - 9 = 0$.

968. On the axis Ox, find a point equidistant from the two planes $12x - 16y + 15z + 1 = 0, 2x + 2y - z - 1 = 0.$

969. Derive the equation of the locus of points whose departure from the plane $4x - 4y - 2z + 3 = 0$ is equal to 2.

970. Derive the equation of the locus of points whose departure from the plane $6x + 3y + 2z - 10 = 0$ is equal to -3.

971. Write the equations of the planes parallel to the plane $2x - 2y - z - 3 = 0$ and situated at a distance $d = 5$ from it.

972. In each of the following, find the equation of the locus of points equidistant from the two parallel planes:

1) $4x - y - 2z - 3 = 0, 4x - y - 2z - 5 = 0;$

2) $3x + 2y - z + 3 = 0,\ 3x + 2y - z - 1 = 0;$

3) $5x - 3y + z + 3 = 0, 10x - 6y + 2z + 7 = 0.$

973. In each of the following, write the equations of the planes which bisect the dihedral angles formed by the two intersecting planes :

1) $x - 3y + 2z - 5 = 0,\quad 3x - 2y - z + 3 = 0;$

2) $5x - 5y - 2z - 3 = 0,\quad x + 7y - 2z + 1 = 0;$

(3) $2x - y + 5z + 3 = 0, 2x - 10y + 4z - 2 = 0.$

974. In each of the following, determine whether the point $M(2, -1, 3)$ and the origin lie inside one dihedral angle, or in two complementary angles, or in two vertical dihedral angles (formed by the two intersecting planes)

1) $2x - y + 3z - 5 = 0,\quad 3x + 2y - z + 3 = 0;$

2) $2x + 3y - 5z - 15 = 0, 5x - y - 3z - 7 = 0;$

3) $x + 5y - z + 1 = 0,\ 2x + 17y + z + 2 = 0.$

975. In each of the following, determine whether the points $M\,(2, -1, 1)$ and $N\,(1, 2, -3)$ lie inside one dihedral angle, or in two complementary angles, or in two vertical dihedral angles (formed by the two intersecting planes):

1) $3x - y + 2z - 3 = 0,\quad x - 2y - z + 4 = 0;$

2) $2x - y + 5z - 1 = 0,\ 3x - 2y + 6z - 1 = 0.$

976. Determine whether the origin lies inside the acute or the obtuse angle formed by the two planes $x - 2y + 3z - 5 = 0, 2x - y - z + 3 = 0.$

977. Determine whether the point $M\,(3, 2, -1)$ lies inside the acute or the obtuse angle formed by the two planes $5x - y + z + 3 = 0,$ $4x - 3y + 2z + 5 = 0.$

978. Write the equation of the plane bisecting that dihedral angle between the two planes $2x - 14y + 6z - 1 = 0,\ 3x + 5y - 5z + 3 = 0$ which contains the origin.

979. Write the equation of the plane bisecting that dihedral angle between the two planes $2x-y+2z-3=0$, $3x+2y-6z-1=0$ which contains the point $(1,2,-3)$.

980. Write the equation of the plane which bisects the acute dihedral angle formed by the two planes $2x-3y-4z-3=0$, $4x-3y-2z-3=0$.

981. Write the equation of the plane which bisects the obtuse dihedral angle formed by the two planes $3x-4y-z+5=0$, $4x-3y+z+5=0$.

41. The Equations of a Straight Line

A straight line is determined (as the intersection of two planes) by two simultaneously considered equations of the first degree,

$$\begin{cases} A_1x+B_1y+C_1z+D_1=0, \\ A_2x+B_2y+C_2z+D_2=0, \end{cases} \tag{1}$$

provided that the coefficients A_1, B_1, C_1 of the first equation are not proportional to the coefficients A_2, B_2, C_2 of the second (otherwise these equations will represent parallel or coincident planes).

Let a straight line a be represented by equation (1), and let α and β be any numbers, not both zero; then the equation

$$\alpha(A_1x+B_1y+C_1z+D_1)+\beta(A_2x+B_2y+C_2z+D_2)=0 \tag{2}$$

represents a plane through the straight line a.

By an appropriate choice of the numbers α, β, an equation of the form (2) can be made to represent every plane passing through the line a.

The totality of planes passing through the same straight line is called a pencil of planes. An equation of the form (2) is called the equation of a pencil of planes.

If $\alpha \neq 0$, then equation (2) can be reduced, by letting $\frac{\beta}{\alpha}=\lambda$, to the form

$$A_1x+B_1y+C_1z+D_1+\lambda(A_2x+B_2y+C_2z+D_2)=0. \tag{3}$$

This form of the equation of a pencil of planes is used more widely than equation (2); note, however, that equation (3) can be made to represent every plane of the pencil except the plane corresponding to $\alpha=0$, that is, except the plane $A_2x+B_2y+C_2z+D_2=0$.

982. Write the equations of the straight lines in which the plane $5x-7y+2z-3=0$ intersects the coordinate planes.

983. Write the equations of the straight line in which the plane $3x-y-7z+9=0$ intersects the plane passing through the axis Ox and the point $E(3,2,-5)$.

984. Find the points in which the line $\begin{cases} 2x + y - z - 3 = 0, \\ x + y + z - 1 = 0 \end{cases}$

pierces the coordinate planes.

985. Prove that the line

$$\begin{cases} 2x - 3y + 5z - 6 = 0, \\ x + 5y - 7z + 10 = 0 \end{cases}$$

meets the axis Oy.

986. Determine the value of D for which the line

$$\begin{cases} 2x + 3y - z + D = 0, \\ 3x - 2y + 2z - 6 = 0 \end{cases}$$

meets: 1) the axis Ox; 2) the axis Oy; 3) the axis Oz.

987. Find the conditions which must be satisfied by the coefficients of the equations

$$\begin{cases} A_1x + B_1y + C_1z + D_1 = 0, \\ A_2x + B_2y + C_2z + D_2 = 0 \end{cases}$$

of a straight line in order that the line should be parallel:
1) to the axis Ox; 2) to the axis Oy; 3) to the axis Oz.

988. Find the conditions which must be satisfied by the coefficients of the equations

$$\begin{cases} A_1x + B_1y + C_1z + D_1 = 0, \\ A_2x + B_2y + C_2z + D_2 = 0 \end{cases}$$

of a straight line in order that the line should: 1) meet the x-axis; 2) meet the y-axis; 3) meet the z-axis; 4) coincide with the x-axis; 5) coincide with the y-axis; 6) coincide with the z-axis.

989. In the pencil of planes

$2x - 3y + z - 3 + \lambda(x + 3y + 2z + 1) = 0$, find the plane which 1) passes through the point $M_1(1, -2, 3)$; 2) is parallel to the axis Ox; 3) is parallel to the axis Oy; 4) is parallel to the axis Oz.

990. Write the equation of the plane which passes through the line of intersection of the planes $3x - y + 2z + 9 = 0$, $x + z - 3 = 0$ and : 1) passes through the point $M_1(4, -2, -3)$; 2) is parallel to the axis Ox; 3) is parallel to the axis Oy; 4) is parallel to the axis Oz.

991. Write the equation of the plane which passes through the line of intersection of the planes $2x - y + 3z - 5 = 0$, $x + 2y - z + 2 = 0$ and is parallel to the vector $l = \{2, -1, -2\}$.

992. Write the equation of the plane which passes through the line of intersection of the planes $5x - 2y - z - 3 = 0$, $x + 3y - 2z + 5 = 0$ and is parallel to the vector $l = \{7, 9, 17\}$.

993. Write the equation of the plane which passes through the line of intersection of the planes $3x-2y+z-3=0$, $x-2z=0$ and is perpendicular to the plane $x-2y+z+5=0$.

994. Find the equation of the plane which passes through the line

$$\begin{cases} 5x-y-2z-3=0, \\ 3x-2y-5z+2=0 \end{cases}$$

and is perpendicular to the plane $x+19y-7z-11=0$.

995. Find the equation of the plane which passes through the line of intersection of the planes $2x+y-z+1=0$, $x+y+2z+1=0$ and is parallel to the segment bounded by the points $M_1(2,5,-3)$ and $M_2(3,-2,2)$.

996. Find the equation of the plane belonging to the pencil of planes

$$\alpha(3x-4y+z+6)+\beta(2x-3y+z+2)=0$$

and equidistant from the points $M_1(3,-4,-6)$, $M_2(1,2,2)$.

997. Determine whether the plane $4x-8y+17z-8=0$ belongs to the pencil of planes $\alpha(5x-y+4z-1)+\beta(2x+2y-3z+2)=0$.

998. Determine whether the plane $5x-9y-2z+12=0$ belongs to the pencil of planes $\alpha(2x-3y+z-5)+\beta(x-2y-z-7)=0$.

999. Determine the values of l and m for which the plane $5x+ly+4z+m=0$ belongs to the pencil of planes $\alpha(3x-7y+z-3)+\beta(x-9y-2z+5)=0$.

1000. Write the equation of the plane which belongs to the pencil of planes

$$\alpha(x-3y+7z+36)+\beta(2x+y-z-15)=0$$

and is situated at a distance $p=3$ from the origin.

1001. Write the equation of the plane which belongs to the pencil of planes $\alpha(10x-8y-15z+56)+\beta(4x+y+3z-1)=0$ and is situated at a distance $d=7$ from the point $C(3,-2,-3)$.

1002. Find the equation of the plane which belongs to the pencil of planes

$$\alpha(4x+13y-2z-60)+\beta(4x+3y+3z-30)=0$$

and cuts a triangle of area 6 from the quadrant Oxy.

1003. Find the equations of the plane which project the line

$$\begin{cases} 2x-y+2z-3=0, \\ x+2y-z-1=0 \end{cases}$$

on the coordinate planes.

1004. Write the equations of the projections of the line

$$\begin{cases} x+2y-3z-5=0, \\ 2x-y+z+2=0 \end{cases}$$

on the coordinate planes.

1005. Write the equation of the plane which projects the line

$$\begin{cases} 3x + 2y - z - 1 = 0, \\ 2x - 3y + 2z - 2 = 0 \end{cases}$$

on the plane $x + 2y + 3z - 5 = 0$.

1006. Write the equations of the projections of the line

$$\begin{cases} 5x - 4y - 2z - 5 = 0, \\ x + 2z - 2 = 0 \end{cases}$$

on the plane $2x - y + z - 1 = 0$.

42. The Direction Vector of a Straight Line. The Canonical Equations of a Straight Line. The Parametric Equations of a Straight Line

Every non-zero vector lying on a given straight line or parallel to it is called the direction vector of that straight line.

In what follows, the direction vector of an arbitrary straight line will be denoted by the letter a, and its coordinates by l, m, n

$\boldsymbol{a} = \{l, m, n\}$.

Given, one point $M_0(x_0, y_0, z_0)$ of a straight line and its direction vector $\boldsymbol{a} = \{l, m, n\}$, the line can be represented by (two) equations of the form

$$\frac{x - x_0}{l} = \frac{y - y_0}{m} = \frac{z - z_0}{n}. \tag{1}$$

Equations of a straight line written in this form are called its canonical equations.

The canonical equations of the straight line passing through two given points $M_1(x_1, y_1, z_1)$ and $M_2(x_2, y_2, z_2)$ have the form

$$\frac{x - x_1}{x_2 - x_1} = \frac{y - y_1}{y_2 - y_1} = \frac{z - z_1}{z_2 - z_1}. \tag{2}$$

Denoting by t each of the equal ratios in the canonical equations (1), we obtain

$$\frac{x - x_0}{l} = \frac{y - y_0}{m} = \frac{z - z_0}{n} = t.$$

Hence,

$$\begin{cases} x = x_0 + lt, \\ y = y_0 + mt, \\ z = z_0 + nt. \end{cases} \tag{3}$$

These are the parametric equations of the straight line passing through the point $M_0(x_0, y_0, z_0)$ in the direction of the vector $\boldsymbol{a} = \{l, m, n\}$. In equation (3), t is regarded as an arbitrarily varying parameter, and x, y, z as functions of t; the quantities x, y, z vary with t so that the point $M(x, y, z)$ moves along the given straight line.

If the parameter t is taken as the variable time, and equations (3) are considered to be the equations of motion of the variable point M, then these equations will determine the uniform rectilinear motion of the point M. When $t = 0$, the point M coincides with the point M_0. The speed v of the point M is a constant given by the formula

$$v = \sqrt{l^2 + m^2 + n^2}.$$

1007. Write the canonical equations of the straight line passing through the point $M_1(2, 0, -3)$ and parallel:

1) to the vector $\boldsymbol{a} = \{2, -3, 5\}$;

2) to the straight line $\frac{x-1}{5} = \frac{y+2}{2} = \frac{z+1}{-1}$;

3) to the axis Ox; (4) to the axis Oy; 5) to the axis Oz.

1008. Write the canonical equations of the straight line passing through the two given points:

1) $(1, -2, 1)$, $(3, 1, -1)$; 2) $(3, -1, 0)$, $(1, 0, -3)$;

3) $(0, -2, 3)$, $(3, -2, 1)$; 4) $(1, 2, -4)$, $(-1, 2, -4)$.

1009. Write the parametric equations of the straight line passing through the point $M_1(1, -1, -3)$ and parallel:

(1) to the vector $\boldsymbol{a} = \{2, -3, 4\}$;

(2) to the line $\frac{x-1}{2} = \frac{y+2}{5} = \frac{z-1}{0}$

(3) to the line $x = 3t - 1$, $y = -2t + 3$, $z = 5t + 2$.

1010. Write the parametric equations of the straight line passing through the two given points:

1) $(3, -1, 2)$, $(2, 1, 1)$; 2) $(1, 1, -2)$, $(3, -1, 0)$; 3) $(0, 0, 1)$, $(0, 1, -2)$.

1011. A straight line is drawn through the points $M_1(-6, 6, -5)$ and $M_2(12, -6, 1)$. Find the points in which this line pierces the coordinate planes.

1012. Given the vertices $A(3, 6, -7)$, $B(-5, 2, 3)$, $C(4, -7, -2)$ of a triangle. Write the parametric equations of the median drawn from the vertex C.

1013. Given the vertices $A(3, -1, -1)$, $B(1, 2, -7)$, $C(-5, 14, -3)$ of a triangle. Find the canonical equations of the bisector of the interior angle at the vertex B.

1014. Given the vertices $A(2, -1, -3)$, $B(5, 2, -7)$, $C(-7, 11, 6)$ of a triangle. Find the canonical equations of the bisector of the exterior angle at the vertex A.

1015. Given the vertices $A(1,-2,-4)$, $B(3,1,-3)$, $C(5,1,-7)$ of a triangle. Find the parametric equations of the altitude drawn from the vertex B.

1016. Given the line $\begin{cases} 2x-5y+z-3=0, \\ x+2y-z+2=0. \end{cases}$

Calculate the projections on the coordinate axes of a direction vector $\boldsymbol{a}$ of this line. Find the general expression for the projection on the coordinate axes of any direction vector of this line.

1017. Given the line

$$\begin{cases} 2x-y+3z+1=0, \\ 3x+y-z-2=0. \end{cases}$$

Find the resolution (with respect to the basis $\boldsymbol{i}, \boldsymbol{j}, \boldsymbol{k}$) of a direction vector $\boldsymbol{a}$ of this line. Write the general expression for the resolution (with respect to the basis $\boldsymbol{i}, \boldsymbol{j}, \boldsymbol{k}$) of any direction vector of this line.

1018. Write the canonical equations of the straight line through the point $M_1(2,3,-5)$ and parallel to the line

$$\begin{cases} 3x-y+2z-7=0, \\ x+3y-2z+3=0. \end{cases}$$

1019. Write the canonical equations of the following lines:

1) $\begin{cases} x-2y+3z-4=0, \\ 3x+2y-5z-4=0; \end{cases}$ 2) $\begin{cases} 5x+y+z=0, \\ 2x+3y-2z+5=0; \end{cases}$

3) $\begin{cases} x-2y+3z+1=0, \\ 2x+y-4z-8=0. \end{cases}$

1020. Write the parametric equations of the following lines:

1) $\begin{cases} 2x+3y-z-4=0, \\ 3x-5y+2z+1=0, \end{cases}$ 2) $\begin{cases} x+2y-z-6=0, \\ 2x-y+z+1=0. \end{cases}$

1021. In each of the following, prove that the given lines are parallel:

1) $\dfrac{x+2}{3}=\dfrac{y-1}{-2}=\dfrac{z}{1}$ and $\begin{cases} x+y-z=0, \\ x-y-5z-8=0; \end{cases}$

2) $x=2t+5$, $y=-t+2$, $z=t-7$ and $\begin{cases} x+3y+z+2=0, \\ x-y-3z-2=0; \end{cases}$

3) $\begin{cases} x+y-3z+1=0, \\ x-y+z+3=0 \end{cases}$ and $\begin{cases} x+2y-5z-1=0, \\ x-2y+3z-9=0. \end{cases}$

1022. In each of the following, prove that the given lines are mutually perpendicular:

1) $\dfrac{x}{1}=\dfrac{y-1}{-2}=\dfrac{z}{3}$ and $\begin{cases} 3x+y-5z+1=0, \\ 2x+3y-8z+3=0; \end{cases}$

2) $x = 2t + 1, y = 3t - 2, z = -6t + 1$ and $\begin{cases} 2x + y - 4z + 2 = 0, \\ 4x - y - 5z + 4 = 0; \end{cases}$

3) $\begin{cases} x + y - 3z - 1 = 0, \\ 2x - y - 9z - 2 = 0 \end{cases}$ and $\begin{cases} 2x + y + 2z + 5 = 0, \\ 2x - 2y - z + 2 = 0. \end{cases}$

1023. Find the acute angle between the lines
$\frac{x-3}{1} = \frac{y+2}{-1} = \frac{z}{\sqrt{2}}, \frac{x+2}{1} = \frac{y-3}{1} = \frac{z+5}{\sqrt{2}}.$

1024. Find the obtuse angle between the lines
$x = 3t - 2, y = 0, z = -t + 3; x = 2t - 1, y = 0, z = t - 3.$

1025. Determine the cosine of the angle between the lines
$\begin{cases} x - y - 4z - 5 = 0, \\ 2x + y - 2z - 4 = 0; \end{cases} \begin{cases} x - 6y - 6z + 2 = 0, \\ 2x + 2y + 9z - 1 = 0. \end{cases}$

1026. Prove that the lines represented by the parametric equations $x = 2t - 3,\ y = 3t - 2,\ z = -4t + 6$ and $x = t + 5,\ y = -4t - 1,\ z = t - 4$ intersect.

1027. Given the lines
$\frac{x+2}{2} = \frac{y}{-3} = \frac{z-1}{4}, \frac{x-3}{l} = \frac{y-1}{4} = \frac{z-7}{2};$
find the value of l for which they intersect.

1028. Prove that the condition for the two lines
$\frac{x-a_1}{l_1} = \frac{y-b_1}{m_1} = \frac{z-c_1}{n_1}$ and $\frac{x-a_2}{l_2} = \frac{y-b_2}{m_2} = \frac{z-c_2}{n_2}$
to lie in the same plane can be written in the form

$$\begin{vmatrix} a_2 - a_1 & b_2 - b_1 & c_2 - c_1 \\ l_1 & m_1 & n_1 \\ l_2 & m_2 & n_2 \end{vmatrix} = 0.$$

1029. Write the equations of the straight line which passes through the point $M_1(-1, 2, -3)$, is perpendicular to the vector $\boldsymbol{a} = \{6, -2, -3\}$ and cuts the line $\frac{x-1}{3} = \frac{y+1}{2} = \frac{z-3}{-5}$.

1030. Write the equations of the straight line which passes through the point $M_1(-4, -5, 3)$ and cuts the two lines
$\frac{x+1}{3} = \frac{y+3}{-2} = \frac{z-2}{-1}, \frac{x-2}{2} = \frac{y+1}{3} = \frac{z-1}{-5}.$

1031. Write the parametric equations of the common perpendicular to the two straight lines

$$x = 3t - 7, y = -2t + 4, z = 3t + 4$$

and

$$x = t + 1, y = 2t - 9, z = -t - 12.$$

1032. Given the equations of motion of the point $M(x, y, z)$:

$$x = 3 - 4t,\ y = 5 + 3t,\ z = -2 + 12t.$$

Determine the speed v of M.

1033. Given the equations of motion of the point $M(x, y, z)$:

$$x = 5 - 2t,\ y = -3 + 2t,\ z = 5 - t.$$

Determine the distance d covered by M in the interval of time from $t_1 = 0$ to $t_2 = 7$.

1034. Write the equations of motion of the point $M(x, y, z)$ which starts from $M_0(3, -1, -5)$ and moves rectilinearly and uniformly in the direction of the vector $\boldsymbol{s} = \{-2, 6, 3\}$, with a speed $v = 21$.

1035. Write the equations of motion of the point $M(x, y, z)$ which is in uniform rectilinear motion and has covered the distance from $M_1(-7, 12, 5)$ to $M_2(9, -4, -3)$ in the interval of time from $t_1 = 0$ to $t_2 = 4$.

1036. The point $M(x, y, z)$ starts from $M_0(20, -18, -32)$ and moves rectilinearly and uniformly in the direction opposite to that of the vector $\boldsymbol{s} = \{3, -4, -12\}$, with a speed $v = 26$. Write the equations of motion of M and find the position of M at the instant $t = 3$.

1037. The points $M(x, y, z)$ and $N(x, y, z)$ are in uniform rectilinear motion: M starts from $M_0(-5, 4, -5)$ and moves with a speed $v_M = 14$ in the direction of the vector $\boldsymbol{s} = \{3, -6, 2\}$, and N starts from $N_0(-5, 16, -6)$ and moves with a speed $v_N = 13$ in the direction opposite to that of the vector $\mathbf{r} = \{-4, 12, -3\}$. Write the equations of motion for each of these two points, show that their paths intersect, and find:

(1) the point of intersection P of their paths;

(2) the time in which the point M travels from M_0 to P;

(3) the time in which the point N travels from N_0 to P;

(4) the lengths of the segments M_0P and N_0P.

43. Miscellaneous Problems Involving the Equation of a Plane and the Equations of a Straight Line

1038. Prove that the line $x = 3t - 2,\ y = -4t + 1,\ z = 4t - 5$ is parallel to the plane $4x - 3y - 6z - 5 = 0$.

1039. Prove that the line

$$\begin{cases} 5x - 3y + 2z - 5 = 0, \\ 2x - y - z - 1 = 0 \end{cases}$$

lies in the plane $4x - 3y + 7z - 7 = 0$.

1040. In each of the following, find the point of intersection of the given line and the given plane:

1) $\frac{x-1}{1}=\frac{y+1}{-2}=\frac{z}{6}, 2x+3y+z-1=0;$

2) $\frac{x+3}{3}=\frac{y-2}{-1}=\frac{z+1}{-5}, x-2y+z-15=0;$

3) $\frac{x+2}{-2}=\frac{y-1}{3}=\frac{z-3}{2}, x+2y-2z+6=0.$

1041. Write the canonical equations of the straight line passing through the point $M_0(2,-4,-1)$ and the mid point of that segment of the line

$$\begin{cases} 3x+4y+5z-26=0, \\ 3x-3y-2z-5=0 \end{cases}$$

intercept by the planes $5x+3y-4z+11=0, 5x+3y-4z-41=0.$

1042. Find the equations of the straight line passing through the point $M_0(2,-3,-5)$ and perpendicular to the plane $6x-3y-5z+2=0.$

1043. Find the equation of the plane passing through the point $M_0(1,-1,-1)$ and perpendicular to the straight line

$$\frac{x+3}{2}=\frac{y-1}{-3}=\frac{z+2}{4}.$$

1044. Find the equation of the plane passing through the point $M_0(1,-2,1)$ and perpendicular to the straight line

$$\begin{cases} x-2y+z-3=0, \\ x+y-z+2=0. \end{cases}$$

1045. Find the value of m for which the line

$$\frac{x+1}{3}=\frac{y-2}{m}=\frac{z+3}{-2}$$

is parallel to the plane $x-3y+6z+7=0.$

1046. What is the value of C for which the line

$$\begin{cases} 3x-2y+z+3=0, \\ 4x-3y+4z+1=0 \end{cases}$$

is parallel to the plane

$$2x-y+Cz-2=0?$$

1047. What are the values of A and D for which the line

$$x=3+4t, y=1-4t, z=-3+t$$

lies in the plane $Ax+2y-4z+D=0$?

1048. What are the values of A and B for which the plane $Ax+By+3z-5=0$ is perpendicular to the line

$$x=3+2t, y=5-3t, z=-2-2t?$$

1049. What are the values of l and C for which the line $\frac{x-2}{l}=\frac{y+1}{4}=\frac{z-5}{-3}$ is perpendicular to the plane $3x-2y+Cz+1=0$?

1050. Find the projection of the point $P(2,-1,3)$ on the line

$$x=3t,\ y=5t-7,\ z=2t+2.$$

1051. Find the point Q symmetric to the point $P(4,1,6)$ with respect to the line

$$\begin{cases} x-y-4z+12=0, \\ 2x+y-2z+3=0. \end{cases}$$

1052. Find the point Q symmetric to the point $P(2,-5,7)$ with respect to the straight line passing through the points $M_1(5,4,6)$ and $M_2(-2,-17,-8)$.

1053. Find the projection of the point $P(5,2,-1)$ on the plane $2x-y+3z+23=0$.

1054. Find the point Q symmetric to the point $P(1,3,-4)$ with respect to the plane $3x+y-2z=0$.

1055. In the plane Oxy, find a point P such that the sum of its distances from the points $A(-1,2,5)$ and $B(11,-16,10)$ will have the least value.

1056. In the plane Oxz, find a point P such that the difference of its distances from the points $M_1(3,2,-5)$ and $M_2(8,-4,-13)$ will have the greatest value.

1057. In the plane $2x-3y+3z-17=0$, find a point P such that the sum of its distances from the points $A(3,-4,7)$ and $B(-5.-14,17)$ will have the least value.

1058. In the plane $2x+3y-4z-15=0$, find a point P such that the difference of its distances from the points $M_1(5,2,-7)$ and $M_2(7,-25,10)$ will have the greatest value.

1059. The point $M(x,y,z)$ starts from $M_0(15,-24,-16)$ and moves rectilinearly and uniformly, with a speed $v=12$, in the direction of the vector $\mathbf{s}=\{-2,2,1\}$. Show that the path of M intersects the plane $3x+4y+7z-17=0$, and find.

1) their point of intersection P;

2) the time in which the point M travels from M_0 to P;

3) the length of the segment M_0P.

1060. The point $M(x, y, z)$ starts from $M_0(28, -30, -27)$ and moves rectilinearly and uniformly, with a speed $v = 12.5$, along the perpendicular dropped from the point M_0 to the plane $15x - 16y - 12z + 26 = 0$. Write the equations of motion of M and find:

1) the point P in which the path of M pierces this plane;

2) the time in which the point M travels from M_0 to P;

3) the length of the segment M_0P.

1061. The point $M(x, y, z)$ starts from $M_0(11, -21, 20)$ and moves rectilinearly and uniformly, with a speed $v = 12$, in the direction of the vector $\mathbf{s} = \{-1, 2, -2\}$. Find the time in which M traces that segment of its path which is intercepted by the parallel planes

$2x + 3y + 5z - 41 = 0, 2x + 3y + 5z + 31 = 0.$

1062. Find the distance d of the point $P(1, -1, -2)$ from the line

$$\frac{x+3}{3} = \frac{y+2}{2} = \frac{z-8}{-2}.$$

1063. Find the distance d from the point $P(2, 3, -1)$ to the following lines:

1) $\frac{x-5}{3} = \frac{y}{2} = \frac{z+25}{-2}$; 2) $x = t + 1;\ y = t + 2, z = 4t + 13$;

3) $\begin{cases} 2x - 2y + z + 3 = 0, \\ 3x - 2y + 2z + 17 = 0. \end{cases}$

1064. Show that the lines

$$\begin{cases} 2x + 2y - z - 10 = 0, \\ x - y - z - 22 = 0, \end{cases} \quad \frac{x+7}{3} = \frac{y-5}{-1} = \frac{z-9}{4}$$

are parallel and calculate the distance d between them.

1065. Write the equation of the plane which passes through the point $M_1(1, 2, -3)$ and is parallel to the lines

$$\frac{x-1}{2} = \frac{y+1}{-3} = \frac{z-7}{3},$$

$$\frac{x+5}{3} = \frac{y-2}{-2} = \frac{z+3}{-1}.$$

1066. Prove that the equation of the plane which passes through the point $M_0(x_0, y_0, z_0)$ and is parallel to the lines

$$\frac{x-a_1}{l_1} = \frac{y-b_1}{m_1} = \frac{z-c_1}{n_1}, \frac{x-a_2}{l_2} = \frac{y-b_2}{m_2} = \frac{z-c_2}{n_2}$$

can be written in the form

$$\begin{vmatrix} x-x_0 & y-y_0 & z-z_0 \\ l_1 & m_1 & n_1 \\ l_2 & m_2 & n_2 \end{vmatrix} = 0.$$

1067. Prove that the equation of the plane which passes through the points $M_1(x_1, y_1, z_1)$, $M_2(x_2, y_2, z_2)$ and is parallel to the line

$$\frac{x-a}{l} = \frac{y-b}{m} = \frac{z-c}{n}$$

can be written in the form $\begin{vmatrix} x-x_1 & y-y_1 & z-z_1 \\ x_2-x_1 & y_2-y_1 & z_2-z_1 \\ l & m & n \end{vmatrix} = 0.$

1068. Find the equation of the plane passing through the line $x = 2t+1, y = -3t+2, z = 2t-3$ and through the point $M_1(2, -2, 1)$.

1069. Prove that the equation of the plane passing through the line

$$x = x_0 + lt, y = y_0 + mt, z = z_0 + nt$$

and through the point $M_1(x_1, y_1, z_1)$ can be put in the form

$$\begin{vmatrix} x-x_1 & y-y_1 & z-z_1 \\ x_1-x_0 & y_1-y_0 & z_1-z_0 \\ l & m & n \end{vmatrix} = 0.$$

1070. Prove that the lines

$\frac{x-1}{2} = \frac{y+2}{-3} = \frac{z-5}{4}$ and $x = 3t+7, y = 2t+2, z = -2t+1$

lie in the same plane, and find the equation of this plane.

1071. Prove that, if the two lines

$\frac{x-a_1}{l_1} = \frac{y-b_1}{m_1} = \frac{z-c_1}{n_1}, \frac{x-a_2}{l_2} = \frac{y-b_2}{m_2} = \frac{z-c_2}{n_2}$

intersect, the equation of the plane in which they lie can be written in the form

$$\begin{vmatrix} x-a_1 & y-b_1 & z-c_1 \\ l_1 & m_1 & n_1 \\ l_2 & m_2 & n_2 \end{vmatrix} = 0.$$

1072. Find the equation of the plane passing through the two parallel lines

$\frac{x-2}{3} = \frac{y+1}{2} = \frac{z-3}{-2}, \frac{x-1}{3} = \frac{y-2}{2} = \frac{z+3}{-2}.$

1073. Prove that the equation of the plane passing through the two parallel lines

$x = a_1 + lt,\ y = b_1 + mt,\ z = c_1 + nt$ and $x = a_2 + lt,\ y = b_2 + mt,\ z = c_2 + nt$, can be written in the form

$$\begin{vmatrix} x-a_1 & y-b_1 & z-c_1 \\ a_2-a_1 & b_2-b_1 & c_2-c_1 \\ l & m & n \end{vmatrix} = 0.$$

1074. Find the projection of the point $C(3, -4, -2)$ on the plane passing through the parallel lines

$$\frac{x-5}{13} = \frac{y-6}{1} = \frac{z+3}{-4}, \frac{x-2}{13} = \frac{y-3}{1} = \frac{z+3}{-4}.$$

1075. Find the point Q symmetric to the point $P(3, -4, -6)$ with respect to the plane which passes through $M_1(-6, 1, -5), M_2(7, -2, -1)$ and $M_3(10, -7, 1)$.

1076. Find the point Q symmetric to the point $P(-3, 2, 5)$ with respect to the plane passing through the lines

$$\begin{cases} x - 2y + 3z - 5 = 0, \\ x - 2y - 4z + 3 = 0; \end{cases} \begin{cases} 3x + y + 3z + 7 = 0, \\ 5x - 3y + 2z + 5 = 0. \end{cases}$$

1077. Write the equation of the plane passing through the line $x = 3t + 1, y = 2t + 3, z = -t - 2$

and parallel to the line $\begin{cases} 2x - y + z - 3 = 0, \\ x + 2y - z - 5 = 0. \end{cases}$

1078. Prove that the equation of the plane passing through the line $\frac{x-x_1}{l_1} = \frac{y-y_1}{m_1} = \frac{z-z_1}{n_1}$ and parallel to the line

$x = x_0 + lt, y = y_0 + mt, z = z_0 + nt,$

can be written in the form $\begin{vmatrix} x-x_1 & y-y_1 & z-z_1 \\ l & m & n \\ l_1 & m_1 & n_1 \end{vmatrix} = 0.$

1079. Write the equation of the plane passing through the line $\frac{x-1}{2} = \frac{y+2}{-3} = \frac{z-2}{2}$ and perpendicular to the plane $3x + 2y - z - 5 = 0$.

1080. Prove that the equation of the plane passing through the line $x = x_0 + lt, y = y_0 + mt, z = z_0 + nt$ and perpendicular to the plane $Ax + By + Cz + D = 0$

can be written in the form $\begin{vmatrix} x-x_0 & y-y_0 & z-z_0 \\ l & m & n \\ A & B & C \end{vmatrix} = 0.$

1081. Find the canonical equations of the straight line which passes through the point $M_0(3, -2, -4)$, is parallel to the plane $3x - 2y - 3z - 7 = 0$, and cuts the line

$$\frac{x-2}{3} = \frac{y+4}{-2} = \frac{z-1}{2}.$$

1082. Find the parametric equations of the line parallel to the planes $3x + 12y - 3z - 5 = 0, 3x - 4y + 9z + 7 = 0$ and cutting the lines $\frac{x+5}{2} = \frac{y-3}{-4} = \frac{z+1}{3}, \frac{x-3}{-2} = \frac{y+1}{3} = \frac{z-2}{4}$.

1083. In each of the following, calculate the shortest distance between the two given lines:

1) $\frac{x+7}{3} = \frac{y+4}{4} = \frac{z+3}{-2}; \frac{x-21}{6} = \frac{y+5}{-4} = \frac{z-2}{-1}$;

2) $x = 2t - 4, y = -t + 4, z = -2t - 1; x = 4t - 5, y = -3t + 5, z = -5t + 5$;

3) $\frac{x+5}{3} = \frac{y+5}{2} = \frac{z-1}{-2}; x = 6t + 9, y = -2t, z = -t + 2$.

44. The Sphere

In rectangular cartesian coordinates, a sphere with centre $C(\alpha, \beta, \gamma)$ and radius r is represented by the equation $(x-\alpha)^2 + (y-\beta)^2 + (z-\gamma)^2 = r^2$. A sphere of radius r and with centre at the origin has the equation $x^2 + y^2 + z^2 = r^2$.

1084. In each of the following, find the equation of the sphere determined by the stated conditions:

1) the sphere, of radius $r = 9$, has its centre at $C(0, 0, 0)$;
2) the sphere, of radius $r = 2$, has its centre at $C(5, -3, 7)$;
3) the sphere passes through the origin and has its centre at $C(4, -4, -2)$;
4) the sphere passes through the point $A(2, -1, -3)$ and has its centre at $C(3, -2, 1)$;
5) the points $A(2, -3, 5)$ and $B(4, 1, -3)$ are the extremities of a diameter of the sphere;
6) the centre of the sphere is at the origin, and the plane $16x - 15y - 12z + 75 = 0$ is tangent to the sphere;
7) the centre of the sphere is at $C(3, -5, -2)$ and, the plane $2x - y - 3z + 11 = 0$ is tangent to the sphere;
8) the sphere passes through the three points $M_1(3, 1, -3)$, $M_2(-2, 4, 1)$, $M_3(-5, 0, 0)$, and its centre lies in the plane $2x + y - z + 3 = 0$;
9) the sphere passes through the four points $M_1(1, -2, -1)$, $M_2(-5, 10, -1)$, $M_3(4, 1, 11)$ and $M_4(-8, -2, 2)$.

1085. Write the equation of a sphere of radius $r = 3$ which touches the plane $x + 2y + 2z + 3 = 0$ at the point $M_1(1, 1, -3)$.

1086. Calculate the radius R of the sphere which touches the planes $3x + 2y - 6z - 15 = 0, 3x + 2y - 6z + 55 = 0$.

1087. A sphere has its centre on the line

$$\begin{cases} 2x + 4y - z - 7 = 0, \\ 4x + 5y + z - 14 = 0 \end{cases}$$

and touches the planes $x + 2y - 2z - 2 = 0$, $x + 2y - 2z + 4 = 0$. Find the equation of the sphere.

1088. Find the equation of a sphere tangent to the two parallel planes. $6x - 3y - 2z - 35 = 0$, $6x - 3y - 2z + 63 = 0$, if $M_1(5, -1, -1)$ is its point of contact with one of these planes.

1089. Write the equation of the sphere with centre $C(2, 3, -1)$ which cuts off a chord of length 16 on the line

$$\begin{cases} 5x - 4y + 3z + 20 = 0, \\ 3x - 4y + z - 8 = 0. \end{cases}$$

1090. In each of the following, determine the coordinates of the centre C and the radius r of the given sphere:

1) $(x - 3)^2 + (y + 2)^2 + (z - 5)^2 = 16$;
2) $(x + 1)^2 + (y - 3)^2 + z^2 = 9$;
3) $x^2 + y^2 + z^2 - 4x - 2y + 2z - 19 = 0$;
4) $x^2 + y^2 + z^2 - 6z = 0$;
5) $x^2 + y^2 + z^2 + 20y = 0$.

1091. Write the parametric equations of that diameter of the sphere $x^2 + y^2 + z^2 + 2x - 6y + z - 11 = 0$ which is perpendicular to the plane $5x - y + 2z - 17 = 0$.

1092. Write the canonical equations of that diameter of the sphere $x^2 + y^2 + z^2 - x + 3y + z - 13 = 0$ which is parallel to the line $x = 2t - 1, y = -3t + 5, z = 4t + 7$.

1093. In each of the following, determine whether the point $A(2, -1, 3)$ is inside, on, or outside the given sphere

1) $(x - 3)^2 + (y + 1)^2 + (z - 1)^2 = 4$;
2) $(x + 14)^2 + (y - 11)^2 + (z + 12)^2 = 625$;
3) $(x - 6)^2 + (y - 1)^2 + (z - 2)^2 = 25$;
4) $x^2 + y^2 + z^2 - 4x + 6y - 8z + 22 = 0$;
5) $x^2 + y^2 + z^2 - x + 3y - 2z - 3 = 0$.

1094. In each of the following, calculate the shortest distance from the point A to the given sphere:

1) $A(-2, 6, -3)$, $x^2 + y^2 + z^2 = 4$;
2) $A(9, -4, -3)$, $x^2 + y^2 + z^2 + 14x - 16y - 24z + 241 = 0$;
3) $A(1, -1, 3)$, $x^2 + y^2 + z^2 - 6x + 4y - 10z - 62 = 0$.

1095. In each of the following, determine whether the given plane cuts, touches or passes outside the given sphere

1) $z = 3, x^2 + y^2 + z^2 - 6x + 2y - 10z + 22 = 0$;

2) $y = 1, x^2 + y^2 + z^2 + 4x - 2y - 6z + 14 = 0$;

3) $x = 5, x^2 + y^2 + z^2 - 2x + 4y - 2z - 4 = 0$.

1096. In each of the following, determine whether the given line intersects, touches or passes outside the given sphere

1) $x = -2t + 2, y = 3t - \frac{7}{2}, z = t - 2, x^2 + y^2 + z^2 + x - 4y - 3z + \frac{1}{2} = 0$;

2) $\frac{x-5}{3} = \frac{x}{2} = \frac{z+25}{-2}, x^2 + y^2 + z^2 - 4x - 6y + 2z - 67 = 0$;

3) $\begin{cases} 2x - y + 2z - 12 = 0, \\ 2x - 4y - z + 6 = 0, \end{cases} x^2 + y^2 + z^2 - 2x + 2y + 4z - 43 = 0.$

1097. On the sphere $(x-1)^2 + (y+2)^2 + (z-3)^2 = 25$, find the point M_1 nearest to the plane $3x - 4z + 19 = 0$, and calculate the distance d of the point M_1 from this plane.

1098. Find the centre C and the radius R of the circle

$$\begin{cases} (x-3)^2 + (y+2)^2 + (z-1)^2 = 100, \\ 2x - 2y - z + 9 = 0. \end{cases}$$

1099. The points $A(3, -2, 5)$ and $B(-1, 6, -3)$ are the extremities of a diameter of a circle. Write the equations of this circle, given that it passes through the point $C(1, -4, 1)$.

1100. The point $C(1, -1, -2)$ is the centre of the circle which cuts off a chord of length 8 on the line

$$\begin{cases} 2x - y + 2z - 12 = 0, \\ 4x - 7y - z + 6 = 0. \end{cases}$$

Write the equations of the circle.

1101. Find the equations of the circle through the three points $M_1(3, -1, -2), M_2(1, 1, -2)$ and $M_3(-1, 3, 0)$.

1102. Given the two spheres $(x - m_1)^2 + (y - n_1)^2 + (z - p_1)^2 = R_1^2$, $(x - m_2)^2 + (y - n_2)^2 + (z - p_2)^2 = R_2^2$, which intersect in a circle lying in a plane τ. Prove that an equation of the form

$\alpha[(x - m_1)^2 + (y - n_1)^2 + (z - p_1)^2 - R_1^2] +$

$\beta[(x - m_2)^2 + (y - n_2)^2 + (z - p_2)^2 - R_2^2] = 0$

can be made, by an appropriate choice of the numbers α and β, to represent the plane τ as well as every sphere passing through the circle of intersection of the given spheres.

1103. Write the equation of the plane passing through the curve of intersection of the two spheres
$2x^2 + 2y^2 + 2z^2 + 3x - 2y + z - 5 = 0, x^2 + y^2 + z^2 - x + 3y - 2z + 1 = 0.$

1104. Write the equation of the sphere passing through the origin and through the circle
$$\begin{cases} x^2 + y^2 + z^2 = 25, \\ 2x - 3y + 5z - 5 = 0. \end{cases}$$

1105. Find the equation of the sphere passing through the circle
$$\begin{cases} x^2 + y^2 + z^2 - 2x + 3y - 6z - 5 = 0, \\ 5x + 2y - z - 3 = 0 \end{cases}$$
and through the point $A(2, -1, 1)$.

1106. Find the equation of the sphere passing through the two circles
$$\begin{cases} x^2 + z^2 = 25, \\ y = 2, \end{cases} \quad \begin{cases} x^2 + z^2 = 16, \\ y = 3. \end{cases}$$

1107. Find the equation of the tangent plane to the sphere $x^2 + y^2 + z^2 = 49$ at the point $M_1(6, -3, -2)$.

1108. Prove that the plane $2x - 6y + 3z - 49 = 0$ is tangent to the sphere $x^2 + y^2 + z^2 = 49$. Calculate the coordinates of their point of contact.

1109. Find the values of a for which the plane
$x + y + z = a$ is tangent to the sphere $x^2 + y^2 + z^2 = 12$.

1110. Find the equation of the tangent plane to the sphere $(x-3)^2 + (y-1)^2 + (z+2)^2 = 24$ at the point $M_1(-1, 3, 0)$.

1111. The point $M_1(x_1, y_1, z_1)$ lies on the sphere $x^2 + y^2 + z^2 = r^2$. Write the equation of the tangent plane at M_1 to the sphere.

1112. Find the condition that the plane $Ax + By + Cz + D = 0$ should be tangent to the sphere $x^2 + y^2 + z^2 = R^2$.

1113. The point $M_1(x_1, y_1, z_1)$ lies on the sphere
$(x-\alpha)^2 + (y-\beta)^2 + (z-\gamma)^2 = r^2$. Write the equation of the tangent plane at M_1 to the sphere.

1114. Through the points of intersection of the line
$x = 3t - 5, y = 5t - 11, z = -4t + 9$ and the sphere
$(x+2)^2 + (y-1)^2 + (z+5)^2 = 49$, tangent planes are passed to the sphere. Write the equations of these planes.

1115. Write the equations of the tangent planes to the sphere
$x^2 + y^2 + z^2 = 9$ which are parallel to the plane
$x + 2y - 2z + 15 = 0$.

1116. Write the equations of the tangent planes to the sphere

$$(x-3)^2+(y+2)^2+(z-1)^2=25$$

which are parallel to the plane $4x+3z-17=0$.

1117. Write the equations of the tangent planes to the sphere

$$x^2+y^2+z^2-10x+2y+26z-113=0$$

which are parallel to the lines

$$\frac{x+5}{2}=\frac{y-1}{-3}=\frac{z-13}{2},\ \frac{x+7}{3}=\frac{y+1}{-2}=\frac{z-8}{0}.$$

1118. Prove that two planes tangent to the sphere $x^2+y^2+z^2+2x-6y+4z-15=0$ can be passed through the line

$$\begin{cases} 8x-11y+8z-30=0, \\ x-y-2z=0; \end{cases}$$

write the equations of these planes.

1119. Prove that no tangent plane to the sphere $x^2+y^2+z^2-4x+2y-4z+4=0$ can be passed through the line $\frac{x+6}{2}=y+3=z+1$.

1120. Prove that only one tangent plane to the sphere $x^2+y^2+z^2-2x+6y+2z+8=0$ can be passed through the line $x=4t+4,\ y=3t+1,\ z=t+1$; write the equation of this plane.

45. The Equations of the Plane, Straight Line and Sphere in Vector Notation

As used below, the symbol $M(\boldsymbol{r})$ means that $\boldsymbol{r}$ is the radius vector of the point M.

1121. Find the equation of the plane α which passes through the point $M_0(\boldsymbol{r}_0)$ and has $\boldsymbol{n}$ as its normal vector.

Solution *: Let $M(\boldsymbol{r})$ be an arbitrary point. It lies in the plane α if, and only if, the vector $\overline{M_0M}$ is perpendicular to $\boldsymbol{n}$. The perpendicularity condition for vectors is that their scalar product should be zero. Hence, $\overline{M_0M}\perp\boldsymbol{n}$ if, and only if,

$$\overline{M_0M}\cdot\boldsymbol{n}=0. \tag{1}$$

Let us express the vector $\overline{M_0M}$ in terms of the radii vectors of its terminal and initial points : $\overline{M_0M}=\boldsymbol{r}-\boldsymbol{r}_0$.

Hence, from (1), we find $(\boldsymbol{r}-\boldsymbol{r}_0)\boldsymbol{n}=0$. (2)

This is the equation of the plane α in vector notation; it is satisfied by the radius vector $\boldsymbol{r}$ of a point M if, and only if, M lies in the plane α; [$\boldsymbol{r}$ is called the current radius vector of equation (2).]

1122. Prove that the equation $\boldsymbol{rn} + D = 0$ represents a plane perpendicular to the vector $\boldsymbol{n}$. Write the equation of this plane in terms of coordinates, if $\boldsymbol{n} = \{A, B, C\}$.

1123. Given a unit vector $\boldsymbol{n}^0$ and a number $p > 0$. Prove that the equation $\boldsymbol{rn}^0 - p = 0$ represents a plane perpendicular to the vector $\boldsymbol{n}^0$ and that p is the distance from the origin to the plane. Write the equation of this plane in terms of coordinates, if the vector $\boldsymbol{n}^0$ makes angles α, β, γ with the coordinate axes.

1124. Calculate the distance d from the point $M_1(\mathbf{r}_1)$ to the plane $\boldsymbol{rn}^0 - p = 0$. Also, express the distance d in terms of coordinates, if $r_1 = \{x_1, y_1, z_1\}, \boldsymbol{n}^0 = \{\cos\alpha, \cos\beta, \cos\gamma\}$.

1125. Given the two points $M_1(r_1)$ and $M_2(\boldsymbol{r}_2)$. Write the equation of the plane through the point M_1 and perpendicular to the vector $\overline{M_1M_2}$. Also, write the equation of this plane in terms of coordinates, if

$r_1 = \{x_1, y_1, z_1\}$,

$r_2 = \{x_2, y_2, z_2\}$.

1126. Write the equation of the plane through the point $M_0(r_0)$ and parallel to the vectors $\boldsymbol{a}_1$ and $\boldsymbol{a}_2$. Also, write the equation of this plane in terms of coordinates, if

$r_0 = \{x_0, y_0, z_0\}$,

$\boldsymbol{a}_1 = \{l_1, m_1, n_1\}$,

$\boldsymbol{a}_2 = \{l_2, m_2, n_2\}$.

1127. Write the equation of the plane through three points $M_1(\mathbf{r}_1), M_2(\mathbf{r}_2)$, $M_3(\mathbf{r}_3)$. Also, write the equation of this plane in terms of coordinates, if

$$r_1 = \{x_1, y_1, z_1\}, r_2 = \{x_2, y_2, z_2\}, r_3 = \{x_3, y_3, z_3\}.$$

1128. Write the equation of the plane through the point $M_0(\boldsymbol{r}_0)$ and perpendicular to the planes

$$\boldsymbol{rn}_1 + D_1 = 0, \boldsymbol{rn}_2 + D_2 = 0.$$

Also, write the equation of this plane in terms of coordinates, given that

$$\boldsymbol{r}_0 = \{x_0, y_0, z_0\}, \boldsymbol{n}_1 = \{A_1, B_1, C_1\},$$
$$\boldsymbol{n}_2 = \{A_2, B_2, C_2\}.$$

* Problems 1121 and 1129, whose solutions are given here, are essential for a correct understanding of the remaining problems of this section.

1129. Prove that the equation

$$[(\boldsymbol{r} - \boldsymbol{r}_0)\boldsymbol{a}] = 0$$

represents the straight line through the point $M_0(\boldsymbol{r}_0)$ and parallel to the vector $\boldsymbol{a}$, i.e., that this equation is satisfied by the radius vector $\boldsymbol{r}$ of a point $M(\mathbf{r})$ if, and only if, M lies on the indicated straight line.

Proof. Consider an arbitrary point $M(\boldsymbol{r})$. Let $\boldsymbol{r}$ satisfy the given equation; by the rule for subtraction of vectors, $r - r_0 == \overline{M_0M}$; since $[(\boldsymbol{r} - r_0)a] = 0$, it follows that $[\overline{M_0M}\ \boldsymbol{a}] = 0$; hence, the vector $\overline{M_0M}$ is collinear with the vector $\boldsymbol{a}$. This means that the point M actually lies on the straight line which passes through M_0 in the direction of the vector $\boldsymbol{a}$. Conversely, let M lie on this line. Then $\overline{M_0M}$ is collinear with a. It follows that $[\overline{M_0M}\ \boldsymbol{a}] = 0$; but $\overline{M_0M} = \boldsymbol{r} - \boldsymbol{r}_0$; hence $[(\boldsymbol{r} - \boldsymbol{r}_0)\,a] = 0$. Thus, the given equation is satisfied by the radius vector r of a point M if, and only if, M lies on the indicated straight line (r is called the current radius vector of the equation).

1130. Prove that the equation $[\boldsymbol{ra}] = \boldsymbol{m}$

represents a straiaght line parallel to the vector $\boldsymbol{a}$.

1131. Prove that the parametric equation

$$\boldsymbol{r} = r_0 + \boldsymbol{at},$$

where t is a variable parameter, represents a straight line passing through the point $M_0(r_0)$ (i.e., as t varies, the point $M(\boldsymbol{r})$ moves along this line). Write the canonical equations of the line in terms of coordinates, if

$$r_0 = \{x_0, y_0, z_0\}, \boldsymbol{a} = \{l, m, n\}.$$

1132. A straight line passes through two points $M_1(r_1)$ and $M_2(r_2)$. Write its equations in the form indicated in Problems 1129, 1130, 1131.

1133. Write the equation of the plane through the point $M_1(r_1)$ and perpendicular to the line $\boldsymbol{r} = r_0 + \mathbf{a}t$. Also, write the equation of this plane in terms of coordinates, given that

$$r_1 = \{x_1, y_1, z_1\}, \boldsymbol{a} = \{l, m, n\}.$$

1134. Write the equation of the plane through the point $M_0(r_0)$ and parallel to the lines $[\boldsymbol{ra}_1] = \mathbf{m}_1$, $[\boldsymbol{ra}_2] = \boldsymbol{m}_2$.

1135. Write the equation of the plane through the point $M_0(r_0)$ and perpendicular to the planes

$$\boldsymbol{rn}_1 + D_1 = 0, \boldsymbol{rn}_2 + D_2 = 0.$$

1136. A straight line passes through the point $M_0(r_0)$ and is perpendicular to the plane $\boldsymbol{rn} + D = 0$. Write its equation in parametric form. Also, write its canonical equations in terms of coordinates, if

$$r_0 = \{x_0, y_0, z_0\}, \boldsymbol{n} = \{A, B, C\}.$$

1137. A straight line passes through the point $M_0(\boldsymbol{r}_0)$ and is parallel to the planes $\boldsymbol{rn}_1 + D_1 = 0$, $\boldsymbol{rn}_2 + D_2 = 0$. Write its equation in parametric form. Also, write its canonical equations in terms of coordinates, if

$$\boldsymbol{r}_0 = \{x_0, y_0, z_0\}, \boldsymbol{n}_1 = \{A_1, B_1, C_1\}, \boldsymbol{n}_2 = \{A_2, B_2, C_2\}.$$

1138. Find the condition for the line $\boldsymbol{r} = \boldsymbol{r}_0 + \boldsymbol{a}t$ to lie in the plane $\boldsymbol{rn} + D = 0$. Also, write this condition in terms of coordinates, if

$$\boldsymbol{r}_0 = \{x_0, y_0, z_0\}, \boldsymbol{a} = \{l, m, n\}, \boldsymbol{n} = \{A, B, C\}.$$

1139. Find the equation of the plane passing through the line $\boldsymbol{r} = \boldsymbol{r}_0 + \boldsymbol{a}_1 t$ and parallel to the line

$$[\boldsymbol{ra}_2] = \boldsymbol{m}.$$

1140. Find the condition for the two lines

$\boldsymbol{r} = \boldsymbol{r}_1 + \boldsymbol{a}_1 t$ and $\boldsymbol{r} = \boldsymbol{r}_2 + \boldsymbol{a}_2 t$ to lie in the same plane.

1141. Find the radius vector of the point of intersection of the line $\boldsymbol{r} = \boldsymbol{r}_0 + \boldsymbol{a}t$ and the plane $\boldsymbol{rn} + D = 0$. Calculate the coordinates x, y, z of the point of intersection, given that

$$\boldsymbol{r}_0 = \{x_0, y_0, z_0\}, \boldsymbol{a} = \{l, m, n\}, \boldsymbol{n} = \{A, B, C\}.$$

1142. Find the radius vector of the projection of $M_1(\boldsymbol{r}_1)$ on the plane $\boldsymbol{rn} + D = 0$. Calculate the coordinates x, y, z of this projection, if

$$\boldsymbol{r}_1 = \{x_1, y_1, z_1\}, \boldsymbol{n} = \{A, B, C\}.$$

1143. Find the radius vector of the projection of the point $M_1(\boldsymbol{r}_1)$ on the line $\boldsymbol{r} = \boldsymbol{r}_0 + \boldsymbol{a}t$. Calculate the coordinates x, y, z of this projection, if

$$\boldsymbol{r}_1 = \{x_1, y_1, z_1,\} \boldsymbol{r}_0 = \{x_0, y_0, z_0\}, \boldsymbol{a} = \{l, m, n\}.$$

1144. Calculate the distance d of the point $M_1(\boldsymbol{r}_1)$ from the line $\boldsymbol{r} = \boldsymbol{r}_0 + \boldsymbol{a}t$. Also, express the distance d in terms of coordinates, if

$$\boldsymbol{r}_1 = \{x_1, y_1, z_1\}, \boldsymbol{r}_0 = \{x_0, y_0, z_0\}, \boldsymbol{a} = \{l, m, n\}.$$

1145. Calculate the shortest distance d between the two skew lines

$$\boldsymbol{r} = \boldsymbol{r}_1 + \boldsymbol{a}_1 t \text{ and } \boldsymbol{r} = \boldsymbol{r}_2 + \boldsymbol{a}_2 t.$$

Also, express this distance d in terms of coordinates, if

$$\boldsymbol{r}_1 = \{x_1, y_1, z_1\}, \boldsymbol{r}_2 = \{x_2, y_2, z_2\},$$
$$\boldsymbol{a}_1 = \{l_1, m_1, n_1\}, \boldsymbol{a}_2 = \{l_2, m_2, n_2\}.$$

1146. Prove that the equation $(\boldsymbol{r} - \boldsymbol{r}_0)^2 = R^2$

represents a sphere with centre $C(\boldsymbol{r}_0)$ and with radius equal to R (i.e., that this equation is satisfied by the radius vector $\boldsymbol{r}$ of a point M if, and only if, M lies on the indicated sphere).

1147. Find the radii vectors of the points of intersection of the line

$$\boldsymbol{r} = \boldsymbol{a}t \text{ and the sphere } \boldsymbol{r}^2 = R^2.$$

Calculate the coordinates of these points of intersection, given that

$$\boldsymbol{a} = \{l, m, n\}.$$

1148. Find the radii vectors of the points of intersection of the line

$$\boldsymbol{r} = \boldsymbol{r}_0 + \boldsymbol{a}t \text{ and the sphere } (\boldsymbol{r} - \boldsymbol{r}_0)^2 = R^2.$$

Calculate the coordinates of these points of intersection, given that

$$\boldsymbol{r}_0 = \{x_0, y_0, z_0\}, \boldsymbol{a} = \{l, m, n\}.$$

1149. The point $M_1(\boldsymbol{r}_1)$ lies on the sphere $(\boldsymbol{r} - \boldsymbol{r}_0)^2 = R^2$.

Find the equation of the tangent plane at M_1 to the sphere.

1150. Write the equation of the sphere which has its centre at $C(\boldsymbol{r}_1)$ and is tangent to the plane $\boldsymbol{rn} + D = 0$. Also, write the equation of this sphere in terms of coordinates, given that

$$\boldsymbol{r}_1 = \{x_1, y_1, z_1\}, \boldsymbol{n} = \{A, B, C\}.$$

1151. Write the equations of the planes which touch the sphere $\boldsymbol{r}^2 = R^2$ and are parallel to the plane $\boldsymbol{rn} + D = 0$.

Also, write the equations of these tangent planes in terms of coordinates, if $\boldsymbol{n} = \{A, B, C\}$.

1152. Through the points of intersection of the line

$$\boldsymbol{r} = \boldsymbol{r}_0 + \mathbf{a}t \text{ and the sphere } (\boldsymbol{r} - \boldsymbol{r}_0)^2 = R^2,$$

tangent planes are drawn to the sphere. Write the equations of these planes. Also, write their equations in terms of coordinates, given that

$$\boldsymbol{r}_0 = \{x_0, y_0, z_0\}, \boldsymbol{a} = \{l, m, n\}.$$

46. Quadric Surfaces

The ellipsoid is defined as the surface represented, in a rectangular cartesian coordinate system, by the equation

$$\frac{x^2}{a^2} + \frac{y^2}{b^2} + \frac{z^2}{c^2} = 1. \quad (1)$$

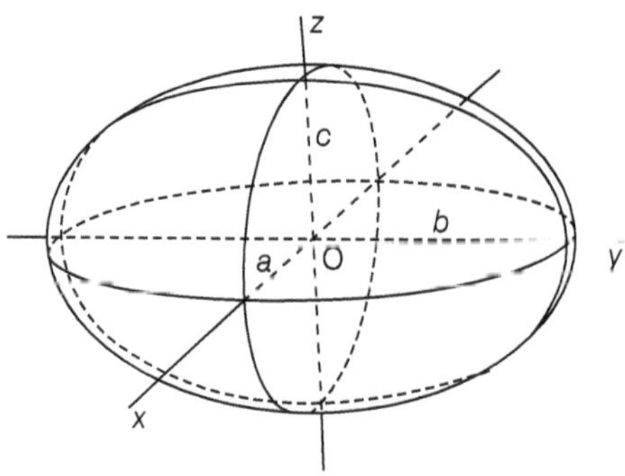

Fig. 47.

Equation (1) is called the canonical equation of an ellipsoid.The quantities a, b, c are the semi-axes of an ellipsoid (see Fig.47). If the semi-axes are all of a different length, the ellipsoid is referred to as triaxial; if any two of them are equal, the ellipsoid is a surface of revolution. For example, when $a = b$, Oz will be the axis of revolution. When $a = b < c$, the ellipsoid is called a prolate ellipsoid of revolution; when $a = b > c$, it is called an oblate ellipsoid of revolution. In the case $a = b = c$, the ellipsoid is a sphere.

The hyperboloids are defined as the surfaces represented, in a rectangular cartesian coordinate system, by the equations

$$\frac{x^2}{a^2} + \frac{y^2}{b^2} - \frac{z^2}{c^2} = 1, \quad (2) \qquad\qquad \frac{x^2}{a^2} + \frac{y^2}{b^2} - \frac{z^2}{c^2} = -1. \quad (3)$$

The hyperboloid represented by equation (2) is referred to as a hyperboloid of one sheet (Fig. 48), and that represented by equation (3), as a hyperboloid of two sheets (Fig. 49); equations (2) and (3) are said to be the canonical equations of the respective hyper boloids. The quantities a, b, c are called the semi-axes of a hyperboloid. Fig. 48 shows the semi-axes a and b of the hyperboloid of one sheet represented by equation (2). In the case of the hyperboloid of two sheets represented by equation (3), one of the semi-axes (namely, c) is shown in Fig. 49. When $a = b$, the hyperboloids represented by equations (2) and (3) are surfaces of revolution.

The paraboloids are the surfaces represented, in some rectangular cartesian coordinate system, by the equations

$$\frac{x^2}{p} + \frac{y^2}{q} = 2z, \quad (4) \qquad\qquad \frac{x^2}{p} - \frac{y^2}{q} = 2z, \quad (5)$$

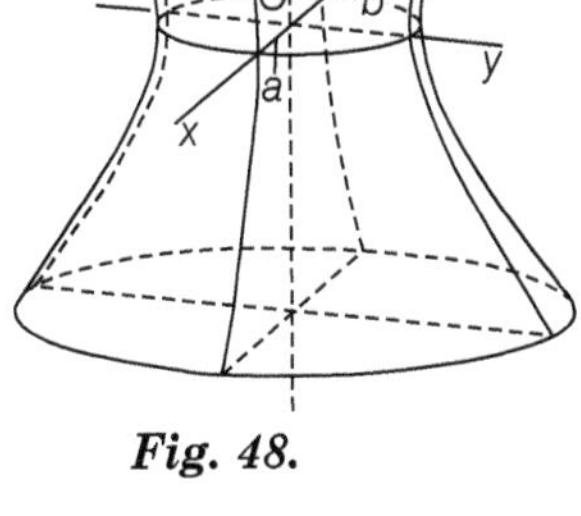

Fig. 48.

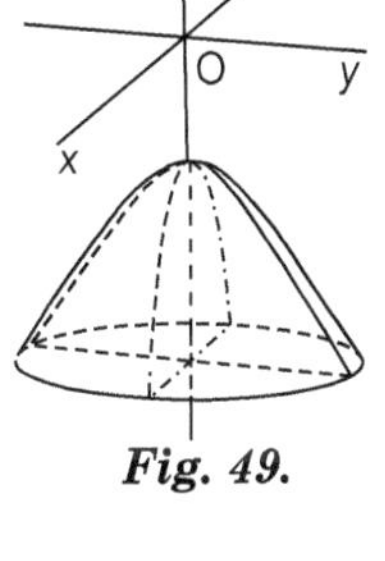

Fig. 49.

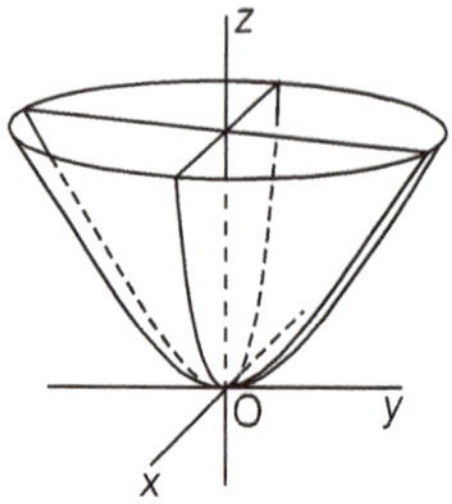

Fig. 50.

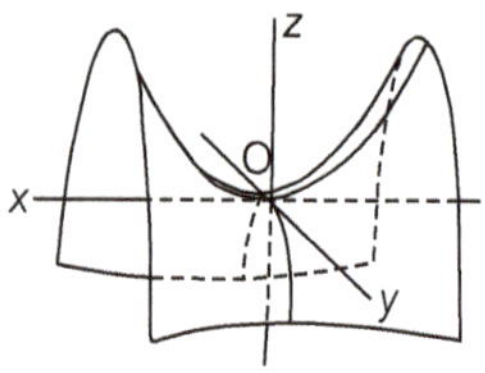

Fig. 51.

where p and q are positive numbers (called the parameters of a paraboloid). The paraboloid represented by equation (4) is referred to as an elliptic paraboloid (Fig. 50), and that represented by equation (5), as a hyperbolic paraboloid (Fig. 51). Equations (4) and (5) are said to be the canonical equations of the respective paraboloids. In the case $p = q$, the paraboloid represented by equation (4) is a surface of revolution (about Oz).

Let us now consider the transformation of space referred to as uniform compression (or uniform elongation).

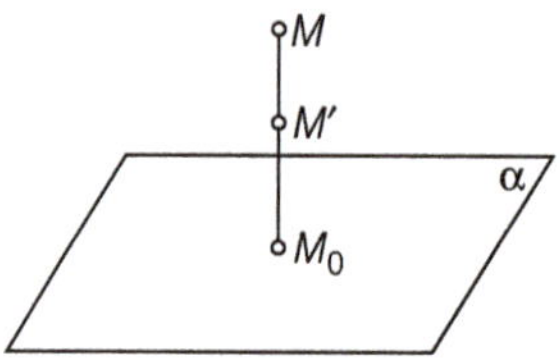

Fig. 52.

Take any plane and denote it by the letter α; also, choose some positive number q. Let M be an arbitrary point (in space) not lying in the plane α, and M_0 the foot of the perpendicular dropped from the point M onto the plane α. Next, move the point M along the line MM_0 to a new position M' such that the condition

$$M_0M' = q \cdot M_0M$$

will be satisfied and the point M will remain on the same side of the plane α as before the motion (Fig. 52). Let all points in space not lying in the plane α be subjected to this procedure; let the points situated in the plane α remain in their original positions. Thus, all points in space (except those lying in the plane α) will undergo a shift such that the distance of each point from the plane α will change to q times its former value. This motion of points in space is referred to as a uniform compression of space towards the plane α; the number q is called the coefficient of compression.

Let there be given a surface F; under a uniform compression of space, the points of which F is made up will move to new positions so as to make up a surface F'. We shall agree to say that the surface F' has been obtained from F by a uniform compression of space.Many quadric surfaces (in fact, all except the hyperbolic paraboloid) can be obtained from surfaces of revolution by uniform compressions.

Example. Prove that an arbitrary triaxial ellipsoid.

$$\frac{x^2}{a^2}+\frac{y^2}{b^2}+\frac{z^2}{c^2}=1,$$

can be obtained from the sphere

$$x^2+y^2+z^2=a^2$$

by two consecutive uniform compressions of space towards the coordinate planes : towards the plane Oxy with coefficient of compression $q_1=\frac{c}{a}$, and towards the plane Oxz with coefficient of compression $q_2=\frac{b}{a}$.

Proof : Let us perform a uniform compression of space towards the plane Oxy (with coefficient of compression $q_1=\frac{c}{a}$), and let this compression carry a point $M(x, y, z)$ into $M'(x', y', z')$. Let us express the coordinates x', y', z' of the point M' in terms of the coordinates x, y, z of M. Since the line $M'M$ is perpendicular to the plane Oxy, it follows that $x'=x$, $y'=y$. On the other hand, since the distance from the point M' to the plane Oxy is equal to the distance from the point M to Oxy multiplied by the number $q_1=\frac{c}{a}$, it follows that $z'=\frac{c}{a}z$. We thus obtain the required expressions :

$$x'=x, y'=y, z'=\frac{c}{a}z, \tag{6}$$

or

$$x=x', y=y', z=\frac{a}{c}z'. \tag{7}$$

Suppose that $M(x, y, z)$ is an arbitrary point of the sphere $x^2+y^2+z^2=a^2$.

Replacing x, y, z by their expressions (7), we get

$$x'^2+y'^2+\frac{a^2}{c^2}z'^2=a^2, \text{ whence } \quad \frac{x'^2}{a^2}+\frac{y'^2}{a^2}+\frac{z'^2}{c^2}=1.$$

Consequently, the point $M'(x', y', z')$ lies on an ellipsoid of revolution. If we now perform, in an analogous manner, a compression of space towards the plane Oxz according to the formulas

$$x'=x'', y'=\frac{a}{b}y'', z'=z'',$$

we shall obtain a triaxial ellipsoid, namely, the one whose equation has been given.

Note also that the hyperboloid of one sheet and the hyperbolic paraboloid are ruled surfaces, that is, surfaces made up of straight lines; these straight lines are called the rectilinear generators of the respective surfaces.

The hyperboloid of one sheet :

$$\frac{x^2}{a^2}+\frac{y^2}{b^2}-\frac{z^2}{c^2}=1$$

has two systems of rectilinear generators, represented by the equations

$$\begin{cases}\alpha\left(\dfrac{x}{a}+\dfrac{z}{c}\right)=\beta\left(1+\dfrac{y}{b}\right),\\ \beta\left(\dfrac{x}{a}-\dfrac{z}{c}\right)=\alpha\left(1-\dfrac{y}{b}\right),\end{cases}\quad \begin{cases}\alpha\left(\dfrac{x}{a}+\dfrac{z}{c}\right)=\beta\left(1-\dfrac{y}{b}\right)\\ \beta\left(\dfrac{x}{a}-\dfrac{z}{c}\right)=\alpha\left(1+\dfrac{y}{b}\right),\end{cases}$$

where α and β are some numbers, not both zero. The hyperbolic paraboloid

$$\frac{x^2}{p}-\frac{y^2}{q}=2z$$

also has two systems of rectilinear generators, represented by the equations

$$\begin{cases}\alpha\left(\dfrac{x}{\sqrt{\mu}}+\dfrac{y}{\sqrt{q}}\right)=2\beta z,\\ \beta\left(\dfrac{x}{\sqrt{p}}-\dfrac{y}{\sqrt{q}}\right)=\alpha,\end{cases}\quad \begin{cases}\alpha\left(\dfrac{x}{\sqrt{p}}-\dfrac{y}{\sqrt{q}}\right)=2\beta z,\\ \beta\left(\dfrac{x}{\sqrt{p}}+\dfrac{y}{\sqrt{q}}\right)=\alpha.\end{cases}$$

A conical surface, or cone, is the surface generated by a moving straight line (the generator) which passes through a fixed point S and always intersects a given curve L. The point S is called the vertex of the cone; the curve L is called its directing curve.

A cylindrical surface, or cylinder, is the surface generated by a moving straight line (the generator) which has a fixed direction and always intersects a given curve L (the directing curve).

1153. Show that the plane $x-2=0$ intersects the ellipsoid $\dfrac{x^2}{16}+\dfrac{y^2}{12}+\dfrac{z^2}{4}=1$ in an ellipse; find the semi-axes and vertices of this ellipse.

1154. Show that the plane $z+1=0$ intersects the hyperboloid of one sheet $\dfrac{x^2}{32}-\dfrac{y^2}{18}+\dfrac{z^2}{2}=1$ in a hyporbola; find the semi-axes and vertices of the hyperbola.

1155. Show that the plane $y+6=0$ intersects the hyperbolic paraboloid

$$\frac{x^2}{5}-\frac{y^2}{4}=6z$$

in a parabola; find the parameter and vertex of this parabola.

1156. Find the equations of the projections (on the coordinate planes) of the section of the elliptic paraboloid

$$y^2+z^2=x \text{ by the plane } x+2y-z=0.$$

1157. Identify the curve which is the section of the ellipsoid $\frac{x^2}{12}+\frac{y^2}{4}+\frac{z^2}{3}=1$ by the plane $2x-3y+4z-11=0$, and find the centre of the curve.

1158. Identify the curve which is the section of the hyperbolic paraboloid $\frac{x^2}{2}-\frac{z^2}{3}=y$ by the plane $3x-3y+4z+2=0$, and find the centre of the curve.

1159. In each of the following, identify the curve represented by the given equation and find its centre:

1) $\begin{cases} \frac{x^2}{3}+\frac{y^2}{6}=2z, \\ 3x-y+6z-14=0; \end{cases}$ 2) $\begin{cases} \frac{x^2}{4}-\frac{y^2}{3}=2z, \\ x-2y+2=0; \end{cases}$

3) $\begin{cases} \frac{x^2}{4}+\frac{y^2}{9}-\frac{z^2}{36}=1, \\ 9x-6y+2z-28=0. \end{cases}$

1160. Find the values of m for which the plane $x+mz-1=0$ intersects the hyperboloid of two sheets $x^2+y^2-z^2=-1$

1) in an ellipse;

2) in a hyperbola.

1161. Find the values of m for which the plane $x+my-2=0$ intersects the elliptic paraboloid

$$\frac{x^2}{2}+\frac{z^2}{3}=y$$

1) in an ellipse;

2) in a parabola.

1162. Prove that the elliptic paraboloid $\frac{x^2}{9}+\frac{z^2}{4}=2y$ has a point in common with the plane $2x-2y-z-10=0$, and find the coordinates of that point.

1163. Prove that the hyperboloid of two sheets

$$\frac{x^2}{3}+\frac{y^2}{4}-\frac{z^2}{25}=-1$$

has a point in common with the plane

$$5x+2z+5=0,$$

and find the coordinates of that point.

1164. Prove that the ellipsoid $\frac{x^2}{81}+\frac{y^2}{36}+\frac{z^2}{9}=1$

has a point in common with the plane

$$4x-3y+12z-54=0,$$

and find the coordinates of that point.

1165. Determine the value of m for which the plane

$$x-2y-2z+m=0$$

is tangent to the ellipsoid

$$\frac{x^2}{144}+\frac{y^2}{36}+\frac{z^2}{9}=1.$$

1166. Write the equation of the plane which is perpendicular to the vector $\boldsymbol{n}=\{2,-1,-2\}$ and touches the elliptic paraboloid

$$\frac{x^2}{3}+\frac{y^2}{4}=2z.$$

1167. Find the equations of the tangent planes to the ellipsoid

$$4x^2+16y^2+8z^2=1$$

which are parallel to the plane

$$x-2y+2z+17=0;$$

calculate the distance between these tangent planes.

1168. Find the equation of the surface into which the sphere

$$x^2+y^2+z^2=25$$

is transformed by a uniform compression of space towards the plane Oyz, if the coefficient of compression is equal to $\frac{3}{5}$.

1169. Find the equation of the surface into which the ellipsoid

$$\frac{x^2}{64}+\frac{y^2}{25}+\frac{z^2}{16}=1$$

is transformed by three consecutive uniform compressions of space towards the coordinate planes Oxy, Oxz, and Oyz, if the respective coefficients of compression are $\frac{3}{4}$, $\frac{4}{5}$ and $\frac{3}{4}$.

1170. Determine the coefficients q_1 and q_2 of two consecutive uniform compressions of space towards the coordinate planes Oxy, Oxz, which transform the sphere

$$x^2 + y^2 + z^2 = 25$$

into the ellipsoid $\frac{x^2}{25} + \frac{y^2}{16} + \frac{z^2}{4} = 1.$

1171. Find the equation of the surface generated by revolving the ellipse

$$\begin{cases} \frac{y^2}{b^2} + \frac{z^2}{c^2} = 1, \\ x = 0 \end{cases}$$

about the axis Oy.

Solution *. Let $M(x, y, z)$ be an arbitrary point in space, and let C denote the foot of the perpendicular dropped from the point M to the axis Oy (Fig.53). By revolving this perpendicular about the axis Oy, the point M can be moved to the plane Oyz; in this position, we shall denote it by $N(0, Y, Z)$. Since $CM = CN$ and $CM = \sqrt{x^2 + z^2}$, $CN = |Z|$, it follows that

$$|Z| = \sqrt{x^2 + z^2} \tag{1}$$

It is also evident that

$$Y = y \tag{2}$$

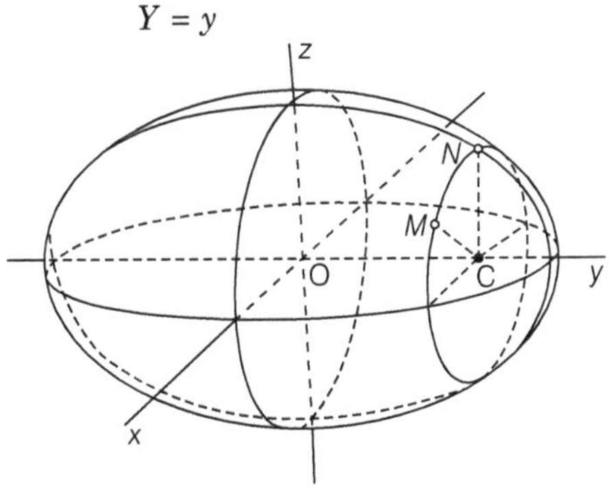

Fig. 53.

The point M lies on the required surface of revolution if, and only if, N lies on the given ellipse, that is, if

$$\frac{Y^2}{b^2} + \frac{Z^2}{c^2} = 1. \tag{3}$$

Hence, by (1) and (2), we obtain the equation for the coordinates of the point M:

$$\frac{y^2}{b^2} + \frac{x^2 + z^2}{c^2} = 1. \tag{4}$$

From the foregoing, it is clear that this equation is satisfied if, and only if, the point M lies on the required surface of revolution. Consequently, equation (4) is the desired equation of this surface.

1172. Find the equation of the surface generated by revolving the ellipse

$$\begin{cases} \dfrac{x^2}{a^2} + \dfrac{y^2}{b^2} = 1, \\ z = 0 \end{cases}$$

about the axis Ox.

1173. Find the equation of the surface generated by revolving the hyperbola

$$\begin{cases} \dfrac{x^2}{a^2} - \dfrac{z^2}{c^2} = 1, \\ y = 0 \end{cases}$$

about the axis Oz.

1174. Prove that the triaxial ellipsoid represented by the equation

$$\frac{x^2}{a^2} + \frac{y^2}{b^2} + \frac{z^2}{c^2} = 1$$

can be obtained by revolving the ellipse

$$\frac{x^2}{a^2} + \frac{y^2}{b^2} = 1, z = 0$$

about the axis Ox and by a subsequent uniform compression of space towards the plane Oxy.

1175. Prove that the hyperboloid of one sheet represented by the equation

$$\frac{x^2}{a^2} + \frac{y^2}{b^2} - \frac{z^2}{c^2} = 1$$

can be obtained by revolving the hyperbola

$$\begin{cases} \dfrac{x^2}{a^2} - \dfrac{z^2}{c^2} = 1, \\ y = 0 \end{cases}$$

about the axis Oz and by a subsequent uniform compression of space towards the plane Oxz.

1176. Prove that the hyperboloid of two sheets represented by the equation

$$\frac{x^2}{a^2} + \frac{y^2}{b^2} - \frac{z^2}{c^2} = -1$$

can be obtained by revolving the hyperbola $\begin{cases} \dfrac{z^2}{c^2} - \dfrac{x^2}{a^2} = 1, \\ y = 0 \end{cases}$

about the axis Oz and by a subsequent uniform compression of space towards the plane Oxz.

1177. Prove that the elliptic paraboloid represented by the equation

$$\frac{x^2}{p} + \frac{y^2}{q} = 2z$$

can be obtained by revolving the parabola $\begin{cases} x^2 = 2pz, \\ y = 0 \end{cases}$

about the axis Oz and by a subsequent uniform compression of space towards the plane Oxz.

1178. Find the equation of the surface generated by a parabola which moves so that the plane of the parabola is always perpendicular to the axis Oy, the axis of the parabola does not change its direction, and the vertex of the parabola slides along another parabola, represented by the equations

$$\begin{cases} y^2 = -2qz, \\ x = 0. \end{cases}$$

The moving parabola is represented, in one of its positions, by the equations

$$\begin{cases} x^2 = 2pz, \\ y = 0. \end{cases}$$

1179. Prove that the equation

$$z = xy$$

represents a hyperbolic paraboloid.

1180. In each of the following, find the points of intersection of the surface and the straight line:

1) $\dfrac{x^2}{81} + \dfrac{y^2}{36} + \dfrac{z^2}{9} = 1$ and $\dfrac{x-3}{3} = \dfrac{y-4}{-6} = \dfrac{z+2}{4}$;

2) $\dfrac{x^2}{16} + \dfrac{y^2}{9} - \dfrac{z^2}{4} = 1$ and $\dfrac{x}{4} = \dfrac{y}{-3} = \dfrac{z+2}{4}$;

3) $\dfrac{x^2}{5} + \dfrac{y^2}{3} = z$ and $\dfrac{x+1}{2} = \dfrac{y-2}{-1} = \dfrac{z+3}{-2}$;

4) $\dfrac{x^2}{9} - \dfrac{y^2}{4} = z$ and $\dfrac{x}{3} = \dfrac{y-2}{-2} = \dfrac{z+1}{2}$.

1181. Prove that the plane

$$2x - 12y - z + 16 = 0$$

intersects the hyperbolic paraboloid

$$x^2 - 4y^2 = 2z$$

in its rectilinear generators. Write the equations of these rectilinear generators.

1182. Prove that the plane $4x-5y-10z-20=0$ intersects the hyperboloid of one sheet

$$\frac{x^2}{25}+\frac{y^2}{16}-\frac{z^2}{4}=1$$

in its rectilinear generators. Write the equations of these rectilinear generators.

1183. Show that the point $M(1,3,-1)$ lies on the hyperbolic paraboloid

$$4x^2-z^2=y$$

and write the equations of its rectilinear generators passing through M.

1184. Write the equations of those rectilinear generators of the hyperboloid of one sheet

$$\frac{x^2}{4}+\frac{y^2}{9}-\frac{z^2}{16}=1$$

which are parallel to the plane

$$6x+4y+3z-17=0.$$

1185. Show that the point $A(-2,0,1)$ lies on the hyperbolic paraboloid

$$\frac{x^2}{4}-\frac{y^2}{9}=z$$

and determine the acute angle formed by its rectilinear generators passing through A.

1186. Write the equation of the cone whose vertex is at the origin and whose directing curve is given by the equations:

$$1)\ \begin{cases}\dfrac{x^2}{a^2}+\dfrac{y^2}{b^2}=1,\\ z=c;\end{cases}\quad 2)\ \begin{cases}\dfrac{x^2}{a^2}+\dfrac{z^2}{c^2}=1,\\ y=b;\end{cases}\quad 3)\ \begin{cases}\dfrac{y^2}{b^2}+\dfrac{z^2}{c^2}=1,\\ x=a.\end{cases}$$

1187. Prove that the equation $z^2=xy$

represents a cone with vertex at the origin.

1188. Find the equation of the cone whose vertex is at the origin and whose directing curve is given by the equations

$$\begin{cases}x^2-2z+1=0,\\ y-z+1=0.\end{cases}$$

1189. Find the equation of the cone whose vertex is at the point $(0,0,c)$ and whose directing curve is given by the equations

$$\begin{cases}\dfrac{x^2}{a^2}+\dfrac{y^2}{b^2}=1,\\ z=0.\end{cases}$$

1190. Find the equation of the cone whose vertex is at the point $(3, -1, -2)$ and whose directing curve is given by the equations

$$\begin{cases} x^2 + y^2 - z^2 = 1, \\ x - y + z = 0. \end{cases}$$

1191. The axis Oz is the axis of a circular cone with vertex at the origin; the point $M_1(3, -4, 7)$ lies on this cone. Write the equation of the cone.

1192. The axis Oy is the axis of a circular cone with vertex at the origin; the elements of this cone make an angle of $60°$ with the axis Oy. Write the equation of the cone.

1193. The straight line $\dfrac{x-2}{2} = \dfrac{y+1}{-2} = \dfrac{z+1}{-1}$

is the axis of a circular cone whose vertex lies in the plane Oyz; the point $M_1\left(1, 1, -\dfrac{5}{2}\right)$ lies on this cone. Write the equation of the cone.

1194. Write the equation of a circular cone which has the coordinate axes as its elements.

1195. Write the equation of the cone whose vertex is at the point $S(5, 0, 0)$ and whose elements are tangent to the sphere

$$x^2 + y^2 + z^2 = 9.$$

1196. Write the equation of the cone whose vertex is at the origin and whose elements are tangent to the sphere

$$(x+2)^2 + (y-1)^2 + (z-3)^2 = 9.$$

1197. Write the equation of the cone whose vertex is at $S(3, 0, -1)$ and whose elements are tangent to the ellipsoid

$$\frac{x^2}{6} + \frac{y^2}{2} + \frac{z^2}{3} = 1.$$

1198. Find the equation of the cylinder whose elements are parallel to the vector $l = \{2, -3, 4\}$ and whose directing curve is represented by the equations

$$\begin{cases} x^2 + y^2 = 9, \\ z = 1. \end{cases}$$

1199. Find the equation of the cylinder whose directing curve is represented by the equations

$$\begin{cases} x^2 - y^2 = z, \\ x + y + z = 0 \end{cases}$$

and whose elements are perpendicular to the plane of the directing curve.

1200. A cylinder with elements perpendicular to the

$$x + y - 2z - 5 = 0$$

plane is circumscribed about the sphere

$$x^2 + y^2 + z^2 = 1.$$

Find the equation of this cylinder.

1201. A cylinder with elements parallel to the line

$$x = 2t - 3,\ y = -t + 7,\ z = -2t + 5$$

is circumscribed about the sphere

$$x^2 + y^2 + z^2 - 2x + 4y + 2z - 3 = 0.$$

Find the equation of this cylinder.

1202. Find the equation of the circular cylinder which passes through the point $S(2, -1, 1)$ and has as its axis the line

$$x = 3t + 1,\ y = -2t - 2,\ z = t + 2.$$

1203. Find the equation of the cylinder circumscribed about the two spheres

$$(x - 2)^2 + (y - 1)^2 + z^2 = 25, x^2 + y^2 + z^2 = 25.$$

APPENDIX

THE ELEMENTS OF THE THEORY OF DETERMINANTS

1. Determinants of the Second Order and Systems of Two Equations of the First Degree in Two Unknowns

Consider a square array of four numbers a_1, a_2, b_1, b_2.

$$\begin{pmatrix} a_1 & b_1 \\ a_2 & b_2 \end{pmatrix} \tag{1}$$

The number $a_1b_2 - a_2b_1$ is referred to as the determinant of the second order associated with the array (1). This determinant is denoted by the symbol $\begin{vmatrix} a_1 & b_1 \\ a_2 & b_2 \end{vmatrix}$; accordingly,

$$\begin{vmatrix} a_1 & b_1 \\ a_2 & b_2 \end{vmatrix} = a_1b_2 - a_2b_1. \tag{2}$$

The numbers a_1, a_2, b_1, b_2 are called the elements of the determinant. The elements a_1, b_2 are said to lie on the principal diagonal of the determinant, and the elements a_2, b_1, on the secondary diagonal. Thus, a determinant of the second order is equal to the product of the elements on the principal diagonal minus the product of the elements on the secondary diagonal.

For example,

$$\begin{vmatrix} -3 & 2 \\ -1 & 4 \end{vmatrix} = -3\cdot 4 - (-1)\cdot 2 = -10.$$

Consider the system of two equations

$$\begin{cases} a_1x + b_1y = h_1, \\ a_2x + b_2y = h_2 \end{cases} \tag{3}$$

In two unknowns x, y. (The coefficients a_1, b_1, a_2, b_2 and the constant terms h_1, h_2 are assumed to be known.) Let us introduce the notation

$$\Delta = \begin{vmatrix} a_1 & b_1 \\ a_2 & b_2 \end{vmatrix}, \Delta_x = \begin{vmatrix} h_1 & b_1 \\ h_2 & b_2 \end{vmatrix}, \Delta_y = \begin{vmatrix} a_1 & h_1 \\ a_2 & h_2 \end{vmatrix} \quad (4)$$

The determinant Δ formed from the coefficients of the unknowns of the system (3) is called the determinant of the system. The determinant Δ_x is obtained by replacing the elements of the first column of Δ by the constant terms of the system (3); the determinant Δ_y is obtained from Δ by replacing the elements of the second column by the constant terms of the system (3).

If $\Delta \neq 0$, the system (3) has a unique solution, which is determined by the formulas

$$x = \frac{\Delta_x}{\Delta}, y = \frac{\Delta_y}{\Delta}. \quad (5)$$

If $\Delta = 0$, but at least one of the determinants Δ_x, Δ_y is different from zero, the system (3) has no solution at all (and the equations of the system are said to be inconsistent).

Finally, if $\Delta = 0$ and also $\Delta_x = \Delta_y = 0$, the system (3) has infinitely many solutions (in this case, one equation of the system is a consequence of the other).

Let $h_1 = h_2 = 0$ is equations (3); then the system (3) will be of the form

$$\begin{cases} a_1x + b_1y = 0, \\ a_2x + b_2y = 0. \end{cases} \quad (6)$$

A system of equations which has the form (6) is called a homogeneous system; such a system always possesses the zero solution $x = 0$, $y = 0$. If $\Delta \neq 0$, this solution is unique; but if $\Delta = 0$, the system (6) has, in addition, an infinite number of solutions other than the zero solution.

1204. Evaluate the determinants:

1) $\begin{vmatrix} -1 & 4 \\ -5 & 2 \end{vmatrix}$; 2) $\begin{vmatrix} 3 & -4 \\ 1 & 2 \end{vmatrix}$;

3) $\begin{vmatrix} 3 & 6 \\ 5 & 10 \end{vmatrix}$; 4) $\begin{vmatrix} 3 & 16 \\ 5 & 10 \end{vmatrix}$;

5) $\begin{vmatrix} a & 1 \\ a^2 & a \end{vmatrix}$; 6) $\begin{vmatrix} 1 & 1 \\ x_1 & x_2 \end{vmatrix}$;

7) $\begin{vmatrix} a+1 & b-c \\ a^2+a & ab-ac \end{vmatrix}$; 8) $\begin{vmatrix} \cos\alpha & -\sin\alpha \\ \sin\alpha & \cos\alpha \end{vmatrix}$.

1205. Solve the equations:

1) $\begin{vmatrix} 2 & x-4 \\ 1 & 4 \end{vmatrix} = 0$; 2) $\begin{vmatrix} 1 & 4 \\ 3x & x+22 \end{vmatrix} = 0$;

3) $\begin{vmatrix} x & x+1 \\ -4 & x+1 \end{vmatrix} = 0$; 4) $\begin{vmatrix} 3x & -1 \\ x & 2x-3 \end{vmatrix} = \dfrac{3}{2}$;

5) $\begin{vmatrix} x+1 & -5 \\ 1 & x-1 \end{vmatrix} = 0$; 6) $\begin{vmatrix} x^2-4 & -1 \\ x-2 & x+2 \end{vmatrix} = 0$;

7) $\begin{vmatrix} 4\sin x & 1 \\ 1 & \cos x \end{vmatrix} = 0$; 8) $\begin{vmatrix} \cos 8x & -\sin 5x \\ \sin 8x & \cos 5x \end{vmatrix} = 0.$

1206. Solve the inequalities:

1) $\begin{vmatrix} 3x-3 & 2 \\ x & 1 \end{vmatrix} > 0$; 2) $\begin{vmatrix} 1 & x+5 \\ 2 & x \end{vmatrix} < 0$;

3) $\begin{vmatrix} 2x-2 & 1 \\ 7x & 2 \end{vmatrix} > 5$; 4) $\begin{vmatrix} x & 3x \\ 4 & 2x \end{vmatrix} < 14.$

1207. In each of the following, find all solutions of the given system of equations:

1) $\begin{cases} 3x - 5y = 13, \\ 2x + 7y = 81; \end{cases}$ 2) $\begin{cases} 3y - 4x = 1, \\ 3x + 4y = 18; \end{cases}$

3) $\begin{cases} 2x - 3y = 6, \\ 4x - 6y = 5; \end{cases}$ 4) $\begin{cases} x - y\sqrt{3} = 1, \\ x\sqrt{3} - 3y = \sqrt{3}; \end{cases}$

5) $\begin{cases} ax + by = c, \\ bx - ay = d; \end{cases}$ 6) $\begin{cases} x\sqrt{5} - 5y = \sqrt{5}, \\ x - y\sqrt{5} = 5. \end{cases}$

1208. Determine the values of a and b for which the system

$$\begin{cases} 3x - ay = 1, \\ 6x + 4y = b \end{cases}$$

1) has a unique solution; 2) has no solutions;

3) has infinitely many solutions.

1209. Determine the value of a for which the homogeneous system

$$\begin{cases} 13x + 2y = 0, \\ 5x + ay = 0 \end{cases}$$

has a non-zero solution.

2. A Homogeneous System of Two Equations of the First Degree in Three Unknowns

Consider a system of two homogeneous equations,

$$\begin{cases} a_1x + b_1y + c_1z = 0, \\ a_2x + b_2y + c_2z = 0, \end{cases} \qquad (1)$$

in three unknowns x, y, z. Let us introduce the notation

$$\Delta_1 = \begin{vmatrix} b_1 & c_1 \\ b_2 & c_2 \end{vmatrix}, \Delta_2 = \begin{vmatrix} a_1 & c_1 \\ a_2 & c_2 \end{vmatrix}, \Delta_3 = \begin{vmatrix} a_1 & b_1 \\ a_2 & b_2 \end{vmatrix}.$$

If at least one of the determinants $\Delta_1, \Delta_2, \Delta_3$ is different from zero, all solutions of the system (1) will be determined by the formulas

$$x = \Delta_1 t, y = -\Delta_2 t, z = \Delta_3 t,$$

where t is an arbitrary number; each separate solution will be obtained by assigning some definite value to t.

For practical calculations, it will be helpful to observe, that the determinants $\Delta_1, \Delta_2, \Delta_3$ are obtained by striking out, in turn, each column of the array $\begin{pmatrix} a_1 & b_1 & c_1 \\ a_2 & b_2 & c_2 \end{pmatrix}$.

If the three determinants $\Delta_1, \Delta_2, \Delta_3$ are all equal to zero, then the coefficients of the equations of the system (1) are all in proportion. In this case, one equation of the system is a consequence of the other, and the system reduces in fact to a single equation. Such a system naturally has an infinite number of solutions; to obtain one of these, it is necessary to assign arbitrary values to two unknowns and find the third unknown from the equation.

1210. In each of the following, find all solutions of the given system of equations

1) $\begin{cases} 3x - 2y + 5z = 0, \\ x + 2y - 3z = 0; \end{cases}$ 2) $\begin{cases} 3x - 2y + z = 0, \\ 6x - 4y + 3z = 0; \end{cases}$

3) $\begin{cases} x - 3y + z = 0, \\ 2x - 9y + 3z = 0; \end{cases}$ 4) $\begin{cases} 3x - 2y + z = 0, \\ x + 2y - z = 0; \end{cases}$

5) $\begin{cases} 3x - 2y + z = 0, \\ x + 2y - 3z = 0; \end{cases}$ 6) $\begin{cases} 2x - y - 2z = 0, \\ x - 5y + 2z = 0; \end{cases}$

7) $\begin{cases} x + 2y - z = 0, \\ 3x - 5y + 2z = 0; \end{cases}$ 8) $\begin{cases} 3x - 5y + z = 0, \\ x + 2y - z = 0; \end{cases}$

9) $\begin{cases} x + 3y - z = 0, \\ 5x - 3y + z = 0; \end{cases}$ 10) $\begin{cases} ax + y + z = 0, \\ x - y + az = 0; \end{cases}$

11) $\begin{cases} ax + 2y - z = 0, \\ 2x + by - 3z = 0; \end{cases}$ 12) $\begin{cases} x - 3y + az = 0, \\ bx + 6y - z = 0. \end{cases}$

3. Determinants of the Third Order

Consider a square array of nine numbers, $a_1, a_2, a_3, b_1, b_2, b_3, c_1, c_2, c_3$:

$$\begin{pmatrix} a_1 & b_1 & c_1 \\ a_2 & b_2 & c_2 \\ a_3 & b_3 & c_3 \end{pmatrix} \quad (1)$$

The determinant of the third order associated with the array (1) is the number denoted by the symbol

$$\begin{vmatrix} a_1 & b_1 & c_1 \\ a_2 & b_2 & c_2 \\ a_3 & b_3 & c_3 \end{vmatrix}$$

and determined by the relation

$$\begin{vmatrix} a_1 & b_1 & c_1 \\ a_2 & b_2 & c_2 \\ a_3 & b_3 & c_3 \end{vmatrix} = a_1b_2c_3 + b_1c_2a_3 + c_1a_2b_3 - c_1b_2a_3 - b_1a_2c_3 - a_1c_2b_3 \quad (2)$$

The numbers $a_1, a_2, a_3, b_1, b_2, b_3, c_1, c_2, c_3$ are called the elements of the determinant. The diagonal containing the elements a_1, b_2, c_3 is called the principal diagonal of the determinant; the elements a_3, b_2, c_1 form the secondary diagonal. For practical computations, it will be helpful to note that the first three terms in the right-hand

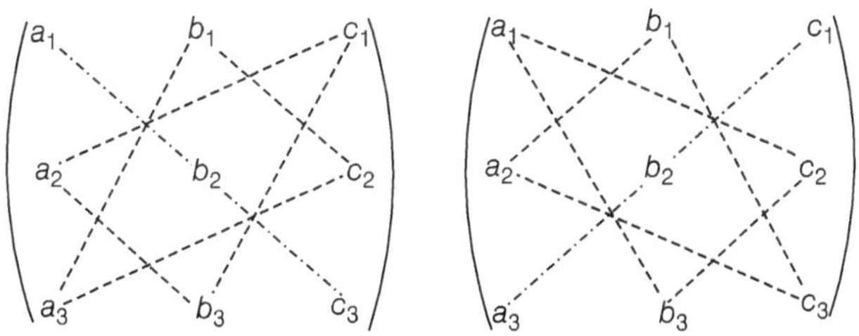

member of (2) are the products of the elements taken three at a time as shown by the various dashed and dotted lines in the left-hand diagram below. The remaining three terms of the right-hand member of (2) are obtained by multiplying the elements three at a time as shown by the various lines in the right-hand diagram, and then changing the sign of each resulting product.

In Problem 1211-1216, evaluate the given determinants of the third order.

1211. $\begin{vmatrix} 3 & -2 & 1 \\ -2 & 1 & 3 \\ 2 & 0 & -2 \end{vmatrix}$ **1212.** $\begin{vmatrix} 1 & 2 & 0 \\ 0 & 1 & 3 \\ 5 & 0 & -1 \end{vmatrix}.$

1213. $\begin{vmatrix} 2 & 0 & 5 \\ 1 & 3 & 16 \\ 0 & -1 & 10 \end{vmatrix}$ **1214.** $\begin{vmatrix} 2 & -1 & 3 \\ -2 & 3 & 2 \\ 0 & 2 & 5 \end{vmatrix}$

1215. $\begin{vmatrix} 2 & 1 & 0 \\ 1 & 0 & 3 \\ 0 & 5 & -1 \end{vmatrix}.$ **1216.** $\begin{vmatrix} 0 & a & a \\ a & 0 & a \\ a & a & 0 \end{vmatrix}.$

4. Properties of Determinants

Property 1. The value of a determinant is unchanged if all its columns are changed into rows so that each row is replaced by the like-numbered column; that is,

$$\begin{vmatrix} a_1 & b_1 & c_1 \\ a_2 & b_2 & c_2 \\ a_3 & b_3 & c_3 \end{vmatrix} = \begin{vmatrix} a_1 & a_2 & a_3 \\ b_1 & b_2 & b_3 \\ c_1 & c_2 & c_3 \end{vmatrix}$$

Property 2. The interchange of two columns or two rows of a determinant is equivalent to multiplying the determinant by -1. For example,

$$\begin{vmatrix} a_1 & b_1 & c_1 \\ a_2 & b_2 & c_2 \\ a_3 & b_3 & c_3 \end{vmatrix} = -\begin{vmatrix} a_1 & c_1 & b_1 \\ a_2 & c_2 & b_2 \\ a_3 & c_3 & b_3 \end{vmatrix}$$

Property 3. If a determinant has two identical columns or two identical rows, the value of the determinant is zero.

Property 4. Multiplying all elements of a column or row by any one number k is equivalent to multiplying the determinant by this number k. For example,

$$\begin{vmatrix} ka_1 & b_1 & c_1 \\ ka_2 & b_2 & c_2 \\ ka_3 & b_3 & c_3 \end{vmatrix} = k\begin{vmatrix} a_1 & b_1 & c_1 \\ a_2 & b_2 & c_2 \\ a_3 & b_3 & c_3 \end{vmatrix}$$

Property 5. If all elements of a column or row are zero, the determinant is zero. This property constitutes a special case (in which $k=0$) of the preceding property.

Property 6. If the corresponding elements of two columns or two rows of a determinant are proportional, the determinant is zero.

Property 7. If each element in the nth column (or the nth row) of a determinant is the sum of two terms, the determinant may be expressed as the sum of two determinants, of which one has in its nth column (or row) the first of the above-mentioned terms, while the other determinant has the second terms; the elements of the remaining columns (or rows) are the same for all the three determinants. For example,

$$\begin{vmatrix} a_1' + a_1'' & b_1 & c_1 \\ a_2' + a_2'' & b_2 & c_2 \\ a_3' + a_3'' & b_3 & c_3 \end{vmatrix} = \begin{vmatrix} a_1' & b_1 & c_1 \\ a_2' & b_2 & c_2 \\ a_3' & b_3 & c_3 \end{vmatrix} + \begin{vmatrix} a_1'' & b_1 & c_1 \\ a_2'' & b_2 & c_2 \\ a_3'' & b_3 & c_3 \end{vmatrix}$$

Property 8. If to the elements of a column (or row) of a determinant are added the corresponding elements of another column (or row), multiplied by any one number, the value of the determinant remains unchanged. For instance,

$$\begin{vmatrix} a_1 + kb_1 & b_1 & c_1 \\ a_2 + kb_2 & b_2 & c_2 \\ a_3 + kb_3 & b_3 & c_3 \end{vmatrix} = \begin{vmatrix} a_1 & b_1 & c_1 \\ a_2 & b_2 & c_2 \\ a_3 & b_3 & c_3 \end{vmatrix}.$$

Further properties of determinants are connected with the concept of cofactors and minors. The determinant obtained from a given determinant by striking out the row and the column, in the intersection of which a particular element lies, is called the minor of that element.

The cofactor of any element of a determinant is equal to the minor of that element taken with its sign unchanged if the sum of the position numbers of the row and column in which the element lies is even, or taken with opposite sign if this sum is odd.

We shall denote the cofactor of an element by the capital letter and subscript corresponding to the letter and subscript of the element.

Property 9. The determinant

$$\Delta = \begin{vmatrix} a_1 & b_1 & c_1 \\ a_2 & b_2 & c_2 \\ a_3 & b_3 & c_3 \end{vmatrix}$$

is equal to the sum of the products of the elements in any column (or row) by their cofactors. In other words, we have the following relations :

$$\Delta = a_1A_1 + a_2A_2 + a_3A_3, \Delta = a_1A_1 + b_1B_1 + c_1C_1,$$
$$\Delta = b_1B_1 + b_2B_2 + b_3B_3, \Delta = a_2A_2 + b_2B_2 + c_2C_2,$$
$$\Delta = c_1C_1 + c_2C_2 + c_3C_3, \Delta = a_3A_3 + b_3B_3 + c_3C_3.$$

In each of Problems 1217-1222, prove the validity of the given relation without expanding the determinants.

1217. $\begin{vmatrix} 3 & 2 & 1 \\ -2 & 3 & 2 \\ 4 & 5 & 3 \end{vmatrix} = \begin{vmatrix} 3 & 2 & 7 \\ -2 & 3 & -2 \\ 4 & 5 & 11 \end{vmatrix}$.

Hint. Use the property 8.

1218. $\begin{vmatrix} 1 & -2 & 3 \\ -2 & 1 & -5 \\ 3 & 2 & 7 \end{vmatrix} = \begin{vmatrix} 1 & 0 & 0 \\ -2 & -3 & 1 \\ 3 & 8 & -2 \end{vmatrix}$

Hint. Use the property 8.

1219. $\begin{vmatrix} a_1 & b_1 & c_1 \\ a_2 & b_2 & c_2 \\ a_1 + \alpha a_2 & b_1 + \alpha b_2 & c_1 + \alpha c_2 \end{vmatrix} = 0.$

Hint. Use the properties 7, 3, 6.

1220. $\begin{vmatrix} \beta b_1 + \gamma c_1 & b_1 & c_1 \\ \beta b_2 + \gamma c_2 & b_2 & c_2 \\ \beta b_3 + \gamma c_3 & b_3 & c_3 \end{vmatrix} = 0.$

Hint. Use the properties 7 and 6.

1221. $\begin{vmatrix} \sin^2\alpha & \cos^2\alpha & \cos 2\alpha \\ \sin^2\beta & \cos^2\beta & \cos 2\beta \\ \sin^2\gamma & \cos^2\gamma & \cos 2\gamma \end{vmatrix} = 0.$

1222. $\begin{vmatrix} 0 & -a & -b \\ a & 0 & -c \\ b & c & 0 \end{vmatrix} = 0.$

In Problems 1223-1227, evaluate the determinants by using the property 9 alone.

1223. $\begin{vmatrix} 1 & 1 & -1 \\ 1 & -1 & 1 \\ -1 & 1 & 1 \end{vmatrix}$. **1224.** $\begin{vmatrix} 1 & 17 & -7 \\ -1 & 13 & 1 \\ 1 & 7 & 1 \end{vmatrix}$.

1225. $\begin{vmatrix} 2 & 0 & 5 \\ 1 & 3 & 16 \\ 0 & -1 & 10 \end{vmatrix}$. **1226.** $\begin{vmatrix} 1 & 2 & 4 \\ -2 & 1 & -3 \\ 3 & -4 & 2 \end{vmatrix}$.

1227. $\begin{vmatrix} 1 & 1 & 1 \\ x & y & z \\ x^2 & y^2 & z^2 \end{vmatrix}$.

1228. By applying the property 8, transform the determinants given in Problems 1223-1227 so as to obtain two zero elements in some column (or row) of each determinant, and then evaluate the determinants by using the property 9.

In each of Problems 1229-1232, evaluate the given determinant.

1229. $\begin{vmatrix} 0 & a & b \\ a & 0 & a \\ b & a & 0 \end{vmatrix}$. **1230.** $\begin{vmatrix} 0 & \sin\alpha & \cot\alpha \\ \sin\alpha & 0 & \sin\alpha \\ \cot\alpha & \sin\alpha & 0 \end{vmatrix}$.

1231. $\begin{vmatrix} x & y & z \\ x^2 & y^2 & z^2 \\ x^3 & y^3 & z^3 \end{vmatrix}$. **1232.** $\begin{vmatrix} a & b & c \\ c & a & b \\ b & c & a \end{vmatrix}$.

1233. Prove the validity of the relations :

1) $\begin{vmatrix} 1 & \sin\alpha & \sin^2\alpha \\ 1 & \sin\beta & \sin^2\beta \\ 1 & \sin\gamma & \sin^2\gamma \end{vmatrix} = (\sin\alpha - \sin\beta)(\sin\beta - \sin\gamma)(\sin\gamma - \sin\alpha)$;

2) $\begin{vmatrix} 1 & 1 & 1 \\ \tan\alpha & \tan\beta & \tan\gamma \\ \tan^2\alpha & \tan^2\beta & \tan^2\gamma \end{vmatrix} = \dfrac{\sin(\alpha-\beta)\sin(\beta-\gamma)\sin(\gamma-\alpha)}{\cos^2\alpha\cos^2\beta\cos^2\gamma}$.

1234. Solve the equations :

1) $\begin{vmatrix} 1 & 3 & x \\ 4 & 5 & -1 \\ 2 & -1 & 5 \end{vmatrix} = 0$, 2) $\begin{vmatrix} 3 & x & -4 \\ 2 & -1 & 3 \\ x+10 & 1 & 1 \end{vmatrix} = 0$.

1235. Solve the inequalities :

1) $\begin{vmatrix} 3 & -2 & 1 \\ 1 & x & -2 \\ -1 & 2 & -1 \end{vmatrix} < 1$; 2) $\begin{vmatrix} 2 & x+2 & -1 \\ 1 & 1 & -2 \\ 5 & -3 & x \end{vmatrix} > 0$.

5. Solution and Analysis of a System of Three First-degree Equations in Three Unknowns

Consider the system of three equations

$$\begin{cases} a_1x + b_1y + c_1z = h_1, \\ a_2x + b_2y + c_2z = h_2, \\ a_3x + b_3y + c_3z = h_3 \end{cases} \qquad (1)$$

In the unknowns x, y, z. (the coefficients $a_1, b_1, \dots, c_3$ and the constant terms h_1, h_2, h_3 are assumed to be known.) Let us use the notation

$$\Delta = \begin{vmatrix} a_1 & b_1 & c_1 \\ a_2 & b_2 & c_2 \\ a_3 & b_3 & c_3 \end{vmatrix}, \Delta_x = \begin{vmatrix} h_1 & b_1 & c_1 \\ h_2 & b_2 & c_2 \\ h_3 & b_3 & c_3 \end{vmatrix}, \Delta_y = \begin{vmatrix} a_1 & h_1 & c_1 \\ a_2 & h_2 & c_2 \\ a_3 & h_3 & c_3 \end{vmatrix},$$

$$\Delta_z = \begin{vmatrix} a_1 & b_1 & h_1 \\ a_2 & b_2 & h_2 \\ a_3 & b_3 & h_3 \end{vmatrix}.$$

The determinant Δ formed from the coefficients of the unknowns of the system (1) is called the determinant of the system.

It will be helpful to note that the determinants Δ_x, Δ_y, and Δ_z are obtained from Δ by replacing its first, second, and third column respectively, by the column of the constant terms of the given system.

If $\Delta \neq 0$, the system (1) has a unique solution, which is determined by the formulas

$$x = \frac{\Delta_x}{\Delta}, \quad y = \frac{\Delta_y}{\Delta}, \quad z = \frac{\Delta_z}{\Delta}.$$

Suppose now that the determinant of the system is zero : $\Delta = 0$. If $\Delta = 0$, but at least one of the determinants $\Delta_x, \Delta_y, \Delta_z$ is different from zero, the system (1) has no solution. If $\Delta = 0$ and also $\Delta_x = 0, \Delta_y = 0, \Delta_z = 0$, the system (1) may, as before, have no solution at all; but if, under these conditions, the system (1) has at least one solution, then the system has an infinite number of different solutions.

A homogeneous system of three first-degree equations in three unknowns is a system of the form

$$\begin{cases} a_1x + b_1y + c_1z = 0, \\ a_2x + b_2y + c_2z = 0, \\ a_3x + b_3y + c_3z = 0, \end{cases} \qquad (2)$$

that is, a system of equations whose constant terms are all equal to zero. Obviously, such a system always possesses the solution $x = 0$, $y = 0$, $z = 0$, which is called the zero solution. If $\Delta \neq 0$, this solution is unique. But if $\Delta = 0$, the homogeneous system (2) has an infinite number of non-zero solutions.

In each of Problems 1236-1243, show that the given system of equations has a unique solution, and find this solution.

1236. $\begin{cases} x + y - z = 36, \\ x + z - y = 13, \\ y + z - x = 7. \end{cases}$

1237. $\begin{cases} x + 2y + z = 4, \\ 3x - 5y + 3z = 1, \\ 2x + 7y - z = 8. \end{cases}$

1238. $\begin{cases} 2x - 4y + 9z = 28, \\ 7x + 3y - 6z = -1, \\ 7x + 9y - 9z = 5. \end{cases}$

1239. $\begin{cases} 2x+y=5, \\ x+3z=16, \\ 5y-z=10. \end{cases}$

1240. $\begin{cases} x+y+z=36, \\ 2x-3z=-17, \\ 6x-5z=7. \end{cases}$

1241. $\begin{cases} 7x+2y+3z=15, \\ 5x-3y+2z=15, \\ 10x-11y+5z=36. \end{cases}$

1242. $\begin{cases} x+y+z=a, \\ x-y+z=b, \\ x+y-z=c. \end{cases}$

1243. $\begin{cases} x-y+z=a, \\ x+y-z=b, \\ y+z-x=c. \end{cases}$

1244. Find all solutions of the system

$$\begin{cases} x+2y-4z=1, \\ 2x+y-5z=-1, \\ x-y-z=-2. \end{cases}$$

1245. Find all solutions of the system

$$\begin{cases} 2x-y+z=-2, \\ x+2y+3z=-1, \\ x-3y-2z=3. \end{cases}$$

1246. Find all solutions of the system

$$\begin{cases} 3x-y+2z=5, \\ 2x-y-z=2, \\ 4x-2y-2z=-3. \end{cases}$$

1247. Determine the values of a and b for which the system

$$\begin{cases} 3x-2y+z=b, \\ 5x-8y+9z=3, \\ 2x+y+az=-1 \end{cases}$$

1) has a unique solution;

2) has no solutions;

3) has infinitely many solutions.

1248. Prove that, if the system

$$\begin{cases} a_1x + b_1y = c_1, \\ a_2x + b_2y = c_2, \\ a_3x + b_3y = c_3 \end{cases}$$

is consistent, then

$$\begin{vmatrix} a_1 & b_1 & c_1 \\ a_2 & b_2 & c_2 \\ a_3 & b_3 & c_3 \end{vmatrix} = 0.$$

1249. Find all solutions of the system

$$\begin{cases} 2x + y - z = 0, \\ x + 2y + z = 0, \\ 2x - y + 3z = 0, \end{cases}$$

1250. Find all solutions of the system

$$\begin{cases} x - y - z = 0, \\ x + 4y + 2z = 0, \\ 3x + 7y + 3z = 0. \end{cases}$$

1251. Determine the value of a for which the homogeneous system

$$\begin{cases} 3x - 2y + z = 0, \\ ax - 14y + 15z = 0, \\ x + 2y - 3z = 0 \end{cases}$$

has a non-zero solution.

6. Determinants of the Fourth Order

The properties of determinants enumerated in 4 hold for determinants of any order. In the present section, these properties should be used in evaluating determinants of the fourth order.

In each of Problems 1252-1260, evaluate the given determinant of the fourth order.

1252. $\begin{vmatrix} -3 & 0 & 0 & 0 \\ 2 & 2 & 0 & 0 \\ 1 & 3 & -1 & 0 \\ -1 & 5 & 3 & 5 \end{vmatrix}.$

1253. $\begin{vmatrix} 2 & -1 & 3 & 4 \\ 0 & -1 & 5 & -3 \\ 0 & 0 & 5 & -3 \\ 0 & 0 & 0 & 2 \end{vmatrix}.$

1254. $\begin{vmatrix} 2 & -1 & 1 & 0 \\ 0 & 1 & 2 & -1 \\ 3 & -1 & 2 & 3 \\ 3 & 1 & 6 & 1 \end{vmatrix}$.

1255. $\begin{vmatrix} 2 & 3 & -3 & 4 \\ 2 & 1 & -1 & 2 \\ 6 & 2 & 1 & 0 \\ 2 & 3 & 0 & -5 \end{vmatrix}$.

1256. $\begin{vmatrix} 8 & 7 & 2 & 0 \\ -8 & 2 & 7 & 10 \\ 4 & 4 & 4 & 5 \\ 0 & 4 & -3 & 2 \end{vmatrix}$.

1257. $\begin{vmatrix} 0 & b & c & d \\ b & 0 & d & c \\ c & d & 0 & b \\ d & c & b & 0 \end{vmatrix}$.

1258. $\begin{vmatrix} a & b & c & d \\ b & a & d & c \\ c & d & a & b \\ d & c & b & a \end{vmatrix}$.

1259. $\begin{vmatrix} a & b & c & d \\ d & a & b & c \\ c & d & a & b \\ b & c & d & a \end{vmatrix}$.

1260. $\begin{vmatrix} 0 & -a & -b & -d \\ a & 0 & -c & -e \\ b & c & 0 & 0 \\ d & e & 0 & 0 \end{vmatrix}$.

1261. Prove that, if the system

$$\begin{cases} A_1x+B_1y+C_1z+D_1=0, \\ A_2x+B_2y+C_2z+D_2=0, \\ A_3x+B_3y+C_3z+D_3=0, \\ A_4x+B_4y+C_4z+D_4=0 \end{cases}$$

is consistent, then

$$\begin{vmatrix} A_1 & B_1 & C_1 & D_1 \\ A_2 & B_2 & C_2 & D_2 \\ A_3 & B_3 & C_3 & D_3 \\ A_4 & B_4 & C_4 & D_4 \end{vmatrix} = 0.$$

ANSWERS AND HINTS

Part One

1. See Fig. 54. 2. *Hint*. The equation $|x| = 2$ is equivalent to the two equations $x = -2$ and $x = 2$; accordingly, we have the two points $A_1(-2)$ and $A_2(2)$ (Fig. 55). The equation $|x-1| = 3$ is equivalent to the two equations $x-1 = -3$ and $x-1 = 3$, whence we find $x = -2$ and $x = 4$ and the points B_1 and B_2 corresponding

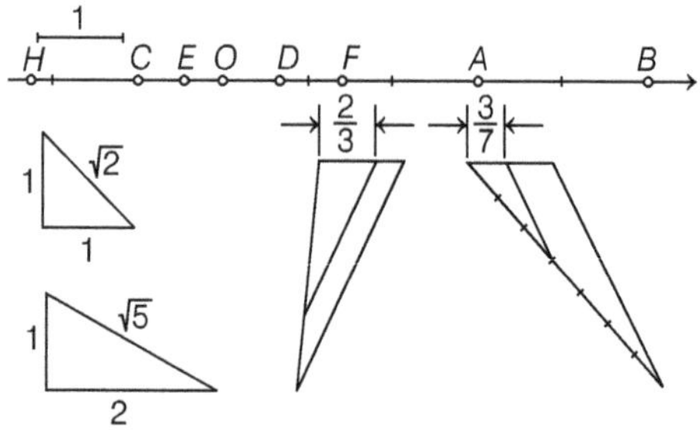

Fig. 54.

to them (Fig. 55). The remaining examples have analogous solutions.

3. The points are situated:

1) to the right of the point $M_1(2)$;

2) to the left of the point M_2 (3) including the point M_2; 3) to the right of the point $M_3(12)$;

4) to the left of the point $M_4\left(\frac{3}{2}\right)$, including the point M_4;

5) to the right of the point $M_5\left(\frac{5}{3}\right)$;

6) inside the segment bounded by the points $M_6(1)$ and $M_2(3)$;

7) inside the segment bounded by the points $M_7(-2)$ and $M_2(3)$, including the points M_7 and M_2;

8) inside the segment bounded by the points A (1) and $B(2)$;

9) outside the segment bounded by the points $P(-1)$ and $Q(2)$;

10) outside the segment bounded by the points $A(1)$ and $B(2)$;

11) inside the segment bounded by the points $P(-1)$) and $Q(2)$;

12) inside the segment bounded by the points $M(3)$ and $N(5)$; including the points M and N;

13) outside the segment bounded by the points $M(3)$ and $N(5)$;

14) outside the segment bounded by the points $P_1(-4)$ and $Q_1(3)$;

15) inside the segment bounded by the points $P_1(-4)$ and $Q_1(3)$ including the points P_1 and Q_1.

4. 1) $AB = 8$, $|AB| = 8$; 2) $AB = -3$, $|AB| = 3$;
3) $AB = 4$, $|AB| = 4$; 4) $AB = 2$, $AB = 2$;
5) $AB = -2$, $AB = 2$;

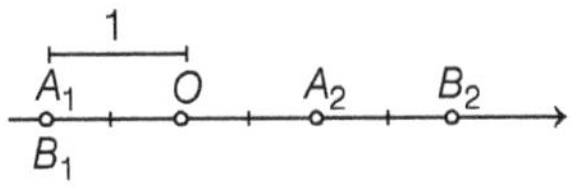

Fig. 55.

6) $AB = 2$, $|AB| = 2$.

5. 1) –2; 2) 5; 3) 1; 4) –8; 5) –2 and 2; 6) –1 and 5; 7) –6 and 4; 8) –7 and –3.

6. The points are situated: 1) Inside the segment bounded by the points $A(-1)$ and $B(1)$, 2) outside the segment bounded by the points $A(-2)$ and $B(2)$, 3) inside the segment bounded by the points $A(-2)$ and $B(2)$, including the points A and B; 4) outside the segment bounded by the points $A(-3)$ and $B(3)$, including the points A and B, 5) inside the segment bounded by the points A (–1) and B 5); 6) inside the segment bounded by the points $A(4)$ and $B(6)$, including the points A and B; 7) outside the segment bounded by the points $A(-1)$ and $B(3)$, including the points A and B, (8) outside the segment bounded by the points A (2) and $B(4)$, including the points A and B;

9) inside the segment bounded by the points $A(-4)$ and $B(2)$;

10) outside the segment bounded by the points $A(-3)$ and $B(-1)$;

11) inside the segment bounded by the points $A(-6)$ and $B(-4)$; including the points A and B;

12) outside the segment bounded by the points $A(-3)$ and $B(1)$, including the points A and B.

7. 1) 1; 2) $-\frac{5}{3}$; 3) 2; 4) $\frac{1}{2}$; 5) $-\frac{10}{3}$.

8. $\lambda_1 = \frac{AB}{BC} = 3$; $\lambda_2 = \frac{CB}{BA} = \frac{1}{3}$; $\lambda_3 = \frac{AC}{CB} = -4$;

$\lambda_4 = \frac{BC}{CA} = -\frac{1}{4}$; $\lambda_5 = \frac{BA}{AC} = -\frac{3}{4}$; $\lambda_6 = \frac{CA}{AB} = -\frac{4}{3}$.

9. $\lambda = \frac{x - x_1}{x_2 - x}$. **10.** $x = \frac{x_1 + \lambda x_2}{1 + \lambda}$. **11.** $x = \frac{x_1 + x_2}{2}$.

12. 1) 4; 2) 2; 3) –2; 4) –2; 5) $-\frac{1}{2}$.

13. 1) $\frac{17}{3}$; 2) $-\frac{13}{4}$; 3) $\frac{1}{3}$; 4) 7; 5) 3; 6) 0.

14. 1) $M(-11)$; 2) $N(13)$.

15. (5) and (12). **16.** $A(7)$ and $B(-41)$.

17. See Fig. 56.

18. $A_x(2,0)$, $B_x(3,0)$, $C_x(-5,0)$, $D_x(-3,0)$, $E_x(-5,0)$.

19. $A_y(0,2)$, $B_y(0,1)$, $C_y(0,-2)$, $D_y(0,1)$, $E_y(0,-2)$.

20. 1) (2, –3); 2) (–3, –2); 3) (–1, 1); 4) (–3, 5); 5) (–4, –6); 6) $(a, -b)$.

21. 1) (1, 2); 2) (–3, –1); 3) (2, –2); 4) (2, 5); 5) (–3,–5); 6) $(-a, b)$.

22. 1) (–3, –3); 2) (–2, 4); 3) (2, –1); 4) (–5, 3); 5) (5, 4); 6) $(-a, -b)$.

23. 1) (3, 2); 2) (–2, 5); 3) (4, –3);

24. 1) (–5, – 3); 2) (–3, 4); 3) (2, – 7).

25. 1) The first and third quadrants; 2) the second and fourth quad-

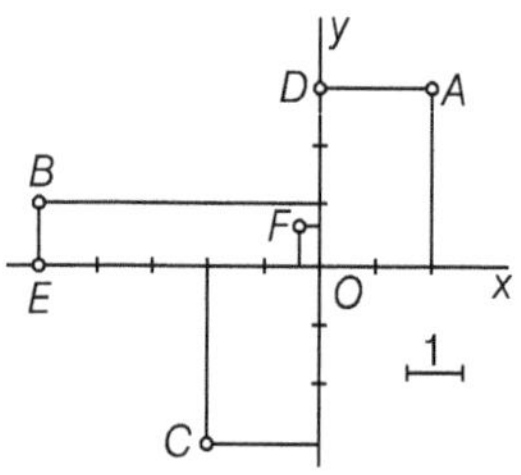

Fig. 56.

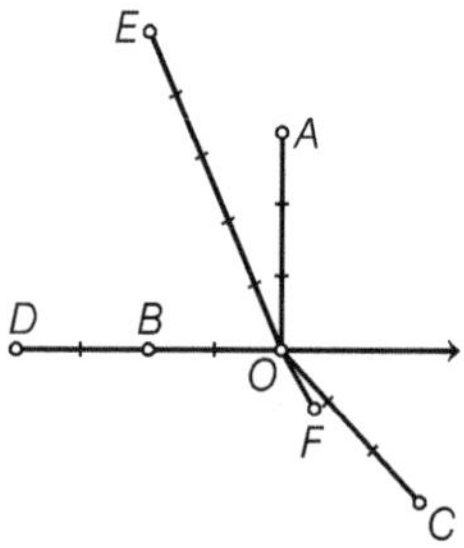

Fig. 57.

rants; 3) the first and third quadrants; 4) the second and fourth quadrants; 5) the first, second, and fourth quadrants; 6) the second, third, and fourth quadrants; 7) the first, third, and fourth quadrants; 8) the first, second, and third quadrants.

26. See Fig. 57.

27. $\left(3, -\frac{\pi}{4}\right), \left(2, \frac{\pi}{2}\right), \left(3, \frac{\pi}{3}\right)$, (1, –2), (5, 1).

28. $\left(1, -\frac{3}{4}\pi\right), (5, -\pi/2), \left(2, \frac{2}{3}\pi\right), \left(4, -\frac{1}{6}\pi\right), (3, \pi - 2)$.

29. $C\left(3, \frac{5}{9}\pi\right)$ and $D\left(5, -\frac{11}{14}\pi\right)$. **30.** $\left(1, -\frac{2\pi}{3}\right)$.

31. $A\left(3, -\frac{\pi}{2}\right), B\left(2, \frac{3}{4}\pi\right), C(1,0), D\left(5, \frac{\pi}{4}\right), E(3, 2-\pi), F(2, \pi - 1)$.

32. $M_1(3,0), M_2\left(1, \frac{\pi}{3}\right), M_3\left(2, -\frac{\pi}{3}\right), M_4\left(5, -\frac{\pi}{12}\right), M_5(3, \pi), M_6\left(1, \frac{7}{12}\pi\right)$.

33. $\left(6, \frac{\pi}{9}\right)$.

34. $d = \sqrt{\rho_1^2 + \rho_2^2 - 2\rho_1\rho_2\cos(\theta_2 - \theta_1)}$.

35. $d = 7$ **36.** $9(17 - 4\sqrt{3})$ square units.

37. $2(13+6\sqrt{2})$ square units. **38.** $28\sqrt{3}$ square units.

39. $S=\frac{1}{2}\rho_1\rho_2\sin(\theta_1-\theta_2)$. **40.** 5 square units.

41. $3(4\sqrt{3}-1)$ square units.

42. $M_1(0,\ 6)$, $M_2(5,\ 0)$, $M_3(\sqrt{2},\sqrt{2})$, $M_4(5,-5\sqrt{3})$, $M_5(-4,4\sqrt{3})$, $M_6(6\sqrt{3},-6)$.

43. $M_1\left(5,\frac{\pi}{2}\right)$, $M_2(3,\pi)$, $M_3\left(2,\frac{\pi}{6}\right)$, $M_4\left(2,-\frac{3}{4}\pi\right)$, $M_5\left(2,-\frac{\pi}{3}\right)$.

44. 1) 3; 2) –3; 3) 0; 4) 5; 5) –5; 6) 2.

47. 1) $X=1, Y=3$; 2) $X=-4, Y=-2$; 3) $X=1, Y=-7$; 4) $X=5, Y=3$.

48. $(3,-1)$.

49. $(-3,2)$.

52. 1) $X=-6, Y=6\sqrt{3}$; 2) $X=3\sqrt{3}, Y=-3$; 3) $X=\sqrt{2},\ y=-\sqrt{2}$.

53. 1) 5; 2) 13; 3) 10.

54. 1) $d=2, \theta=\frac{\pi}{3}$; 2) $d=6, \theta=-\frac{\pi}{4}$; 3) $d=4, \theta=\frac{5}{6}\pi$.

55. 1) $d=\sqrt{2}, \theta=-\frac{3}{4}\pi$; 2) $d=5, \theta=\arctan\frac{4}{3}-\pi$

3) $d=13.\ \theta=\pi-\arctan\frac{12}{5}$; 4) $d=\sqrt{234},\ \theta=-\arctan 5$.

56. 1) 3; 2) –3. **57.** 1) (–9, 3); 2) (–9 – 7).

58. (1) (–15,–12); 2) (1,–12).

59. –2. **60.** $\frac{3\sqrt{3}-4}{2}$.

61. 4. **62.** 1) –5; 2) 5.

63. 1) 5; 2) 10; 3) 5; 4) $\sqrt{5}$; 5) $2\sqrt{2}$; 6) 13.

64. 137 square units. **65.** 34 square units.

66. $8\sqrt{3}$ square units. **67.** 13, 15.

68. 150 square units. **69.** $4\sqrt{2}$.

73. $\measuredangle M_2M_1M_3$ is obtuse.

75. $\measuredangle BAC=45°$, $\measuredangle ABC=45°$, $\measuredangle ACB=90°$.

76. 60°.

Hint. Compute the lengths of the sides of the triangle, and then use the cosine theorem.

77. $M_1(6,0)$ and $M_2(-2, 0)$.

78. $M_1(0, 28)$ and $M_2(0, -2)$.

79. $P_1(1, 0)$ and $P_2(6, 0)$.

80. $C_1(2, 2)$, $R_1=2$, $C_2(10, 10)$. $R_2=10$.

81. $C_1(-3, -5)$, $C_2(5, -5)$.

82. $M_2(3, 0)$.

83. $B(0, 4)$ and $D(-1, -3)$.

84. The conditions of the problem are satisfied by two squares symmetrically situated with respect to the side AB. The points $C_1(-5, 0)$, $D_1(-2, -4)$ are the required vertices of one square;–$C_2(3, 6)$ $D_2(6, 2)$ are the vertices of the other.

85. $C(3,-2)$, $R = 10$. **86.** $(1, -2)$.

87. $Q(4, 6)$.

88. The midpoints of the sides AB, BC, AC are $(2, -4)$, $(-1, 1)$, $(-2, 2)$, respectively.

89. 1) $M(1, 3)$; 2) $N(4, -3)$.

90. $(1, -3)$, $(3, 1)$ and $(-5, 7)$.

91. $D(-3, 1)$. **92.** $(5, -3)$, $(1, -5)$.

93. $D_1(2, 1)$. $D_2(-2, 9)$. $D_3(6, -3)$.

Hint. The fourth vertex of the parallelogram may lie opposite to any one of the given vertices. Hence, three parallelograms actually satisfy the conditions of the problem.

94. 13. **95.** $(2, -1)$ and $(3, 1)$.

96. $\left(\frac{5}{2}, -2\right)$. **97.** $\frac{14}{3}\sqrt{2}$.

98. $(-11, -3)$. **99.** 4 .

100. $\lambda_1 = \frac{AB}{BC} = 2$; $\lambda_2 = \frac{AC}{CB} = -3$; $\lambda_3 = \frac{BA}{AC} = -\frac{2}{3}$.

101. $A(3, -1)$ and $B(0, 8)$.

102. $(3, -1)$. **103.** $(4, -5)$.

104. $(-9, 0)$. **105.** $(0, -3)$.

106. $1 : 3$, starting from the point B.

107. $\left(4\frac{1}{2}, 1\right)$. **108.** $x = \frac{x_1 + x_2 + x_3}{3}$, $y = \frac{y_1 + y_2 + y_3}{3}$

109. $M(-1, 0)$, $C(0, 2)$. **111.** $(5, 5)$.

112. $\left(\frac{5}{12}a, \frac{5}{12}b\right)$. **113.** $\left(\frac{19}{21}a, \frac{19}{21}a\right)$.

114. $x = \frac{mx_1 + nx_2 + px_3}{m + n + p}$, $y = \frac{my_1 + ny_2 + py_3}{m + n + p}$.

115. $(4, 2)$.

Hint. The weight of uniform wire is proportional to its length.

116. 1) 14 square units; 2) 12 square units; 3) 26 square units.

117. 5. **118.** 20 square units.

119. 7.4.

120. $x = -\frac{6}{11}$, $y = 4\frac{1}{11}$. **121.** $x = \frac{7}{17}$, $y = 3\frac{1}{3}$.

122. (0, –8) or (0, –2). **123.** (5, 0) or $\left(-\frac{1}{3}, 0\right)$.

124. (5, 2) or (2, 2).

125. C_1 (–7, –3). D_1(– 6, – 4) or C_2(17, –3), D_2 (18, –4).

126. C_1(–2, 12), D_1(–5, 16) or $C_2\left(-2, \frac{2}{3}\right)$, $D_2\left(-5, \frac{14}{3}\right)$.

127. 1) $x = x' + 3$, $y = y' + 4$; 2) $x = x' - 2$, $y = y' + 1$;
3) $x = x' - 3$, $y = y' + 5$.

128. A (4, – 1), B (0, – 4), C (2, 0).

129. 1) A(0, 0), B (–3, 2), C (–4, 4); 2) A(3, – 2), B(0, 0), C(–1, 2);
3) A (4, –4), B (1, – 2), C (0, 0).

130. 1) (3, 5); 2) (–2, 1); 3) (0, –1); 4) (–5, 0).

131. 1) $x = \frac{x' - y'\sqrt{3}}{2}$, $y = \frac{x'\sqrt{3} + y'}{2}$;

2) $x = \frac{x' + y'}{\sqrt{2}}$; $y = \frac{-x' + y'}{\sqrt{2}}$; 3) $x = -y'$, $y = x'$;

4) $x = y'$, $y = -x'$; 5) $x = -x'$, $y = -y'$.

132. $A(3\sqrt{3}, 1), B\left(\frac{\sqrt{3}}{2}, \frac{3}{2}\right), C(3, -\sqrt{3})$

133. 1) M $(\sqrt{2}, 2\sqrt{2})$, N $(-3\sqrt{2}, 2\sqrt{2})$, P $(-\sqrt{2}, -2\sqrt{2})$; 2) M (1, – 3),
N (5, 1) P(–1, 3); 3) M(–1, 3), N(–5, – 1), P(1, – 3), 4) M(–3, –1),
N(1, –5), P (3, 1).

134. 1) 60°; 2) –30°. **135.** O'(2, – 4).

136. $x = x' + 1$, $y = y' - 3$.

137. $x = \frac{3}{5}x' + \frac{4}{5}y'$, $y = -\frac{4}{5}x' + \frac{3}{5}y'$.

138. M_1(1, 5), M_2(2,0), M_3 (16, –5).

139. A(6, 3), B(0, 0), C(5, –10).

140. 1) O' (3, –2), α = 90°; 2) O' (–1, 3), α = 180°; 3) O'(5, –3) α = – 45°.

141. $x = -\frac{15}{17}x' - \frac{8}{17}y' + 9$, $y = \frac{8}{17}x' - \frac{15}{17}y' - 3$.

142. M_1(1, 9), M_2(4, 2) M_3 (1, –3), $M_4(0, 2 + \sqrt{3})$, M_5 $(1 + \sqrt{3}, 1)$.

143. M_1(0, 5), M_2(3, 0), M_3 (–1, 0), M_4(0, –6), M_5 $(\sqrt{3}, 1)$.

144. $M_1(2, 0)$, $M_2\left(1, -\frac{\pi}{2}\right)$, $M_3\left(3, \frac{\pi}{2}\right)$, $M_4\left(2, -\frac{\pi}{4}\right)$, $M_5\left(2, \frac{\pi}{6}\right)$

145. $M_1\left(\sqrt{2}, \frac{1}{2}\pi\right)$, $M_2\left(2, -\frac{\pi}{2}\right)$, $M_3\left(2, \frac{\pi}{12}\right)$, $M_4\left(2, \frac{7}{12}\pi\right)$, $M_5\left(4, -\frac{5}{12}\pi\right)$.

146. $f(x, y) = 2ax - a^2$.

147. 1) $f(x, y) = 2ax$, 2) $f(x, y) = -2ax - a^2$.

148. $f(x, y) = 4x^2 + 4y^2 + 2a^2$. **149.** $f(x, y) = 4x^2 + 4y^2 - 4ax - 4ay + 4a^2$.

150. $f(x, y) = x^2 + y^2 - 25$. **151.** $f(x, y) = 2xy - 16$.

152. Rotation of the coordinate axes does not affect the expression for this function.

153. (3, 1). **154.** There exists no such point.

155. ± 45° or ± 135°. **156.** 30°, 120°, –60°, –150°.

157. The points M_1, M_4 and M_5 lie on the curve; the points M_2, M_3 and M_6 do not lie on the curve. The equation represents the bisector of the second and fourth quadrants (Fig. 58).

158. a) (0, –5), (0, 5); b) (–3, –4), (–3, 4); c) (5, 0); d) there is no such point on the curve; e) (–4, 3), (4, 3); *f*) (0, –5); g) there is no such point on the curve. The equation represents a circle with centre O (0, 0) and radius 5 (Fig. 59).

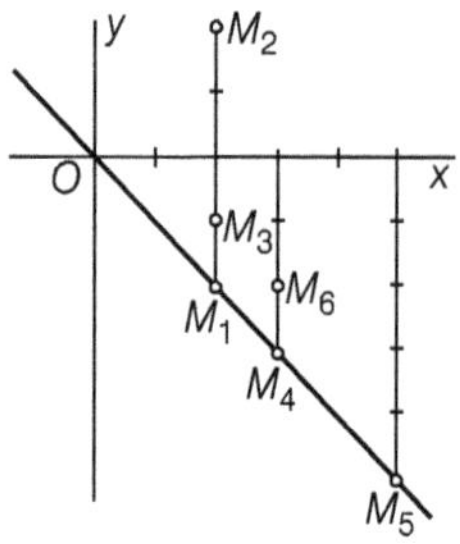

Fig. 58.

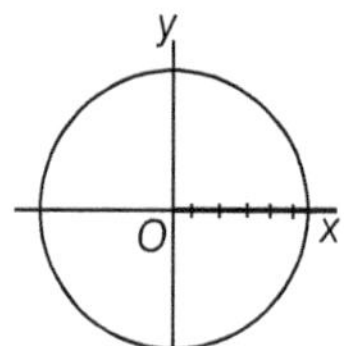

Fig. 59.

159. 1) The bisector of the first and third quadrants; 2) the bisector of the second and fourth quadrants; 3) the straight line parallel to the axis Oy and having an x-intercept of 2 (Fig. 60); 4) the straight line parallel to the axis Oy and having an x-intercept of –3 (Fig. 60): 5) the straight line parallel to the axis Ox and having a y-intercept of 5 (Fig. 60); 6) the straight line parallel to the axis Ox and having a y-intercept of –2 (Fig. 60); 7) the straight line coincident with the y-axis; 8) the straight line coincident with the x-axis, 9) the curve consists of two straight lines, one of which is the bisector of the first and third quadrants, and the other coincides with the y-axis: 10) the curve consists of two straight lines, one of which is the bisector of the second and fourth quadrants, and the other coincides with the x-axis; 11) the curve consists of the two straight lines bisecting the quadrants (Fig. 61); 12) the curve consists of two straight lines, one of which coincides with the x-axis, and the other coincides with the y-axis; 13) the curve consists of two straight lines parallel to the x-axis and whose respective y-intercepts are 3 and –3

(Fig. 62); 14) the curve consists of two straight lines parallel to the y-axis and whose respective x-intercepts are 3 and 5 (Fig. 63); 15) the curve consists of two straight lines parallel to the x-axis and whose respective y-intercepts are –1 and – 4 (Fig. 64): 16) the

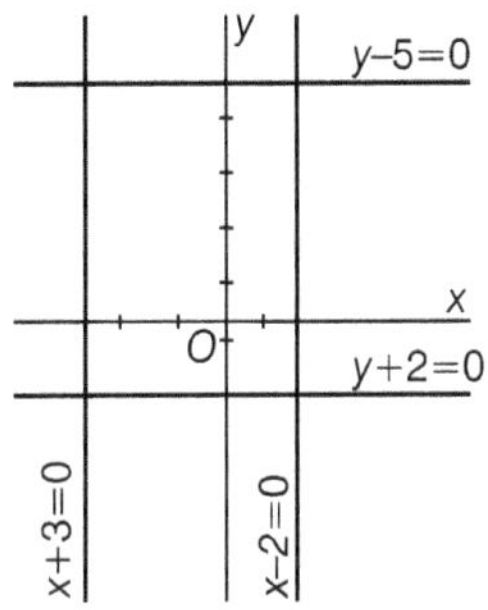

Fig. 60.

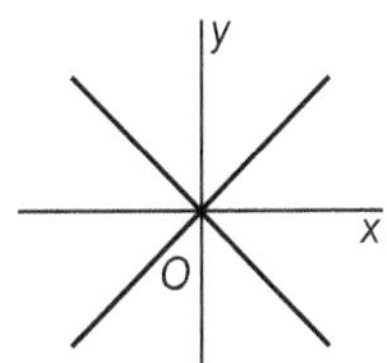

Fig. 61.

curve consists of three straight lines, one of which coincides with the x-axis, and the other two are parallel to the y-axis and have x-intercepts of 2 and 5, respectively; 17) the curve consists of the

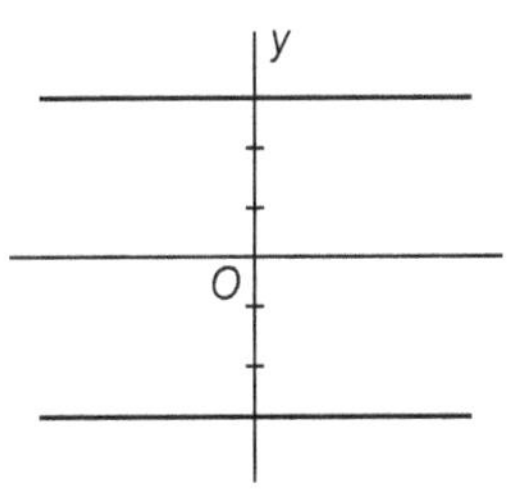

Fig. 62.

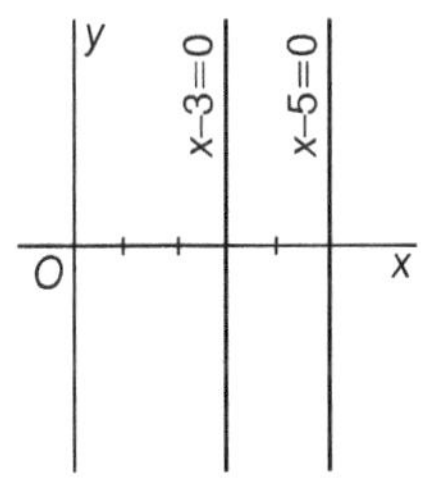

Fig. 63.

two rays bisecting the first and second quadrants (Fig. 65); 18) the curve consists of the two rays bisecting the first and the fourth quadrants (Fig. 66a); 19) the curve consists of the two rays bisecting the third and fourth quadrants (Fig. 66b); 20) the curve consists of

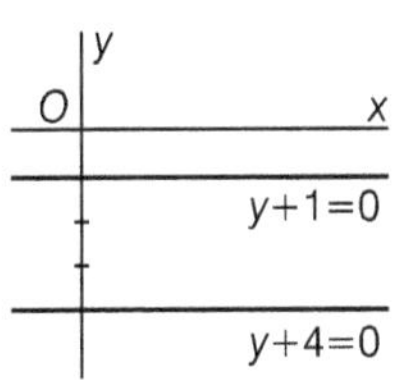

Fig. 64.

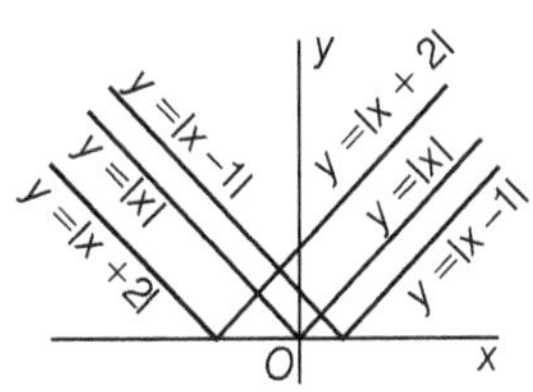

Fig. 65.

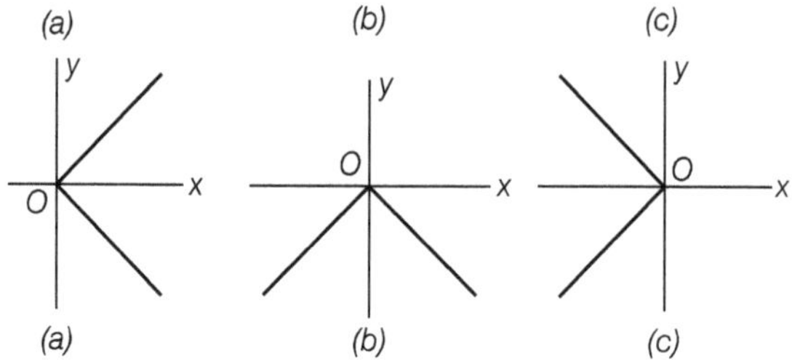

Fig. 66.

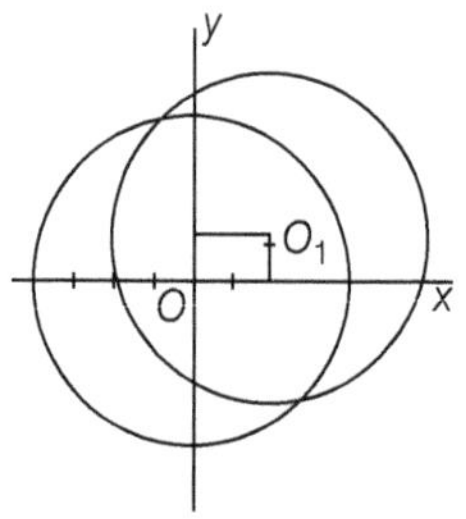

Fig. 67.

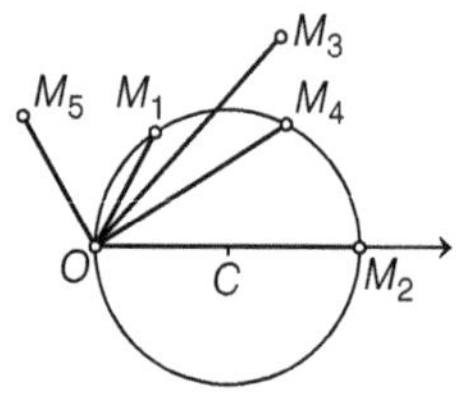

Fig. 68.

the two rays bisecting the second and third quadrants (Fig. 66c); 21) the curve consists of the two rays situated in the upper half-plane and drawn from the point (1, 0) parallel to the bisectors of the quadrants (Fig. 65); 22) the curve consists of the two rays situated in the upper half-plane and drawn from the point (–2, 0) parallel to the bisectors of the quadrants (Fig. 65); 23) the circle with centre at the origin and radius 4 (Fig. 67); 24) the circle with centre at O_1 (2, 1) and radius 4 (Fig. 67); 25) the circle with centre at (– 5, 1) and radius 3; 26) the circle with centre at (1, 0) and radius 2; 27) the circle with centre at (0 – 3) and radius 1; 28) the curve consists of the single point (3, 0) (a degenerate curve); 29) the curve consists of the single point (0, 0) (a degenerate curve); 30) the equation is satisfied by the coordinates of no point (an imaginary curve); 31) the equation is satisfied by the coordinates of no point (an imaginary curve).

160. The curves 1), 2) and 4) pass through the origin.

161. 1) (a) (7, 0), (– 7, 0); b) (0, 7), (0, – 7); 2) a) (0, 0), (6, 0); b) (0, 0), (0, – 8); 3) a) (– 10, 0), (– 2, 0); b) the curve does not intersect the axis *Oy*: 4) the curve does not intersect the coordinate axes; 5) a) (0, 0), (12, 0); b) (0, 0), (0, –16); 6) a) the curve does not intersect the axis *Ox*; b) (0, –1), (0, –7); 7) the curve does not intersect the coordinate axes.

162. 1) (2, 2), (– 2, – 2); 2) (1, – 1), (9, – 9); 3) (3,–4), $\left(1\frac{2}{5}, -4\frac{4}{5}\right)$;

4) the curves do not intersect.

163. The points M_1, M_2 and M_4 lie on the given curve; the points M_3 and M_5 do not lie on the curve. The equation represents a circle (Fig. 68).

164. a) $\left(6, \frac{\pi}{3}\right)$; b) $\left(6, -\frac{\pi}{3}\right)$; c) (3, 0); d) $\left(2\sqrt{3}, \frac{\pi}{6}\right)$; the straight line perpendicular to the polar axis and whose intercept on the polar axis is equal to 3 (Fig. 69).

165. a) $\left(1, \frac{\pi}{2}\right)$; b) $\left(2, \frac{\pi}{6}\right)$ and $\left(2, \frac{5}{6}\pi\right)$; c) $\left(\sqrt{2}, \frac{\pi}{4}\right)$ and $\left(\sqrt{2}, \frac{3}{4}\pi\right)$; the straight line situated in the upper half-plane and drawn parallel to the polar axis one unit above it (Fig. 69).

166. 1) The circle with centre at the pole and radius 5; 2) the ray drawn from the pole and making an angle $\pi/3$ with the polar axis (Fig. 70); 3) the ray drawn from the pole and making an angle $\frac{-\pi}{4}$ with the polar axis (Fig. 70); 4) The straight line perpendicular to the polar axis and making an intercept $a = 2$ on it; 5) the straight line situated in the upper half-plane and drawn parallel to the polar axis one unit above it; 6) the circle with centre $C_1(3, 0)$ and radius 3 (Fig. 71); 7) the circle with centre $C_2\left(5, \frac{\pi}{2}\right)$ and radius 5 (Fig. 71); (8) the curve consists of two rays drawn from the pole and making angle $\frac{\pi}{6}$ and $\frac{5}{6}\pi$, respectively, with the polar axis (Fig. 71);

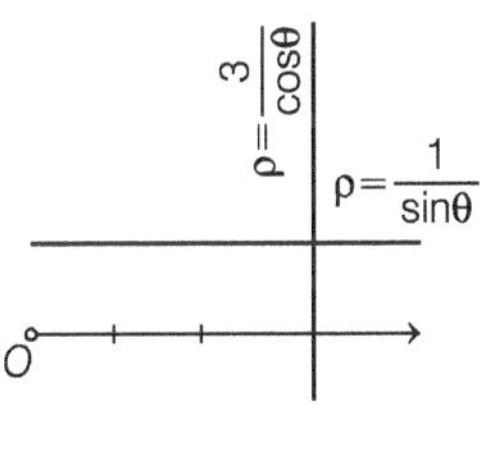

Fig. 69.

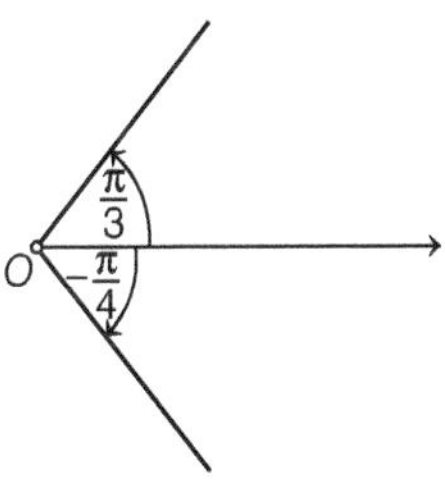

Fig. 70.

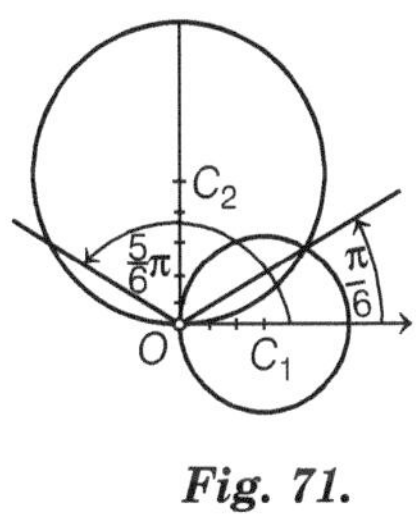

Fig. 71.

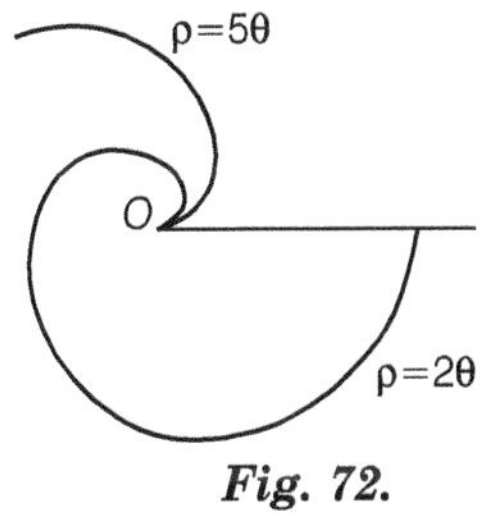

Fig. 72.

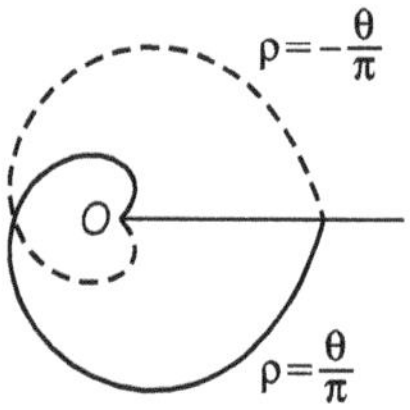

Fig. 73.

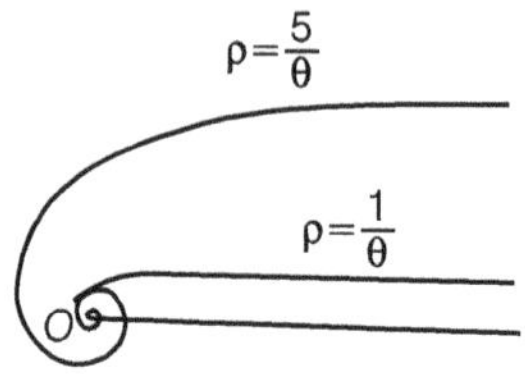

Fig. 74.

9) the curve consists of concentric circles with centre at the pole, the radii r of these circles being determined by the formula $r = (-1)^n \frac{\pi}{6} + \pi n$, where n is any positive integer or zero.

167. Figs. 72 and 73.

168. Figs. 74 and 75.

169. Fig. 76.

170. The segment adjacent to the pole has a length of $\frac{\pi}{2}$; each of the remaining segments has a length of 6π (Fig. 77).

171. Into five parts (Fig. 78).

172. $P\left(12, \frac{1}{2}\right)$ (Fig. 79).

173. Q (81, 4) (Fig. 80).

174. The straight lines $x \pm y = 0$.

175. The straight lines $x \pm a = 0$.

176. The straight lines $y \pm b = 0$.

177. $y + 4 = 0$. **178.** $x - 5 = 0$.

179. 1) The straight line $x - y = 0$; 2) the straight line $x + y = 0$; 3) the straight line $x - 1 = 0$; 4) the straight line $y - 2 = 0$.

180. The straight lines $4ax \pm c = 0$.

181. $x^2 + y^2 = r^2$ **182.** $(x-\alpha)^2 + (y-\beta)^2 = r^2$

183. $x^2 + y^2 = 9$. **184.** $x^2 + y^2 = 16$.

185. $x^2 + y^2 = a^2$. **186.** $(x-4)^2 + y^2 = 16$.

187. $\frac{x^2}{25} + \frac{y^2}{16} = 1$ **188.** $\frac{x^2}{9} - \frac{y^2}{16} = 1$.

189. $y^2 = 12x$.

192. The parabola $y^2 = 2px$.

193. The ellipse $\frac{x^2}{25} + \frac{y^2}{9} = 1$.

194. The hyperbola $\frac{x^2}{16}-\frac{y^2}{9}=1.$

195. The ellipse $\frac{x^2}{25}+\frac{y^2}{16}=1.$

196. The right-hand branch of the hyperbola $\frac{x^2}{64}-\frac{y^2}{36}=1$

197. The parabola $y^2=20x.$

198. $\rho\cos\theta=3.$ **199.** $\theta=\frac{\pi}{3}.$

200. $\tan\theta=1.$

201. $\rho\sin\theta+5=0, \rho\sin\theta-5=0.$

202. $\rho=10\cos\theta.$

203. The conditions of the problem are satisfied by the two circles whose polar equations are $\rho+6\sin\theta=0, \rho-6\sin\theta=0.$

204. $\left.\begin{array}{l} x=a\cos t, \\ y=b\sin t, \end{array}\right\} \frac{x^2}{a^2}+\frac{y^2}{b^2}=1.$

205. $x=\frac{ab\cos t}{\sqrt{a^2\sin^2 t+b^2\cos^2 t}}, y=\frac{ab\sin t}{\sqrt{a^2\sin^2 t+b^2\cos^2 t}}$

206. $x=\frac{ab\cos t}{\sqrt{b^2\cos^2 t-a^2\sin^2 t}}, y=\frac{ab\sin t}{\sqrt{b^2\cos^2 t-a^2\sin^2 t}}.$

207. 1) $x=\frac{t^2}{2p}, y=t;$ 2) $x=2p\cot^2 t, y=2p\cot t;$

3) $x=\frac{p}{2}\cot^2\frac{t}{2}; y==p\cot\frac{t}{2}.$

208. 1) $\left.\begin{array}{l} x=2R\cos^2\theta, \\ y=R\sin 2\theta; \end{array}\right\}$, 2) $\left.\begin{array}{l} x=R\sin 2\theta, \\ y=2R\sin^2\theta; \end{array}\right\}$ 3) $\left.\begin{array}{l} x=2p\cot^2\theta, \\ y=2p\cot\theta. \end{array}\right\}$,

209. 1) $x-y^2=0;$ 2) $x^2+y^2-a^2=0;$

3) $\frac{x^2}{a^2}-\frac{y^2}{b^2}-1=0;$ 4) $\frac{x^2}{a^2}-\frac{y^2}{b^2}-1=0;$

5) $x^2+y^2-2Rx=0;$ 6) $x^2+y^2-2Ry=0;$ 7) $2px-y^2=0.$

210. The points M_1, M_3 and M_4 lie on the given line; the points M_2, M_5 and M_6 do not lie on the line.

211. 3, –3, 0, – 6, and –12.

212. 1, –2, 4, –5, and 7.

213. (6, 0), (0, – 4).

214. (3, – 5).

215. $A(2, -1)$, $B(-1, 3)$, $C(2, 4)$.

216. (1, – 3), (–2, 5), (5, –9), and (8,–17).

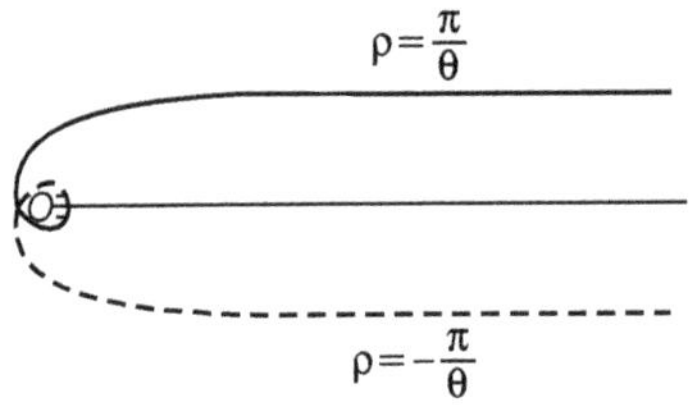

Fig. 75.

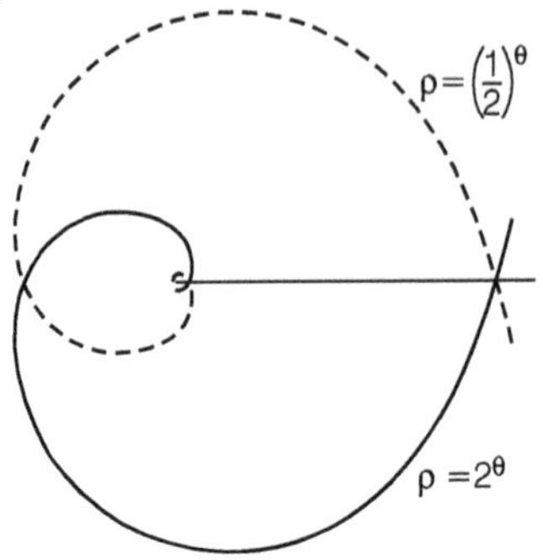

Fig. 76.

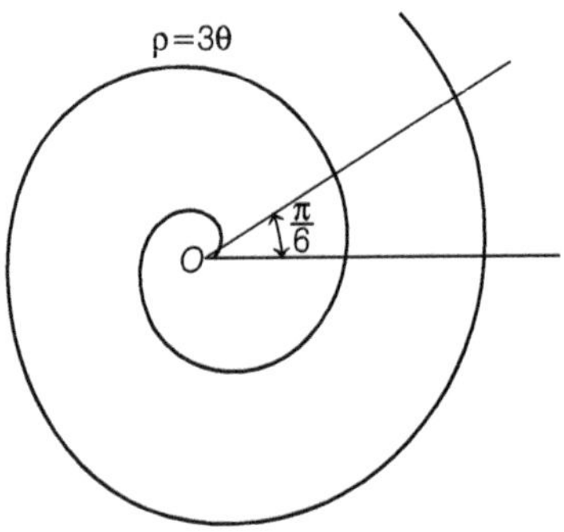

Fig. 77.

217. $S = 17$ square units. **218.** $C_1(-1, 4)$ or $C_2\left(\frac{25}{7}, -\frac{36}{7}\right)$.

219. $C_1(1, -1)$ or $C_2(-2, -10)$. **220.** 1) $2x - 3y + 9 = 0$; 2) $3x - y = 0$; 3) $y + 2 = 0$; 4) $3x + 4y - 12 = 0$; 5) $2x + y + 5 = 0$; 6) $x + 3y - 2 = 0$,

221. 1) $k = 5$, $b = 3$; 2) $k = -\frac{2}{3}$, $b = 2$; 3) $k = -\frac{5}{3}$, $b = -\frac{2}{3}$; 4) $k = -\frac{3}{2}$, $b = 0$; 5) $k = 0$, $b = 3$. **222.** 1)$-\frac{5}{3}$; 2) $\frac{3}{5}$.

223. 1) $2x + 3y - 7 = 0$; 2) $3x - 2y - 4 = 0$.

224. $3x + 2y = 0$, $2x - 3y - 13 = 0$.

225. (2, 1), (4, 2), (–1 , 7), (1, 8)

226. (–2, – 1).

227. $Q(11, -11)$.

228. 1) $3x - 2y - 7 = 0$; 2) $5x + y - 7 = 0$; 3) $8x + 12y + 5 = 0$; 4) $5x + 7y + 9 = 0$; 5) $6x - 30y - 7 = 0$

229. 1) $k=7$; 2) $k=\frac{7}{10}$; 3) $k=-\frac{3}{2}$.

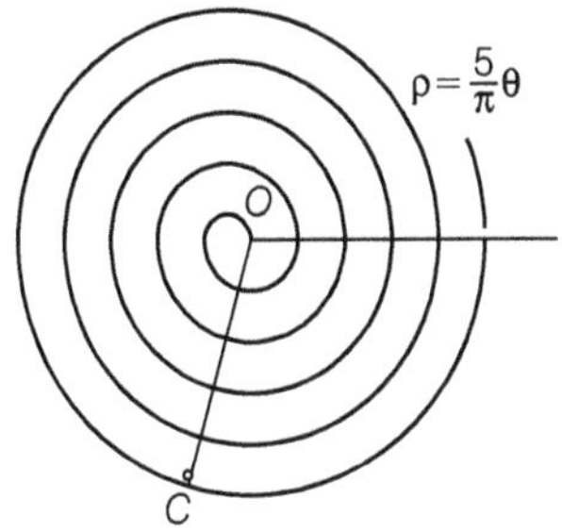

Fig. 78.

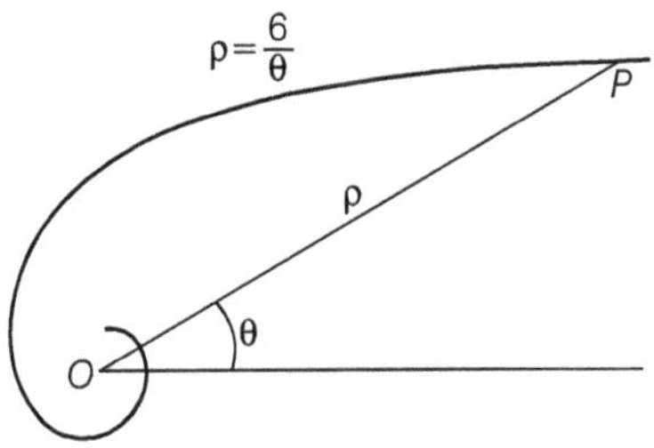

Fig. 79

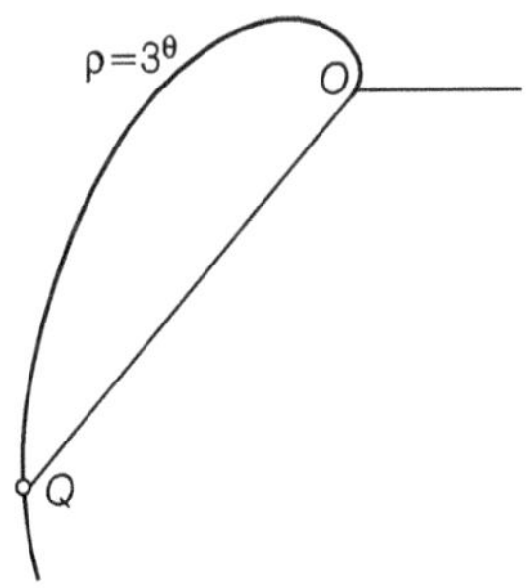

Fig. 80.

230. $5x-2y-33=0,\ x+4y-11=0,\ 7x+6y+33=0.$

231. $7x-2y-12=0, 5x+y-28=0, 2x-3y-18=0.$

232. $x+y+1=0.$ **233.** $2x+3y-13=0.$

234. $4x+3y-11=0, x+y+2=0, 3x+2y-13=0.$

235. (3, 4). **236.** $4x+y-3=0.$ **237.** $x-5=0.$

238. Equation of the side AB: $2x+4y-8=0$; $BC: x+2y-1,=0$, $CA: x-y-1=0$. Equation of the median from the vertex $A: x-3=0$; from the vertex $B: x+y-3=0$, from the vertex $C: y=0$.

239. $(-7, 0), \left(0, +2\frac{1}{3}\right)$.

242. (1, 3) **243.** $3x-5y+4=0$;
$x+7y-16=0$; $3x-5y-22=0$; $x+7y+10=0$.

244. Equations of the sides of the rectangle: $2x-5y+3=0$, $2x-5y-26=0$, equation of its diagonal: $7x-3y-33=0$.

245. The bisector of the interior angle: $5x+y-3=0$; the bisector of the exterior angle: $x-5y-11=0$.

246. $x+y-8=0$, $11x-y-28=0$.

Hint. The conditions of the problem are satisfied by two lines, one of which passes through the point P and bisects the segment joining the points A and B, and the other passes through the point P and is parallel to the segment $\overline{AB}$.

247. (–12, 5). **248.** $M_1(10, -5)$.

249. $P\left(\frac{5}{3}, 0\right)$.

Hint. The problem can be solved by the following procedure: 1) show that the points M and N lie on the same side of the x-axis; 2) find a point symmetric to one of the given points with respect to the X-axis, say, the point N_1, symmetric to the point N; 3) form the equation of the equation of the straight line passing through the points M and N_1; 4) by solving the obtained equation simultaneously with the equation of the x-axis, find the coordinates of the required point.

250. $P(0, 11)$. **251.** $P(2, -1)$.

252. $P(2, 5)$. **253.** 1) $\varphi=\frac{\pi}{4}$; 2) $\varphi=\pi/2$;
3) $\varphi=0$ – the lines are parallel; 4) $\varphi=\arctan\frac{16}{11}$.

254. $x-5y+3=0$ or $5x+y-11=0$.

255. Equations of the sides of the square: $4x+3y+1=0$. $3x-4y+32=0$, $4x+3y-24=0$, $3x-4y+7=0$; equation of its other diagonal: $x+7y-31=0$.

256. $3x-4y+15=0$, $4x+3y-30=0$, $3x-4y-10=0$, $4x+3y-5=0$

257. $2x+y-16=0$, $2x+y+14=0$, $x-2y-18=0$.

258. $3x-y+9=0$, $3x+y+9=0$.

259. $29x-2y+33=0$.

262. 1) $3x-7y-27=0$; 2) $x+9y+25=0$; 3) $2x-3y-13=0$; 4) $x-2=0$; 5) $y+3=0$.

264. The lines 1), 3) and 4) are perpendicular.

266. 1) $\varphi=45°$; 2) $\varphi=60°$; 3) $\varphi=90°$. **267.** $M_3(6,-6)$.

268. $4x-y-13=0$, $x-5=0$, $x+8y+5=0$.

269. $BC: 3x+4y-22=0$; CA; $2x-7y-5=0$, $CN: 3x+5y-23=0$.

270. $x+2y-7=0$; $x-4y-1=0$; $x-y+2=0$.

Hint. The problem can be solved by the following procedure: 1) show that the vertex A lies on neither of the given lines; 2) find the point of intersection of the medians and denote it, say, by M. Since the vertex A and the point M are known, we can now find the equation of the third median; 3) on the line through A and M, lay off the segment $MD = AM$ (Fig. 81). Next, determine the coordinates of the point D, given the midpoint M of the segment AD and one of its end points; A, (4) show that the quadrilateral $BDCM$ is a parallelogram (since its diagonals bisect

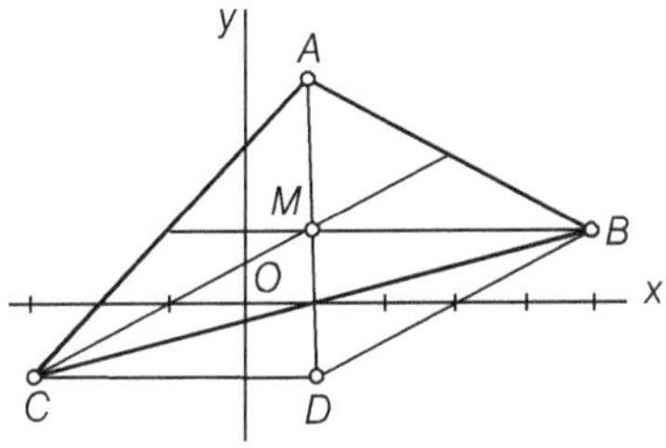

Fig. 81.

each other), and write the equations of the lines DB and DC; 5) calculate the coordinates of the points B and C; 6) now that we know the coordinates of all vertices of the triangle, we can write the equations of its sides.

271. $3x-5y-13=0$, $8x-3y+17=0$, $5x+2y-1=0$.

272. $2x-y+3=0$, $2x+y-7=0$, $x-2y-6=0$. *Hint.* If A is a point on one of the sides of an angle, then the point symmetric to A with respect to the bisector of that angle will lie on the other side of the angle.

273. $4x-3y+10=0$, $7x+y-20=0$, $3x+4y-5=0$.

274. $4x+7y-1=0$, $y-3=0$, $4x+3y-5=0$.

275. $3x+7y-5=0$, $3x+2y-10=0$, $9x+11y+5=0$.

276. $x-3y-23=0$, $7x+9y+19=0$, $4x+3y+13=0$.

277. $x+y-7=0$, $x+7y+5=0$, $x-8y+20=0$.

278. $2x+9y-65=0$, $6x-7y-25=0$, $18x+13y-41=0$.

279. $x+2y=0$, $23x+25y=0$.

280. $8x-y-24=0$.

283. $3x+y=0$, $x-3y=0$.

284. $3x+4y-1=0$, $7x+24y-61=0$.

285. 1) $a=-2$, $5y-33=0$; 2) $a_1=-3$, $x-56=0$; $a_2=3$, $5x+8=0$;
3) $a_1=1$, $3x-8y=0$, $a_2=\frac{5}{3}$, $33x-56y=0$

286. $m=7$, $n=-2$, $y+3=0$. **287.** $m=-4$, $n=2$, $x-5=0$.

288. 1) (5, 6); 2) (3, 2); 3) $\left(\frac{1}{4}, \frac{1}{3}\right)$; 4) $\left(2, -\frac{1}{11}\right)$; 5) $\left(-\frac{5}{3}, 2\right)$.

291. 1) $a \neq 3$; 2) $a = 3$ and $b \neq 2$; 3) $a = 3$ and $b = 2$.

292. 1) $m = -4$, $n \neq 2$ or $m = 4$, $n \neq -2$; 2) $m = -4$, $n = 2$ or $m = 4$, $n = -2$; 3) $m = 0$, n may have any value.

293. $m = \frac{7}{12}$.

294. The conditions of the problem are satisfied by two values of m: $m_1 = 0$, $m_2 = 6$.

295. 1) intersect; 2) do not intersect;

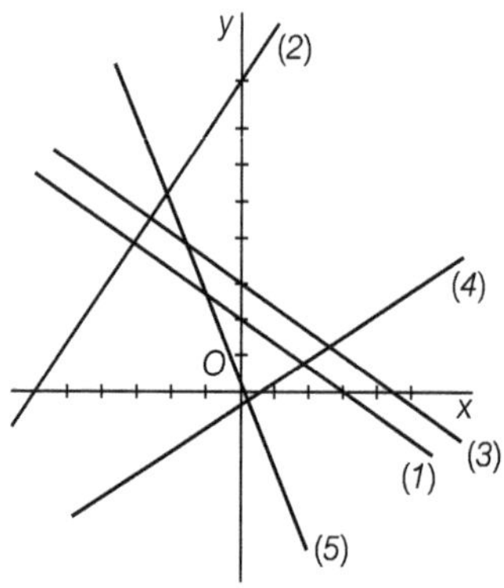

Fig. 82.

3) do not intersect.

298. $a = -7$.

299. 1) $\frac{x}{3} + \frac{y}{2} = 1$; 2) $\frac{x}{-6} + \frac{y}{8} = 1$; 3) $\frac{x}{9/2} + \frac{y}{3} = 1$; 4) $\frac{x}{2/3} + \frac{y}{-2/5} = 1$; 5) $\frac{x}{1/5} + \frac{y}{1/2} = 1$; (Fig. 82).

300. 6 square units.

301. $x + y + 4 = 0$.

302. $x + y - 5 = 0$, $x - y + 1 = 0, 3x - 2y = 0$.

303. *Solution.* Let us write the intercept equation of the required line:

$$\frac{x}{a} + \frac{y}{b} = 1 \tag{1}$$

Our task is to determine the values of the parameters a and b. The point $C(1, 1)$ lies on the required line, and hence its coordinates must satisfy equation (1). Substituting the coordinates of C for the current coordinates in (1) and clearing of fractions, we have

$$a + b = ab. \tag{2}$$

Note now that the area S of the triangle formed by our line and the coordinate axes is determined by the formula $\pm S = \frac{ab}{2}$, where, $+S$ refers to the case when the intercepts a and b have like signs, and $-S$ to the case when the intercepts a and b differ in sign. Hence, by the conditions of the problem we have $ab = \pm 4$ (3)

Solving the system of equations (2) and (3): $\left.\begin{matrix} a+b=4; \\ ab=4; \end{matrix}\right\} \left.\begin{matrix} a+b=-4, \\ ab=-4, \end{matrix}\right\}$

we obtain $a_1 = 2$, $b_1 = 2$; $a_2 = -2+2\sqrt{2}$, $b_2 = -2-2\sqrt{2}$; $a_3 = -2-2\sqrt{2}$, $b_3 = -2+2\sqrt{2}$. Thus, the conditions of the problem are satisfied by three straight lines. Substituting the obtained values of the parameters a and b in 1) gives $\frac{x}{2}+\frac{y}{2}=1$, $\frac{x}{-2+2\sqrt{2}}+\frac{y}{-2-2\sqrt{2}}=1$, $\frac{x}{-2-2\sqrt{2}}+\frac{y}{-2+2\sqrt{2}}=1$. Upon simplifying these equations, we obtain: $x+y-2=0$, $(1+\sqrt{2})x+(1-\sqrt{2})\times y-2=0$, $(1-\sqrt{2})x+(1+\sqrt{2})\,y-2=0$.

304. The conditions of the problem are satisfied by the following three lines:
$(\sqrt{2}+1)x+(\sqrt{2}-1)y-10=0$, $(\sqrt{2}-1)x+(\sqrt{2}+1)y+10=0$, $x-y-10=0$.

305. $3x-2y-12=0$, $3x-8y+24=0$.

306. $x+3y-30=0$, $3x+4y-60=0$, $3x-y-30=0$, $x-12y+60=0$.

307. The conditions of the problem are satisfied by the two lines intersecting the coordinate axes in the points (2,0), (0.– 3) and (– 4, 0), $\left(0, \frac{3}{2}\right)$, respectively,

308. $S \geq 2x_1y_2$.

309. Equations 1), 4), 6) and 8) are in the normal form.

310. 1) $\frac{4}{5}x-\frac{3}{5}y-2=0$; 2) $-\frac{4}{5}x+\frac{3}{5}y-10=0$; 3) $-\frac{12}{13}x+\frac{5}{13}y-1=0$;
4) $-x-2=0$; 5) $\frac{2}{\sqrt{5}}x-\frac{1}{\sqrt{5}}y-1=0$.

311. 1) $\alpha=0$, $p=2$; 2) $\alpha=\pi$, $p=2$; 3) $\alpha=\frac{\pi}{2}$, $p=3$; 4) $\alpha=-\frac{\pi}{\alpha}$, $p=3$;
5) $\alpha=\frac{\pi}{6}$, $p=3$; 6) $\alpha=-\frac{\pi}{4}$, $p=\sqrt{2}$; 7) $\alpha=-\frac{2}{3}\pi$, $p=1$; 8) $\alpha=-\beta$, $p=q$;
9) $\alpha=\beta-\pi$, $p=q$

312. 1) $\delta=-3$, $d=3$. 2) $\delta=1$; $d=1$; 3) $\delta=-4$, $d=4$; 4) $\delta=0$, $d=0$ – the point Q lies on the line.

313. 1) On the same side; 2) on opposite sides; 3) on the same side; 4) on the same side; 5) on opposite sides.

314. 5 square units. **315.** 6 square units.

318. The quadrilateral is convex.

319. The quadrilateral is not convex.

320. 4.

321. 3.

322. 1) $d = 2.5$; 2) $d = 3$; 3) $d = 0.5$; 4) $d = 3.5$.

323. 49 square units.

325. In the ratio 2 : 3, starting from the second line.

326. *Solution.* The problem of drawing straight lines through the point P such that their distance from the point Q will be equal to 5. is equivalent to the problem of drawing through P tangent lines to the circle of radius 5 and with centre at Q. Computing the distance QP gives: $QP = \sqrt{(2-1)^2 + (7-2)^2} = \sqrt{26}$. We see that the distance QP is greater than the radius of the circle; hence, two tangent lines can be drawn from P to the circle. We now proceed to derive their equations.

The equation of every straight line through the point P has the form $y - 7 = k(x-2)$ (1)

or $kx - y + 7 - 2k = 0$, where k is the slope (undetermined as yet). In order to reduce this equation to the normal form, we find the normalizing factor $\mu = \frac{1}{\sqrt{k^2+1}}$. Multiplying (1) by μ, we get the desired normal equation $\frac{kx - y + 7 - 2k}{\pm\sqrt{k^2+1}} = 0.$

Substituting the coordinates of Q in the left-hand member of (2), we have $\frac{|k - 2 + 7 - 2k|}{\sqrt{k^2+1}} = 5$. Solving this equation gives two values of $k: k_1 = -\frac{5}{12}, k_2 = 0$. Substituting these values of the slope in (1), we obtain the required equations $y - 7 = -\frac{5}{12}(x-2)$, or $5x + 12y - 94 = 0$, and $y - 7 = 0$. The problem is solved.

327. $7x + 24y - 134 = 0, x - 2 = 0$. **328.** $3x + 4y - 13 = 0$.

330. $8x - 15y + 9 = 0$.

331. $3x - 4y - 25 = 0, 3x - 4y + 5 = 0$.

332. The conditions of the problem are satisfied by two squares symmetrically situated with respect to the side AB. The equations of the sides of one of the squares are $4x + 3y - 8 = 0$, $4x + 3y + 17 = 0$, $3x - 4y - 6 = 0, 3x + 4y + 19 = 0$.

The equations of the sides of the other square are $4x + 3y - 8 = 0$, $4x + 3y - 33 = 0$, $3x - 4y - 6 = 0$, $3x - 4y + 19 = 0$.

333. The conditions of the problem are satisfied by two squares; the remaining sides of one of the squares lie on the lines $3x+4y-11=0$, $4x-3y-23=0$, $3x+4y-27=0$; the remaining sides of the other square lie on the lines $3x+4y-11=0$, $4x-3y-23=0$, $3x+4y+5=0$.

334. $3x+4y+6=0$, $3x+4y-14=0$ or $3x+4+6=0$, $3x+4y+26=0$.

335. $12x-5y+61=0$, $12x-5y+22=0$ or $12x-5y+61=0$. $12x-5y+100=0$.

336. $M(2, 3)$.

337. $4x+y+5=0$, $y-3=0$.

338. 1) $3x-y+2=0$; 2) $x-2y+5=0$; 3) $20x-8y-9=0$.

339. 1) $4x-4y+3=0$, $2x+2y-7=0$; 2) $4x+1=0$, $8y+13=0$; 3) $14x-8y-3=0$, $64x+112y-23=0$.

340. $x-3y-5=0$, $3x+y-5=0$. *Hint.* The required lines pass through the point P and are perpendicular to the bisectors of the angles formed by the two given lines.

341. 1) By the same angle; 2) by the supplementary angles; 3) by the same angle.

342. 1) By the vertical angles; 2) by the supplementary angles; 3) by the same angle,

343. Inside the triangle

344. Outside the triangle.

345. The acute angle.

346. The obtuse angle,

347. $8x+4y-5=0$.

348. $x+3y-2=0$.

349. $3x-19=0$.

350. $10x-10y-3=0$.

351. $7x+56y-40=0$.

352. $x+y+5=0$.

353. $S(2,-1)$.

354. 1) $3x+2y-7=0$; 2) $2x-y=0$; 3) $y-2=0$; 4) $x-1=0$; 5) $4x+3y-10=0$; 6) $3x-2y+1=0$.

355. $74x+13y+39=0$.

356. $x-y-7=0$.

357. $7x+19y-2=0$.

858. $x-y+1=0$.

359. $4x-5y+22=0$, $4x+y-18=0$, $2x-y+1=0$,

360. $x-5y+13=0$, $5x+y+13=0$.

361. $5x-y-5=0$, (BC), $x-y+3=0$ (AC), $3x-y-1=0$ (CN).

362. $x-5y-7=0$, $5x+y+17=0$, $10x+7y-13=0$

363. $2x+y+8=0$, $x+2y+1=0$.

366. $C=-29$.

367. $a\neq-2$

368. The equations of the sides of the square are $4x+3y-14=0$, $3x-4y+27=0$, $3x-4y+2=0$, $4x+3y+11=0$; the equation of its other diagonal is $7x-y+13=0$.

369. $x+y+5=0$. **370.** $x+y+2=0,\ x-y-4=0,\ 3x+y=0$.

371. $2x+y-6=0, 9x+2y+18=0$.

372. $3x-y+1=0$. **374.** $3x-4y+20=0,\ 4x+3y-15=0$.

375. $x+5y-13=0,\ 5x-y+13=0$.

378. The conditions of the problem are satisfied by the two lines $7x+y-9=0,\ 2x+y+1=0$.

377. $5x-2y-7=0$.

378. $AC: 3x+8y-7=0$, $BD: 8x-3y+7=0$.

379. $4x+y+5=0,\ x-2y-1=0,\ 2x+5y-11=0$.

381. 1) $\rho\sin(\beta-\theta)=p,\ \rho\sin\left(\frac{\pi}{6}-\theta\right)=3$;

2) $\rho\cos(\theta-\alpha)=a\cos\alpha, \rho\cos\left(\theta+\frac{2}{3}\pi\right)=-1$;

3) $\rho\sin(\beta-\theta)=a\sin\beta, \rho\sin\left(\frac{\pi}{6}-\theta\right)=3$.

382. $\rho\sin(\beta-\theta)=\rho_1\sin(\beta-\theta_1)$.

383. $\rho\cos(\theta-\alpha)=\rho_1\cos(\theta_1-\alpha)$.

384. $\dfrac{\rho\sin(\theta-\theta_1)}{\rho_2\sin(\theta_2-\theta_1)}=\dfrac{\sqrt{\rho^2+\rho_1^2-2\rho\rho_1\cos(\theta-\theta_1)}}{\sqrt{\rho_2^2+\rho_1^2-2\rho_2\rho_1\cos(\theta_2-\theta_1)}}$

385. 1) $x^2+y^2=9$; (2) $(x-2)^2+(y+3)^2=49$; 3) $(x-6)^2+(y+8)^2=100$; 4) $(x+1)^2+(y-2)^2=25$; 5) $(x-1)^2+(y-4)^2=8$; 6) $x^2+y^2=16$; 7) $(x-1)^2+(y+1)^2=4$; 8) $(x-2)^2+(y-4)^2=10$; 9) $(x-1)^2+y^2=1$; 10) $(x-2)^2+(y-1)^2=25$.

386. $(x-3)^2+(y+1)^2=38$.

387. $(x-4)^2+(y+1)^2=5$ and $(x-2)^2+(y-3)^2=5$.

388. $(x+2)^2+(y+1)^2=20$.

389. $(x-5)^2+(y+2)^2=20$ and $\left(x-\frac{9}{5}\right)^2+\left(y-\frac{22}{5}\right)^2=20$.

390. $(x-1)^2+(y+2)^2=16$.

391. $(x+6)^2+(y-3)^2=50$ and $(x-29)^2+(y+2)^2=800$.

392. $(x-2)^2+(y-1)^2=5$ and $\left(x-\frac{22}{5}\right)^2+\left(y+\frac{31}{5}\right)^2=\frac{289}{5}$.

393. $(x-2)^2+(y-1)^2=\frac{81}{13}$, $(x+8)^2+(y+7)^2=\frac{25}{13}$.

394. $(x-2)^2+(y-1)^2=25$ and $\left(x+\frac{202}{49}\right)^2+\left(y-\frac{349}{49}\right)^2=\left(\frac{185}{49}\right)^2$.

395. $\left(x+\frac{10}{7}\right)^2+\left(y+\frac{25}{7}\right)^2=1$ and $\left(x-\frac{30}{7}\right)^2+\left(y-\frac{5}{7}\right)^2=1$.

396. $(x-5)^2+y^2=16$, $(x+15)^2+y^2=256$,

$$\left(x-\frac{35}{3}\right)^2+\left(y-\frac{40}{3}\right)^2=\left(\frac{32}{3}\right)^2 \text{ and } \left(x-\frac{35}{3}\right)^2+\left(y+\frac{40}{3}\right)^2=\left(\frac{32}{3}\right)^2$$

397. Equations 1), 2), 4), 5), 8) and 10) represent circles;
1) $C\,5,-2)$, $R=5$; 2) $C(-2,0)$, $R=8$; 3) the equation represents the single point (5, –2); (4) $C(0,\ 5)$, $R=\sqrt{5}$; 5) $C(1,-2)\,R=5$; 6) the equation represents no geometric object in the plane; 7) the equation represents the single point (–2, 1); (8) $C\left(-\frac{1}{2},0\right)$, $R=\frac{1}{2}$;

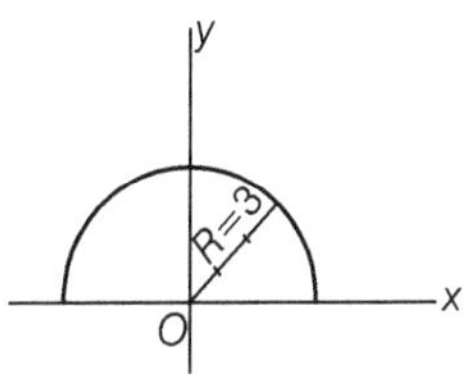

Fig. 83.

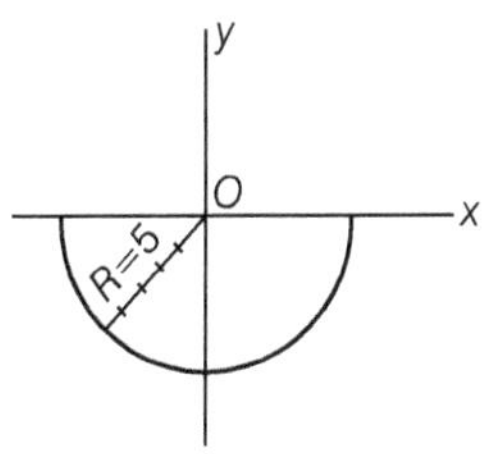

Fig.84.

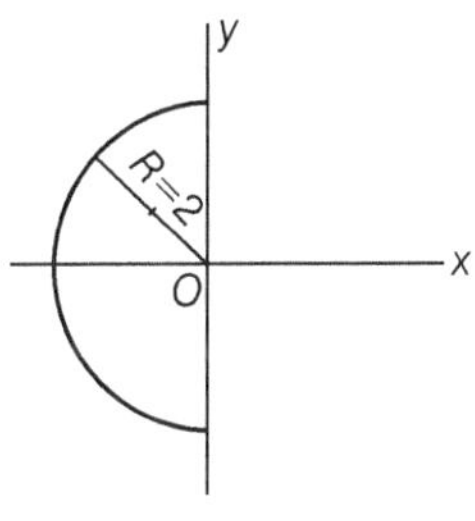

Fig. 85.

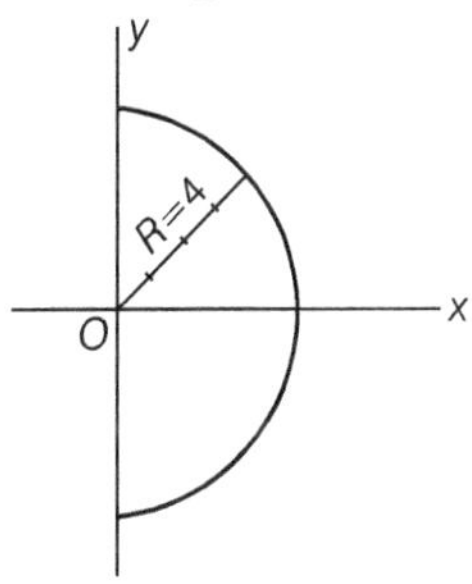

Fig.86.

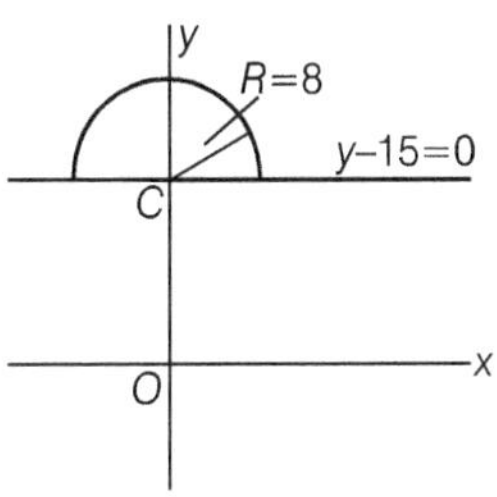

Fig. 87.

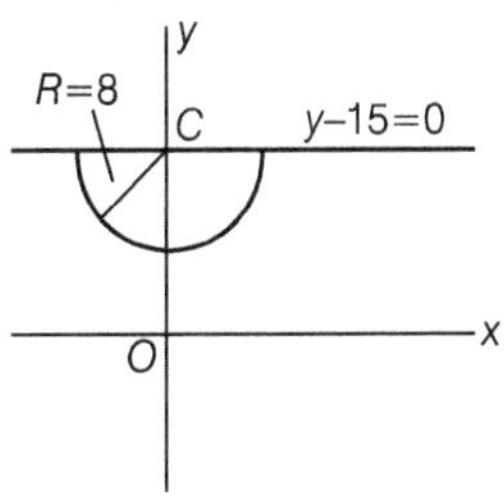

Fig.88.

9) the equation represents no geometric object in the plane;

10) $10C\left(0,-\frac{1}{2}\right)$, $R=\frac{1}{2}$.

398. 1) Semicircle of radius $R = 3$ and with centre at the origin, situated in the upper half-plane (Fig. 83); 2) semicircle of radius $R = 5$ and with centre at the origin, situated in the lower half-plane (Fig. 84); (3) semicircle of radius $R = 2$ and with centre at the origin, situated in the left half-plane (Fig. 85); 4) semicircle of radius $R = 4$ and with centre at the origin, situated in the right half-plane (Fig. 86); 5) semicircle of radius $R = 8$ and with centre C (0, 15). situated above the line $y - 15 = 0$ (Fig. 87); 6) semicircle of radius $R = 8$ and with centre $C(0, 15)$, situated below the line $y - 15 = 0$ (Fig. 88); 7) semicircle of radius $R = 3$ and with centre $C(-2, 0)$,

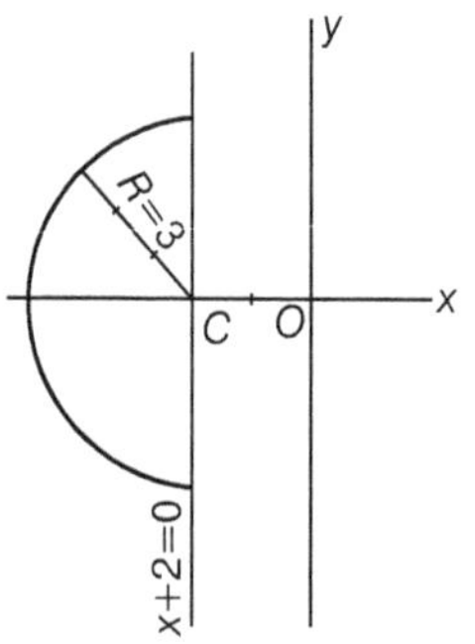

Fig. 89.

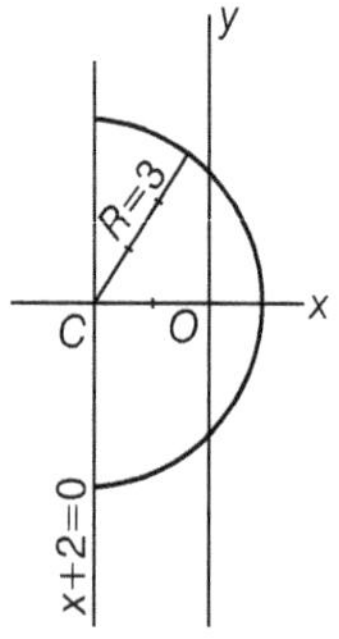

Fig.90.

situated to the left of the line $x + 2 = 0$ (Fig. 89); 8) semicircle of radius $R = 3$ and with centre $C(-2, 0)$, situated to the right of the line $x + 2 = 0$ (Fig. 90); 9) semicircle of radius $R = 5$ and with centre $C(-2, -3)$, situated below the line $y + 3 = 0$ (Fig. 91); 10) semicircle of radius $R = 7$ and with centre $C(-5, -3)$, situated to the right of the line $x + 5 = 0$ (Fig. 92).

399. 1) Outside the circle; 2) on the circle: 3) inside the circle; 4) on the circle; 5) inside the circle.

400. 1) $x + 5y - 3 = 0$; 2) $x + 2 = 0$; 3) $3x - y - 9 = 0$; 4) $y + 1 = 0$.

401. $2x - 5y + 19 = 0$. **402.** 1) 7; 2) 17; 3) 2.

403. $M_1(-1, 5)$ and $M_2(-2, -2)$.

404. 1) Cuts the circle 2) touches the circle; 3) fails to meet the circle.

405. 1) $|k| < \frac{3}{4}$; 2) $k = \pm\frac{3}{4}$ 3) $|k| > \frac{3}{4}$

406. $\dfrac{b^2}{1 + k^2} = R^2$. **407.** $2x + y - 3 = 0$.

408. $11x - 7y - 69 = 0$. **409.** $2\sqrt{5}$.

410. $2x - 3y + 8 = 0$, $3x + 2y - 14 = 0$.

412. $x^2 + y^2 + 6x - 9y - 17 = 0$. **413.** $13x^2 + 13y^2 + 3x + 71y = 0$.

414. $7x-4y=0.$ **415.** 2. **416.** 10.

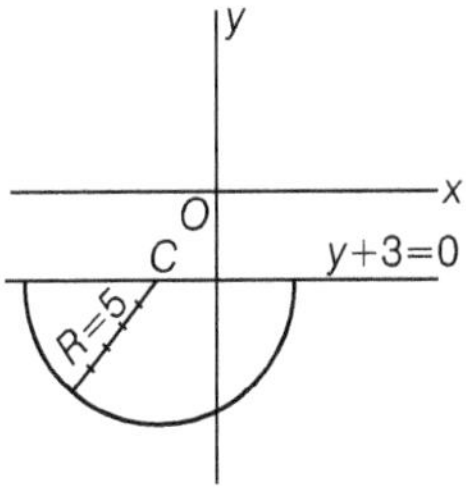

Fig. 91.

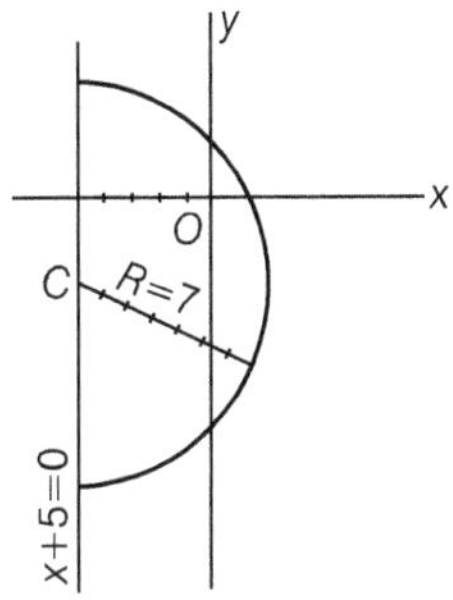

Fig.92.

417. $(x+3)^2+(y-3)^2=10.$

418. $x-2y+5=0.$ **419.** $3x-4y+43=0.$

420. $M_1\left(-\frac{7}{2},\frac{5}{4}\right)$; $d=2\sqrt{5}.$ **421.** $x_1x+y_1y=R^2,$

422. $(x_1-\alpha)\times(x-\alpha)+(y_1-\beta)(y=\beta)=R^2.$

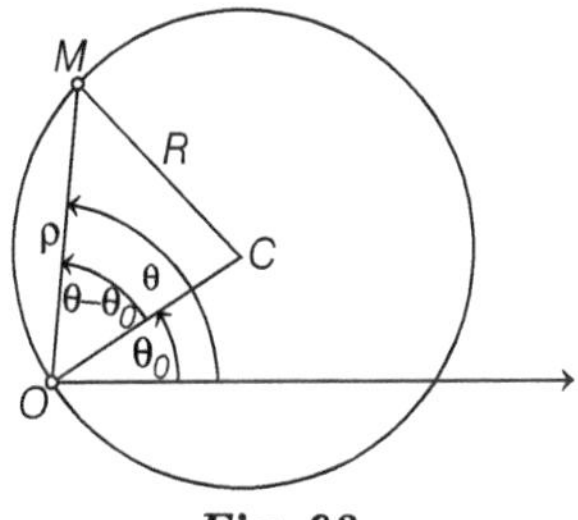

Fig. 93.

423. 45°. **424.** 90°.

425. $(\alpha_1-\alpha_2)^2+(\beta_1-\beta_2)^2=R_1^2+R_2^2.$

427. $x-2y-5=0$ and $2x-y-5=0.$

428. $2x+y-8=0$ and $x-2y+11=0.$

429. $2x+y-5=0, x-2y=0.$

430. 90°. **431.** $x+2y+5=0.$

432. $d=7.5.$ **433.** $d=6.$

434. $d=\sqrt{10}.$ **435.** 3.

436. $2x+y-1=0$ and $2x+y+19=0.$

437. $2x+y-5=0$ and $2x+y+5=0.$

438. $\rho=2R\cos(\theta-\theta_0)$ (Fig. 93).

439. 1) $\rho=2R\cos\theta$ (Fig. 94); 2) $\rho=-2R\cos\theta$ (Fig. 95); 3) $\rho=2R\sin\theta$ (Fig. 96) 4) $\rho=-2R\sin\theta$ (Fig. 97).

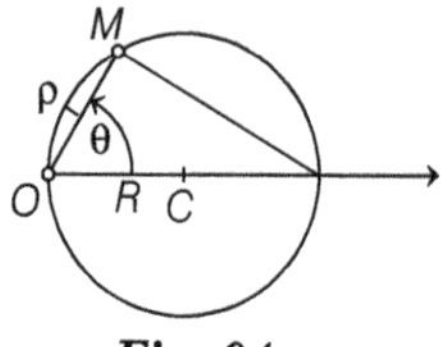

Fig. 94.

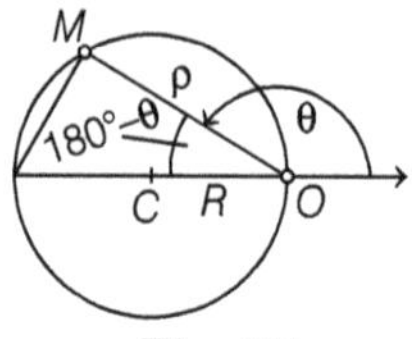

Fig. 95.

440. 1) (2, 0) and $R=-2$; 2) $\left(\frac{3}{2},\frac{\pi}{2}\right)$ and $R=\frac{3}{2}$; 3) $(1,\pi)$ and $R=1$;
4) $\left(\frac{5}{2},-\frac{\pi}{2}\right)$ and $R=\frac{5}{2}$; 5) $\left(3,\frac{\pi}{3}\right)$ and $R=3$; 6) $\left(4,\frac{5}{6}\pi\right)$ and $R=4$;
7) $\left(4,-\frac{\pi}{6}\right)$ and $R=4$.

441. 1) $x^2+y^2-3x=0$; 2) $x^2+y^2+4y=0$; 3) $x^2+y^2-x+y=0$.

442. 1) $\rho=\cos\theta$; 2) $\rho=-3\cos\theta$; 3) $\rho=5\sin\theta$; 4) $\rho=-\sin\theta$; 5) $\rho=\cos\theta+\sin\theta$.

443. $\rho=R\sec(\theta-\theta_0)$

444. 1) $\frac{x^2}{25}+\frac{y^2}{4}=1$; 2) $\frac{x^2}{25}+\frac{y^2}{9}=1$; 3) $\frac{x^2}{169}+\frac{y^2}{144}=1$; 4) $\frac{x^2}{25}+\frac{y^2}{16}=1$;
5) $\frac{x^2}{100}+\frac{y^2}{64}=1$; 6) $\frac{x^2}{169}+\frac{y^2}{25}=1$; 7) $\frac{x^2}{5}+y^2=1$; 8) $\frac{x^2}{16}+\frac{y^2}{12}=1$;
9) $\frac{x^2}{13}+\frac{y^2}{9}=1$; or $\frac{x^2}{\frac{117}{4}}+\frac{y^2}{9}=1$; 10) $\frac{x^2}{64}+\frac{y^2}{48}=1$.

445. 1) $\frac{x^2}{4}+\frac{y^2}{49}=1$; 2) $\frac{x^2}{9}+\frac{y^2}{25}=1$; 3) $\frac{x^2}{25}+\frac{y^2}{169}=1$; 4) $\frac{x^2}{64}+\frac{y^2}{100}=1$;
5) $\frac{x^2}{16}+\frac{y^2}{25}=1$; 6) $\frac{x^2}{7}+\frac{y^2}{16}=1$.

446. 1) 4 and 3; 2) 2 and 1; 3) 5 and 1; 4) $\sqrt{15}$ and $\sqrt{3}$; 5) $\frac{5}{2}$ and $\frac{5}{3}$;
6) $\frac{1}{3}$ and $\frac{1}{5}$; 7) 1 and $\frac{1}{2}$; 8) 1 and 4; 9) $\frac{1}{5}$ and $\frac{1}{3}$; 10) $\frac{1}{3}$ and 1.

447. 1) 5 and 3; 2) $F_1(-4,0)$, $F_2(4,0)$; 3) $\varepsilon=\frac{4}{5}$; 4) $x=\pm\frac{25}{4}$.

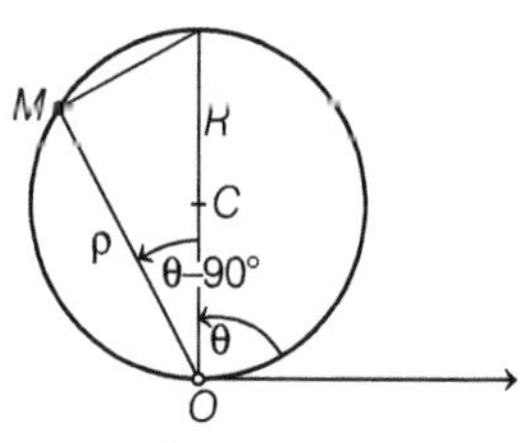

Fig. 96

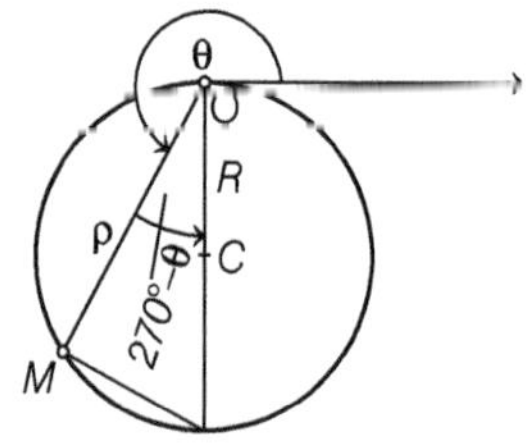

Fig. 97

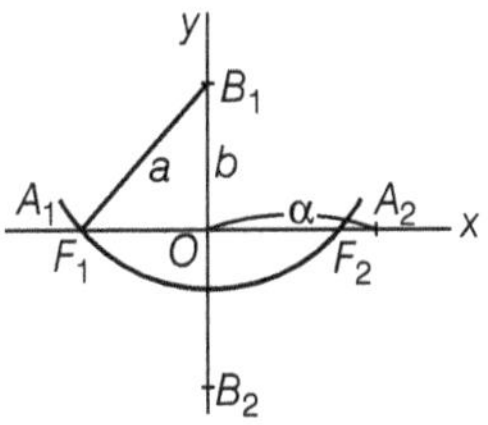

Fig.98

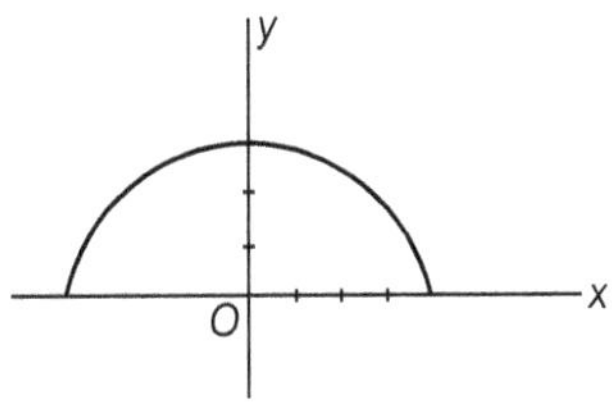

Fig. 99

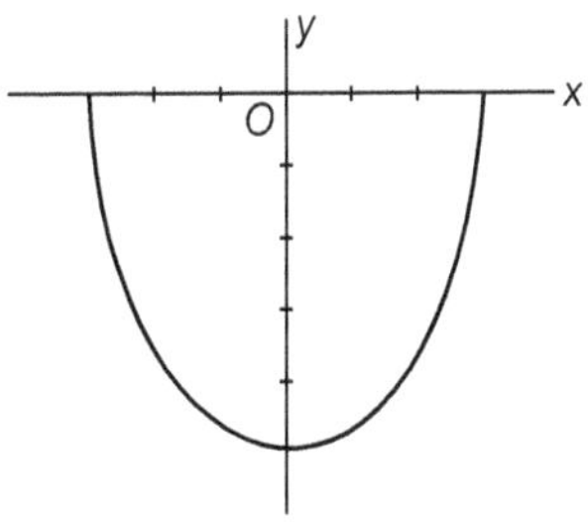

Fig.100.

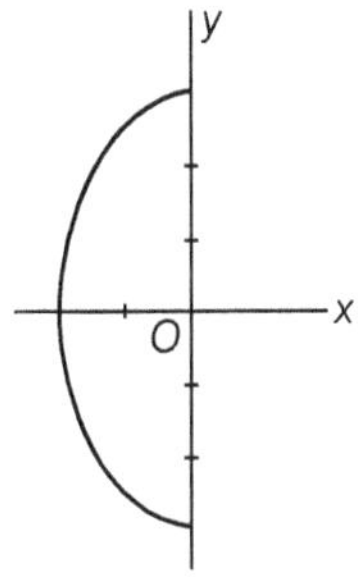

Fig. 101.

448. 16 square units,

449. 1) $\sqrt{5}$ and 3; 2) $F_1(0,-2)$, $F_2(0,2)$; 3) $\varepsilon = \frac{2}{3}$; 4) $y = \pm\frac{9}{2}$.

450. $\frac{4\sqrt{5}}{45}$ square units. **451.** $\frac{b^2}{c}$.

452. See Fig. 98. **453.** $\left(-3, \frac{-8}{5}\right), \left(-3, \frac{8}{5}\right)$.

454. The points A_1 and A_6 lie on the ellipse; A_2, A_4 and A_8 lie inside the ellipse; A_3, A_5, A_7, A_9 and A_{10} lie outside the ellipse.

455. 1) That half of the ellipse $\frac{x^2}{16}+\frac{y^2}{9}=1$ which is situated in the upper half-plane (Fig. 99); 2) that half of the ellipse $\frac{x^2}{9}+\frac{y^2}{25}=1$ which is situated in the lower half-plane (Fig. 100); 3) that half of the ellipse $\frac{x^2}{4}+\frac{y^2}{9}=1$ which is situated in the left half-plane (Fig. 101); 4) that half of the ellipse $x^2+\frac{y^2}{49}=1$ which is situated in the right half-plane (Fig. 102).

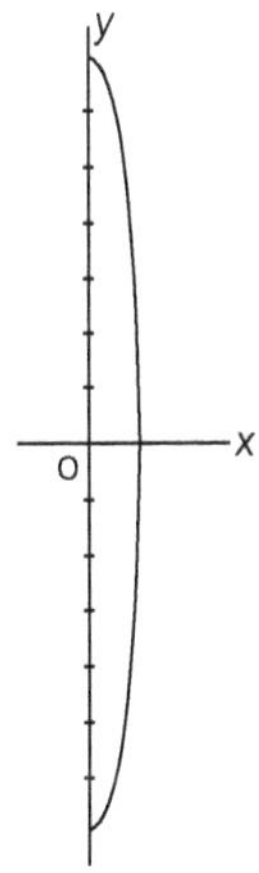

Fig. 102.

456. 15. **457.** 8.

458. $5x+12y+10=0,\ x-2=0.$

459. $r_1=2.6,\ r_2=7.4.$ **460.** 20.

461. 10. **462.** $(-5, 3\sqrt{3})$ and $(-5, -3\sqrt{3})$.

463. $\left(-2, \frac{\sqrt{21}}{2}\right)$ and $\left(-2, -\frac{\sqrt{21}}{2}\right)$.

464. 3 and 7.

465. 1) $\frac{x^2}{36}+\frac{y^2}{9}=1$; 2) $\frac{x^2}{16}+\frac{y^2}{\frac{16}{3}}=1$; 3) $\frac{x^2}{20}+\frac{y^2}{15}=1$; 4) $\frac{x^2}{20}+\frac{y^2}{4}=1$;

5) $\frac{x^2}{9}+\frac{y^2}{5}=1$; 6) $\frac{x^2}{256}+\frac{y^2}{192}=1$; 7) $\frac{x^2}{15}+\frac{y^2}{6}=1$.

466. 1) $\frac{\sqrt{3}}{2}$; 2) $\frac{\sqrt{2}}{2}$; 3) $\frac{\sqrt{3}}{3}$; 4) $\frac{1}{2}$.

467. $\varepsilon=\frac{\sqrt{2}}{2}$.

468. $\frac{(x-x_0)^2}{a^2}+\frac{(y-y_0)^2}{b^2}=1.$ **469.** $\frac{(x-3)^2}{9}+\frac{(y+4)^2}{16}=1.$

470. $\frac{(x+3)^2}{9}+\frac{(y-2)^2}{4}=1.$

471. 1) $C(3,-1)$, semi-axes 3 and $\sqrt{5}$, $\varepsilon=\frac{2}{3}$, equations of the directrices: $2x-15=0, 2x+3=0$; 2) $C(-1,2)$, semi-axes 5 and 4, $\varepsilon=\frac{3}{5}$, equations of the directrices : $3x-22=0, 3x+28=0$; 3) $C(1,-2)$, semi-axes $2\sqrt{3}$ and 4, $\varepsilon=\frac{1}{2}$, equations of the directrices; $y-6=0, y+10=0$.

472. 1) That half of the ellipse $\frac{(x-3)^2}{25}+\frac{(y+7)^2}{4}=1$ which is situated above the line $y+7=0$ (Fig. 103); 2) that half of the ellipse $\frac{(x+3)^2}{9}+\frac{(y-1)^2}{16}=1$ which is situated below the line $y-1=0$ (Fig. 104); 3) that half of the ellipse $\frac{x^2}{16}+\frac{(y+3)^2}{4}=1$ which is situated in the left half-plane (Fig. 105); 4) that half of the ellipse $\frac{(x+5)^2}{4}+\frac{(y-1)^2}{9}=1$ which is situated to the right of the line $x+5=0$ (Fig. 106).

473. 1) $\frac{(x-2)^2}{169}+\frac{y^2}{25}=1$; 2) $2x^2-2xy+2y^2-3=0$;
3) $68x^2+48xy+82y^2-625=0$;
4) $11x^2+2xy+11y^2-48x-48y-24=0$.

474. $5x^2+9y^2+4x-18y-55=0$.

475. $4x^2+3y^2+32x-14y+59=0$.

476. $4x^2+5y^2+14x+40y+81=0$.

477. $7x^2-2xy+7y^2-46x+2y+71=0$.

478. $17x^2+8xy+23y^2+30x-40y-175=0$.

479. $x^2+2y^2-6x+24y+31=0$.

480. $\left(4,\frac{3}{2}\right)$, (3, 2).

481. $\left(3,\frac{8}{5}\right)$ – the line touches the ellipse.

482. The line fails to meet the ellipse.

483. 1) The line cuts the ellipse; 2) the line fails to meet the ellipse; 3) the line touches the ellipse.

484. The line: 1) cuts the ellipse for $|m|<5$; 2) touches the ellipse for $m=\pm5$; 3) passes outside the ellipse for $|m|>5$.

485. $k^2a^2+b^2=m^2$. **486.** $\frac{x_1x}{a^2}+\frac{y_1y}{b^2}=1$.

488. $3x+2y-10=0$ and $3x+2y+10=0$.

489. $x+y-5=0$ and $x+y+5=0$.

490. $2x-y-12=0, 2x-y+12=0$; $d=\frac{24\sqrt{5}}{5}$.

491. $M_1(-3,2)$; $d=\sqrt{13}$.

492. $x+y-5=0$ and $x+4y-10=0$.

493. $4x-5y-10=0$. **494.** $d=18$.

495. $\frac{x^2}{20}+\frac{y^2}{5}=1$ or $\frac{x^2}{80}+\frac{4y^2}{5}=1$. **496.** $\frac{x^2}{40}+\frac{y^2}{10}=1$.

499. $\frac{x^2}{17}+\frac{y^2}{8}=1$. *Hint*. Use the property of the ellipse formulated in Problem 498.

500. $\frac{x^2}{25}+\frac{y^2}{4}=1$. *Hint*. Use the property of the ellipse formulated in Problem 498.

502. $2x+11y-10=0$. *Hint*. Use the property of the ellipse formulated in Problem 501.

503. (3, 2) and (3, −2).

504. $R = \dfrac{mn\sqrt{2}}{\sqrt{m^2 + n^2}}$. **505.** $10.5\sqrt{3}$.

506. $\phi = 60°$. **507.** 16.8.

508. 60°.

509. Into the ellipse $\dfrac{x^2}{25} + \dfrac{y^2}{16} = 1$.

510. $x^2 + y^2 = 9$. **511.** $\dfrac{x^2}{36} + \dfrac{y^2}{16} = 1$.

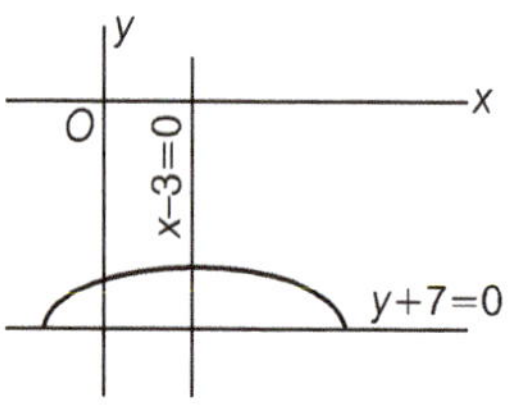

Fig. 103.

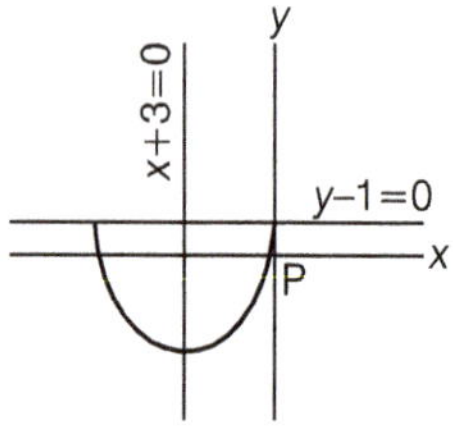

Fig.104.

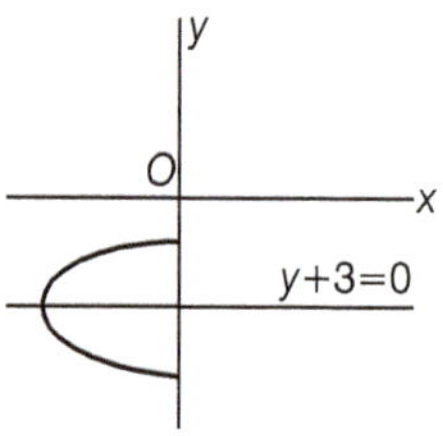

Fig. 105

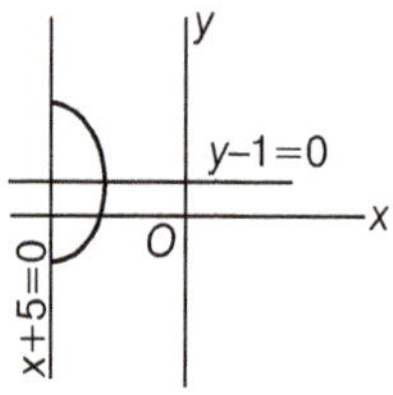

Fig.106

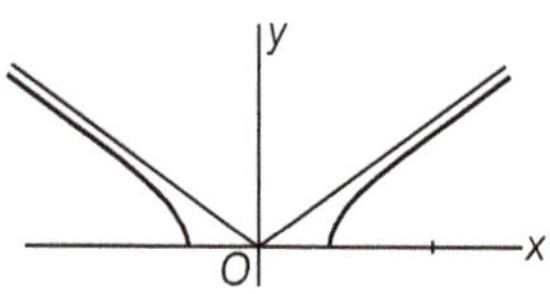

Fig. 107.

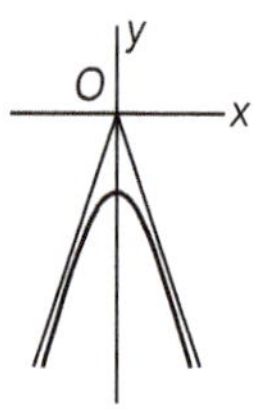

Fig. 180.

512. $q = \dfrac{4}{3}$. **513.** $q = \dfrac{2}{3}$. **514.** $q_1 = \dfrac{4}{3}, q_2 = \dfrac{4}{5}$.

515. 1) $\dfrac{x^2}{25} - \dfrac{y^2}{16} = 1$; 2) $\dfrac{x^2}{9} - \dfrac{y^2}{16} = 1$; 3) $\dfrac{x^2}{4} - \dfrac{y^2}{5} = 1$; 4) $\dfrac{x^2}{64} - \dfrac{y^2}{36} = 1$;
5) $\dfrac{x^2}{36} - \dfrac{y^2}{64} = 1$; 6) $\dfrac{x^2}{144} - \dfrac{y^2}{25} = 1$; 7) $\dfrac{x^2}{16} - \dfrac{y^2}{9} = 1$; 8) $\dfrac{x^2}{4} - \dfrac{y^2}{5} = 1$;
9) $\dfrac{x^2}{64} - \dfrac{y^2}{36} = 1$.

516. 1) $\frac{x^2}{36}-\frac{y^2}{324}=-1$; 2) $\frac{x^2}{16}-\frac{y^2}{9}=-1$ 3) $\frac{x^2}{100}-\frac{y^2}{576}=-1$;
4) $\frac{x^2}{24}-\frac{y^2}{25}=-1$; 5) $\frac{x^2}{9}-\frac{y^2}{16}=-1$.

517. 1) $a=3, b=2$; 2) $a=4, b=1$; 3) $a=4, b=2$; 4) $a=1, b=1$;
5) $a=\frac{5}{2}, b=\frac{5}{3}$, 6) $a=\frac{1}{5}, b=\frac{1}{4}$, 7) $a=\frac{1}{3}, b=\frac{1}{8}$.

518. 1) $a=3, b=4$; 2) $F_1(-5,0), F_2(5,0)$; 3) $\varepsilon=\frac{5}{3}$; 4) $y=\pm\frac{4}{3}x$; 5) $x=\pm\frac{9}{5}$.

519. 1) $a=3, b=4$; 2) $F_1(0,-5), F_2(0,5)$; 3) $\varepsilon=\frac{5}{4}$;
4) $y=\pm\frac{4}{3}x$; 5) $y=\pm\frac{16}{5}$.

520. 12 square units.

521. 1) That portion of the hyperbola $\frac{x^2}{9}-\frac{y^2}{4}=1$ which is situated in the upper half-plane (Fig. 107); 2) that branch of the hyperbola $x^2-\frac{y^2}{9}=-1$ which is situated in the lower half-plane (Fig. 108);
3) that branch of the hyperbola $\frac{x^2}{16}-\frac{y^2}{9}=1$ which is situated in the left half-plane (Fig. 109); 4) that branch of the hyperbola $\frac{x^2}{25}-\frac{y^2}{4}=-1$ which is situated in the upper half-plane (Fig. 110).

522. $x+4\sqrt{5}y+10=0$ and $x-10=0$.

523. $r_1=2\frac{1}{4}$, $r_2=10\frac{1}{4}$.

524. 8. **525.** 12. **526.** 10. **527.** 27.

528. $\left(10, \frac{9}{2}\right)$ and $\left(10, -\frac{9}{2}\right)$. **529.** $(-6, 4\sqrt{3})$ and $(-6, -4\sqrt{3})$.

530. $2\frac{1}{12}$ and $26\frac{1}{12}$. **531.** See Fig. 111.

532. 1) $\frac{x^2}{32}-\frac{y^2}{8}=1$; 2) $x^2-y^2=16$; 3) $\frac{x^2}{4}-\frac{y^2}{5}=1$ or $\frac{x^2}{61/9}-\frac{y^2}{805/16}=1$;
4) $\frac{x^2}{18}-\frac{y^2}{8}=1$; 5) $\frac{x^2}{16}-\frac{y^2}{9}=1$.

533. $\varepsilon=\sqrt{2}$. **534.** $\varepsilon=\sqrt{3}$. **535.** $\frac{x^2}{4}-\frac{y^2}{12}=1$. **536.** $\frac{x^2}{60}-\frac{y^2}{40}=1$.

540. 1) $\frac{(x-x_0)^2}{a^2}-\frac{(y-y_0)^2}{b^2}=1$; 2) $\frac{(x-x_0)^2}{a^2}-\frac{(y-y_0)^2}{b^2}=-1$.

541. 1) $C(2, -3), a = 3,$

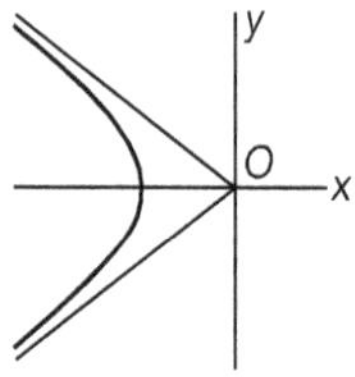

Fig. 109.

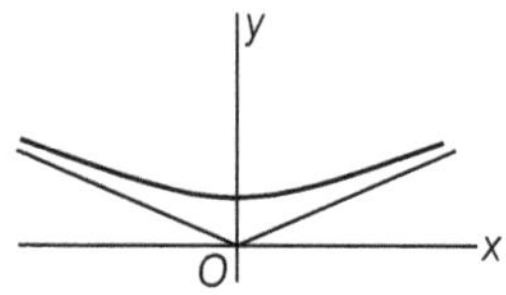

Fig.110.

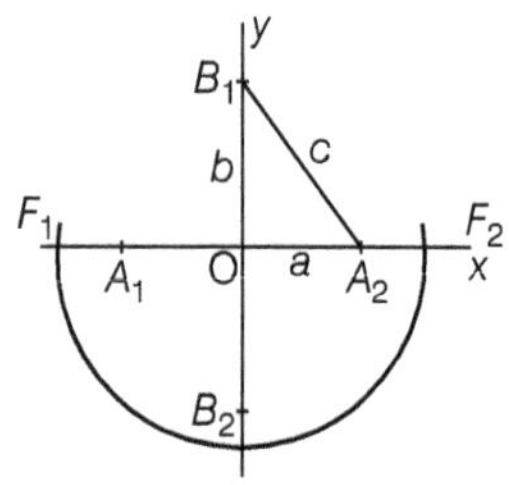

Fig. 111.

$b = 4$, $\varepsilon = \frac{5}{3}$, equations of the directrices, $5x - 1 = 0$, $5x - 19 = 0$, equations of the asymptotes: $4x - 3y - 17 = 0, 4x + 3y + 1 = 0$; 2) $C(-5, 1), a = 8, b = 6, \varepsilon = 1.25$, equations of the directrices: $x = -11.4$ and $x = 1.4$, equations of the asymptotes $3x + 4y + 11 = 0$, $3x - 4y + 19 = 0$; 3) $C(2,-1), a = 3, b = 4, \varepsilon = 1.25$, equations of the directrices: $y = -4.2$, $y = 2.2$, equations of the asymptotes: $4x + 3y - 5 = 0, 4x - 3y - 11 = 0$.

542. That portion of the hyperbola $\frac{(x-2)^2}{9} - \frac{(y+1)^2}{4} = 1$ which is situated above the line $y + 1 = 0$ (Fig. 112); 2) that branch of the hyperbola $\frac{(x-3)^2}{4} - \frac{(y-7)^2}{9} = -1$ which is situated below the line $y - 7 = 0$ (Fig.113); 3) that branch of the hyperbola $\frac{(x-9)^2}{16} - \frac{(y+2)^2}{4} = 1$ which is to the left of the line $x - 9 = 0$ (Fig. 114); 4) that portion of the hyperbola $\frac{(x-5)^2}{9} - \frac{(y+2)^2}{16} = -1$ which is to the left of the line $x - 5 = 0$ (Fig. 115).

543. 1) $\frac{(x-3)^2}{144} - \frac{(y-2)^2}{25} = 1$; 2) $24xy + 7y^2 - 144 = 0$;

3) $2xy + 2x - 2y + 7 = 0$.

544. $\frac{x^2}{16} - \frac{y^2}{9} = 1.$ **545.** $\frac{x^2}{25} - \frac{y^2}{144} = -1.$

546. $x^2 - 4y^2 - 6x - 24y - 47 = 0.$

547. $7x^2 - 6xy - y^2 + 26x - 18y - 17 = 0.$

548. $91x^2 - 100xy + 16y^2 - 136x + 86y - 47 = 0.$

549. $xy = \frac{a^2}{2}$ if the old axes are rotated through an angle of $-45°$; $xy = -\frac{a^2}{2}$ if they are rotated through an angle of $+45°$.

550. 1) C (0,0), $a = b = 6$, equations of the asymptotes: $x = 0$ and $y = 0$;
2) C (0,0), $a = b = 3$, equations of the asymptotes: $x = 0$ and $y = 0$;
3) C (0,0), $a = b = 5$, equations of the asymptotes: $x = 0$ and $y = 0$.

551. (6, 2) and $\left(\frac{14}{3}, -\frac{2}{3}\right)$.

552. $\left(\frac{25}{4}, 3\right)$ – the line touches the hyperbola.

553. The line fails to meet the hyperbola.

554. The line: 1) touches the hyperbola;
2) cuts the hyperbola at two points;
3) fails to meet the hyperbola.

555. The line 1) cuts the hyperbola for $|m| > 4.5$;
2) touches the hyperbola for $m = \pm 4.5$;
3) passes outside the hyperbola for $|m| < 4.5$.

556. $k^2a^2 - b^2 = m^2.$

557. $\frac{x_1 x}{a^2} - \frac{y_1 y}{b^2} = 1.$

559. $3x - 4y - 10 = 0, 3x - 4y + 10 = 0.$

560. $10x - 3y - 32 = 0, 10x - 3y + 32 = 0.$

561. $x + 2y - 4 = 0$. $x + 2y + 4 = 0$; $d = \frac{8\sqrt{5}}{5}.$

562. $M_1(-6, 3)$; $d = \frac{11}{13}\sqrt{13}.$

563. $5x - 3y - 16 = 0, 13x + 5y + 48 = 0.$

564. $2x + 5y - 16 = 0.$

565. $d = \frac{17}{10}\sqrt{10}$

566. $\frac{x^2}{5} - \frac{y^2}{45} = 1, \frac{3x^2}{10} - \frac{4y^2}{45} = 1.$

567. $\frac{x^2}{16} - \frac{y^2}{4} = 1.$

568. $x = -4$, $x = 4$, $y = -1$ and $y = 1$.

572. $\frac{x^2}{5} - \frac{y^2}{4} = 1,$

573. $\frac{x^2}{16} - \frac{y^2}{9} = 1.$

575. $2x + 11y + 6 = 0.$

Hint. Use the property of the hyperbola formulated in Problem 574.

577. $x^2 - y^2 = 16.$

578. $\frac{x^2}{16} - \frac{y^2}{9} = 1.$

579. $\frac{x^2}{25} - \frac{y^2}{4} = 1.$

580. $q = \frac{2}{3}.$

581. $q = 2.$

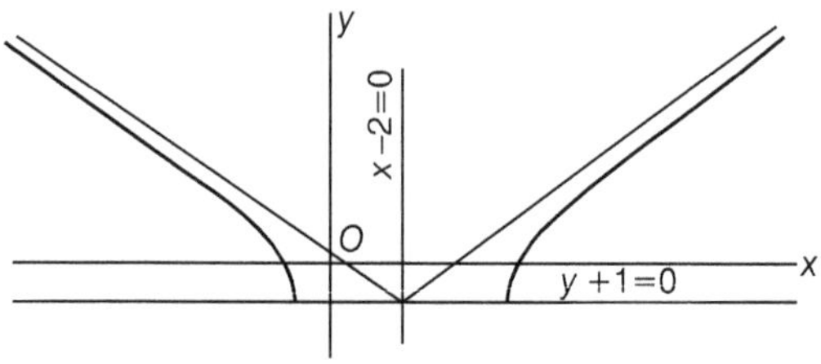

Fig. 112.

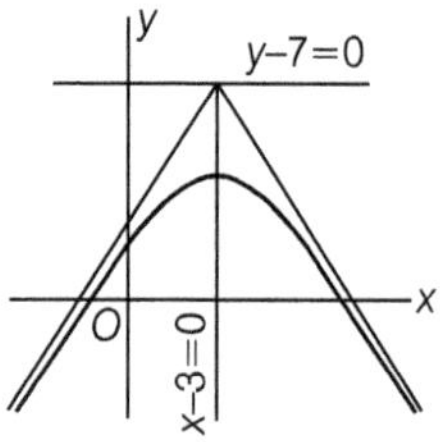

Fig. 113

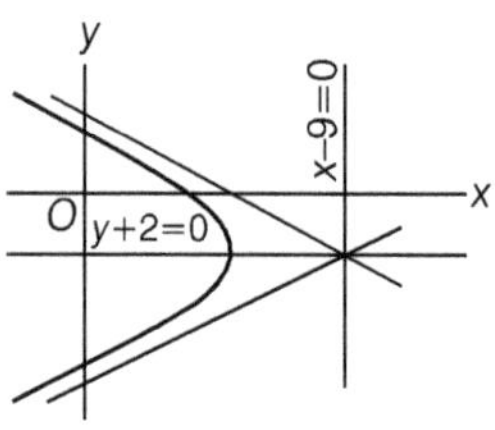

Fig. 114.

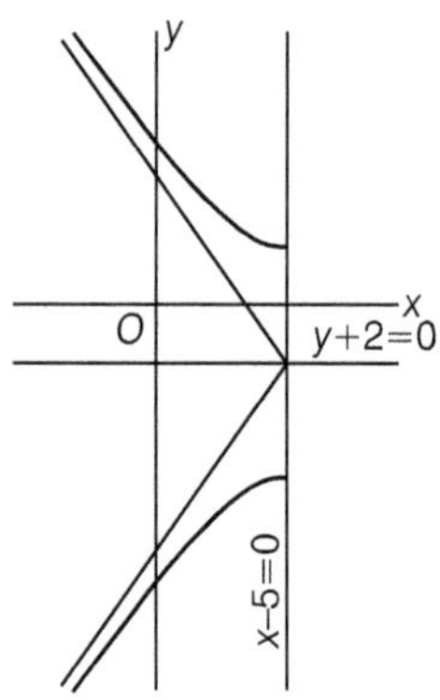

Fig. 115.

582. $q_1 = 2; q_2 = \frac{5}{7}.$

583. 1) $y^2 = 6x;$

2) $y^2 = -x;$

3) $x^2 = \frac{1}{2}y;$ (4) $x^2 = -6y.$

584. 1) $p=3$; situated in the right half-plane (symmetrically with respect to the axis Ox); 2) $p=2.5$; situated in the upper half-plane (symmetrically with respect to the axis Oy); 3) $p=2$; situated in the left half-plane (symmetrically with respect to the axis Ox); 4) $p=\frac{1}{2}$; situated in the lower half-plane (symmetrically with respect to the axis Oy).

585. 1) $y^2=4x$; 2) $y^2=-9x$; 3) $x^2=y$; 4) $x^2=-2y$.

586. 40 cm. **587.** $x^2=-12y$.

588. 1) That portion of the parabola $y^2=4x$ which is situated in the first quadrant (Fig. 116);

2) that portion of the parabola $y^2=-x$ which is situated in the second quadrant (Fig. 117);

3) that portion of the parabola $y^2=-18x$ which is situated in the third quadrant (Fig.118);

4) that portion of the parabola $y^2=4x$ which is situated in the fourth quadrant (Fig. 119);

5) that portion of the parabola $x^2=5y$ which is situated in the first quadrant (Fig. 120);

6) that portion of the parabola $x^2=-25y$ which is situated in the third quadrant (Fig. 121);

7) that portion of the parabola $x^2=3y$ which is situated in the second quadrant (Fig. 122);

8) that portion of the parabola $x^2=-16y$ which is situated in the fourth quadrant (Fig. 123).

589. $F(6,0)$, $x+6=0$. **590.** 12.

591. 6. **592.** (9, 12); (9, – 12).

593. $y^2=-28x$.

594. 1) $(y-\beta)^2=2p(x-\alpha)$; 2) $(y-\beta)^2=-2p(x-\alpha)$.

595. 1) $(x-\alpha)^2=2p(y-\beta)$; 2) $(x-\alpha)^2=-2p(y-\beta)$.

596. 1) $A(2,0), p=2\ x-1=0$; 2) $A\left(\frac{2}{3},0\right), p=3, 6x-13=0$;

3) $A\left(0,-\frac{1}{3}\right), p=3, 6y+11=0$; 4) $A(0,2), p=\frac{1}{2}, 4y-9=0$.

597. 1) $A(-2,1), p=2$;

2) $A(1,3), p=\frac{1}{8}$;

3) $A(6,-1), p=3$.

599. 1) That portion of the parabola $(y-3)^2 = 16(x-1)$ which is situated below the line $y-3=0$ (Fig. 124); 2) that portion of the parabola $(x+4)^2 = 9(y+5)$ which is situated to the right of the line $x+4=0$ (Fig. 125); 3) that portion of the parabola $(x-2)^2 = -2(y-3)$ which is to the left of the line $x-2=0$ (Fig. 126); 4) that portion of the parabola $(y+5)^2 = -3(x+7)$ which is situated below the line $y+5=0$ (Fig. 127).

600. $x = \frac{1}{4}y^2 - y + 7.$

601. $y = \frac{1}{8}x^2 - x + 3.$

602. $x^2 + 2xy + y^2 - 6x + 2y + 9 = 0.$

603. $F(9, -8).$

604. $4x^2 - 4xy + y^2 + 32x + 34y + 89 = 0.$

605. $(2, 1)$ $(-6, 9)$.

606. $(-4, 6)$ – the line touches the parabola.

607. The line and the parabola do not intersect.

608. The line:

1) touches the parabola;

2) cuts the parabola at two points;

3) fails to meet the parabola.

609. 1) $k < \frac{1}{2}$; 2) $k = 1/2$; 3) $k > 1/2$

610. $p = 2bk.$

612. $y_1 y = p(x + x_1).$

613. $x + y + 2 = 0.$

614. $2x - y - 16 = 0.$

615. $d = 2\sqrt{13}.$

616. $M_1(9, -24); d = 10.$

617. $3x - y + 3 = 0$ and $3x - 2y + 12 = 0$

619. $5x - 18y + 25 = 0.$

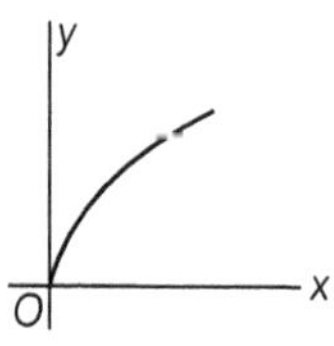

Fig. 116.

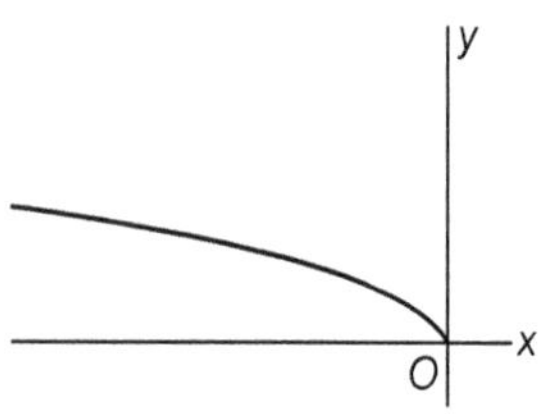

Fig. 117.

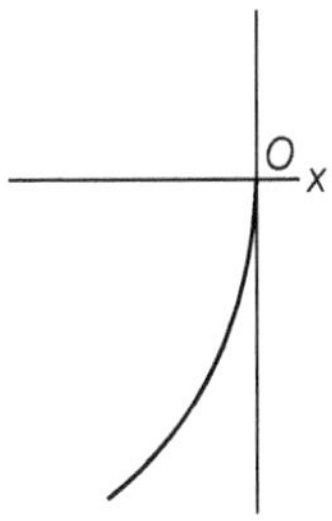

Fig. 118.

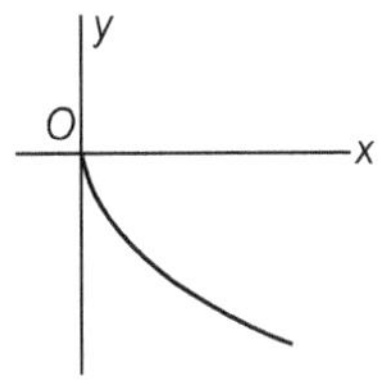

Fig. 119.

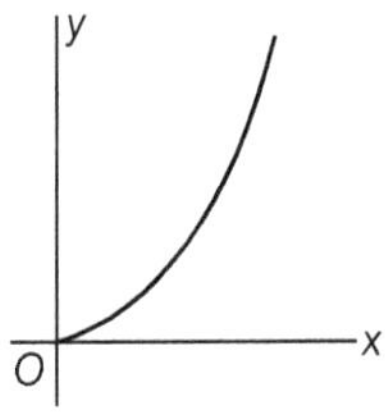

Fig. 120.

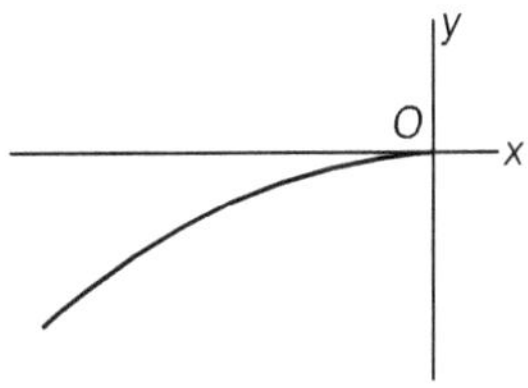

Fig. 121.

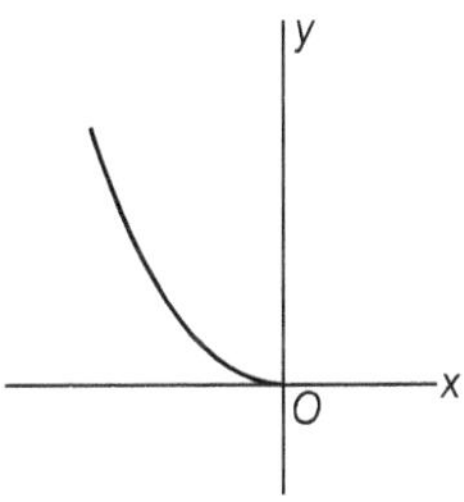

Fig. 122.

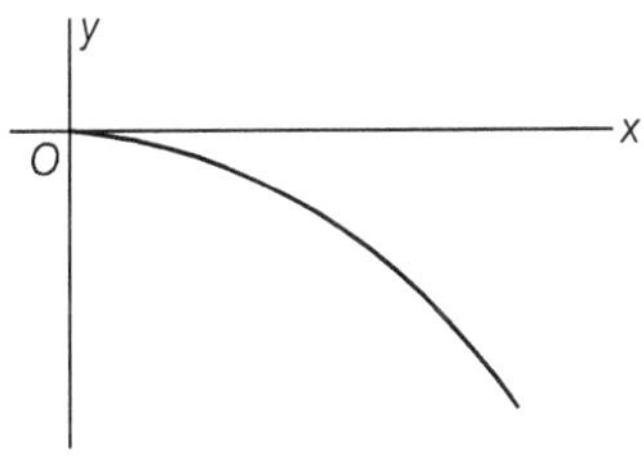

Fig. 123.

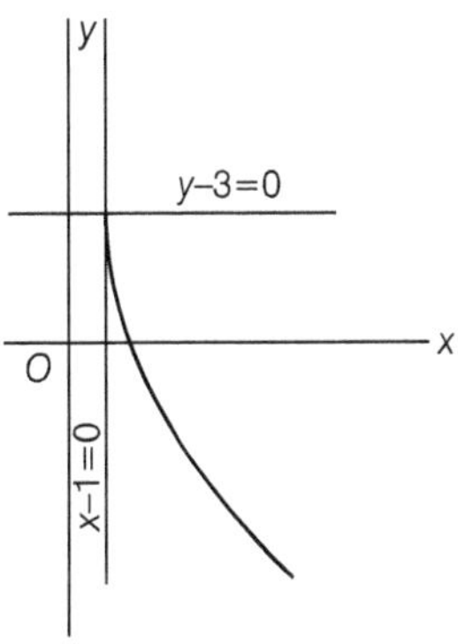

Fig. 124.

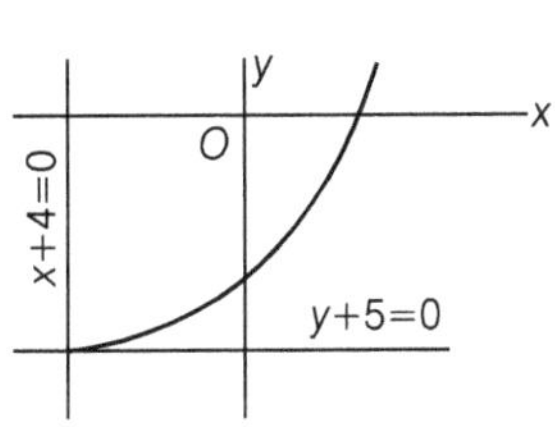

Fig. 125.

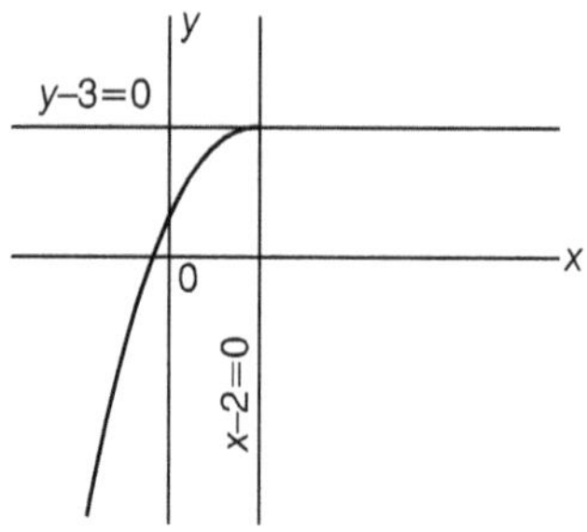

Fig. 126.

620. $d = 13\frac{5}{13}$.

621. (6, 12) and (6, – 12).

622. $(10, \sqrt{30})$, $(10, -\sqrt{30},)$, $(2, \sqrt{6})$, $(2, -\sqrt{6})$.

623. $(2,1)$, $(-1,4)$, $\left(\frac{3+\sqrt{13}}{2}, \frac{7+\sqrt{13}}{2}\right)$ and $\left(\frac{3-\sqrt{13}}{2}, \frac{7-\sqrt{13}}{2}\right)$.

625. $y - 18 = 0$. Hint. Use the property of the parabola formulated in Problem 624

628. 1) $\rho = \frac{16}{5 - 3\cos\theta}$; 2) $\rho = \frac{16}{15 + 3\cos\theta}$.

629. 1) $\rho = \frac{9}{4 - 5\cos\theta}$; 2) $\rho = -\frac{9}{4 - 5\cos\theta}$.

630. 1) $\rho = \frac{144}{5 + 13\cos\theta}$; 2) $\rho = \frac{144}{5 + 13\cos\theta}$.

631. $\rho = \frac{3}{1 - \cos\theta}$.

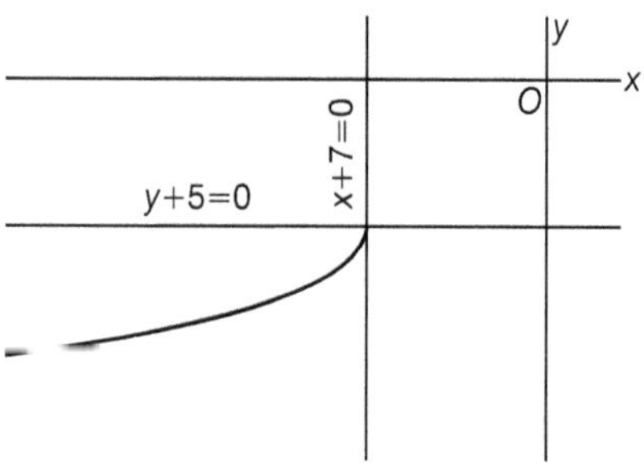

Fig. 127.

632. 1) An ellipse;
2) a parabola;
3) one branch of a hyperbola;
4) an ellipse;
5) one branch of a hyperbola;
6) a parabola.

633. 13, 12.

634. 8, 6.

635. $\rho = -\dfrac{21}{2\cos\theta}, \rho = \dfrac{29}{2\cos\theta}$.

636. The equations of the directrices are
$\rho = -\dfrac{34}{5\cos\theta}, \rho = -\dfrac{16}{5\cos\theta}$; the equations of the asymptotes are
$\rho = \dfrac{20}{3\sin\theta - 4\cos\theta}, \rho = -\dfrac{20}{3\sin\theta + 4\cos\theta}$.

637. $\left(6, \dfrac{\pi}{4}\right), \left(6, -\dfrac{\pi}{4}\right)$.

638. $\left(3, \dfrac{2}{3}\pi\right), \left(3, -\dfrac{2}{3}\pi\right)$.

639. 1) $\left(\dfrac{p}{2}, \pi\right)$; 2) $\left(p, \dfrac{\pi}{2}\right), \left(p, -\dfrac{\pi}{2}\right)$.

640. $\rho^2 = \dfrac{b^2}{1 - \varepsilon^2\cos^2\theta}$.

641. $\rho^2 = \dfrac{b^2}{\varepsilon^2\cos^2\theta - 1}$.

642. $\rho = \dfrac{2p\cos\theta}{\sin^2\theta}$

643. $8x + 25y = 0$.

644. $9x - 32y - 73 = 0$.

645. $x - y = 0, x + 4y = 0$.

646. $x + 2y = 0, 8x - 9y = 0$.

647. $x + 2y = 0, 2x - 3y = 0$.

654. $2x - 5y = 0$.

655. $7x + y - 20 = 0$.

656. $x - 8y = 0, 2x - y = 0$.

657. $x - 2y = 0, 3x - y = 0$; $x + 2y = 0, 3x + y = 0$.

661. $y + 2 = 0$.

662. $2x - y + 1 = 0$

665. Curves 1), 2), 5) and 8) have a single centre; curves 3), 7) have no centre; curves 4), 6) have infinitely many centres.

666. 1) $(3-2)$; 2) $(0,-5)$;
3) $(0,0)$; 4) $(-1,3)$.

667. 1) $x-3y-6=0$; 2) $2x+y-2-0$; 3) $5x-y+4=0$.

668. 1) $9x^2-18xy+6y^2+2=0$; 2) $6x^2+4xy+y^2-7=0$;
3) $4x^2+6xy+y^2-5=0$; 4) $4x^2+2xy+6y^2+1=0$

669. 1) $m \neq 4, n$ may have any value; 2) $m=4, n\neq 6$; 3) $m=4, n=6$.

670. 1) $k=2$; 2) $k_1=-1, k_2=5$;
3) for every $k\neq 2$ which satisfies the inequalities $-1<k<5$;
4) for $k<-1$ and for $k>5$.

671. $x^2-8y^2-4=0$. **672.** $x^2+xy+y^2+3y=0$.

673. 1) Elliptic equation; represents the ellipse $\frac{x'^2}{9}+\frac{y'^2}{4}=1$; the new origin is $O'(5,-2)$; 2) hyperbolic equation; represents the hyperbola $\frac{x'^2}{16}-\frac{y'^2}{9}=1$; the new origin is $O'(3,-2)$; 3) elliptic equation $\frac{x'^2}{4}+\frac{y'^2}{9}=-1$; represents no geometric object (is an equation of an imaginary ellipse); 4) hyperbolic equation; represents a degenerate hyperbola (the pair of intersecting lines $(4x'^2-y'^2=0)$; the new origin is $O'(-1,-1)$;
5) elliptic equation; represents the degenerate ellipse $2x'^2+3y'^2=0$ (a single point).

674. * 1) Hyperbolic equation; represents the hyperbola $\frac{x'^2}{9}-\frac{y'^2}{4}=1$; $\tan\alpha=-2$, $\cos\alpha=\frac{1}{\sqrt{5},}\sin\alpha=-\frac{2}{\sqrt{5}}$; 2) elliptic equation; represents the ellipse $\frac{x'^2}{16}+\frac{y'^2}{4}=1; \alpha=45°$; 3) elliptic equation; represents the degenerate ellipse $x'^2+4y'^2=0$ (a single point); $\tan\alpha=2, \cos\alpha=\frac{1}{\sqrt{5}}, \sin\alpha=\frac{2}{\sqrt{5}}$; 4) hyperbolic equation; represents the degenerate hyperbola $x'^2-y'^2=0$ (a pair of intersecting lines); $\tan\alpha=\frac{2}{3}, \cos\alpha=\frac{3}{\sqrt{13}}, \sin\alpha=\frac{2}{\sqrt{13}}$;
(5) elliptic equation; represents no geometric object (is an equation of an imaginary ellipse); has the form $\frac{x'^2}{4}+y'^2=-1$ in the new coordinates; $\alpha=45°$.

*In Problem 674. (1) (–5), α is the angle from the positive x–axis to the positive x'–axis

675. 1) Hyperbolic; 2) elliptic;
3) parabolic; 4) elliptic;
5) parabolic; 6) hyperbolic.

676. 1) Hyperbolic equation; represents the hyperbola whose equation can be reduced to the form $x'^2 - \frac{y'^2}{4} = 1$ by two consecutive trans formations of coordinates: $x = \tilde{x} + 2, y = \tilde{y} - 1$ and $\tilde{x} = \frac{x' - y'}{\sqrt{2}}, \tilde{y} = \frac{x' + y'}{\sqrt{2}}$ (Fig. 128)

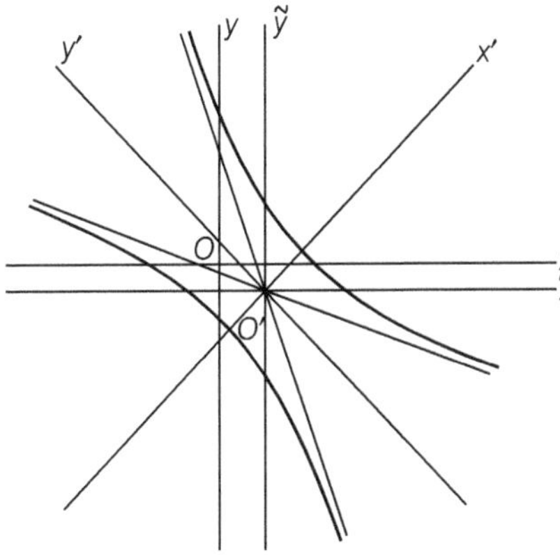

Fig.128.

2) elliptic equation, represents the ellipse $\frac{x'^2}{16} + \frac{y'^2}{9} = 1$, to which form it can be reduced by two consecutive transformations of coordinates: $x = \tilde{x} - 1, y = \tilde{y} + 1$ and $\tilde{x} = \frac{x' - y'}{\sqrt{2}}, \tilde{y} = \frac{x' + y'}{\sqrt{2}}$ (Fig. 129);

3) hyperbolic equation; represents the hyperbola $\frac{x'^2}{9} - \frac{y'^2}{36} = 1$, to which form it can be reduced by two consecutive transformations of coordinates: $x = \tilde{x} + 3, y = \tilde{y} - 4$ and $\tilde{x} = \frac{x' - 2y'}{\sqrt{5}}, \tilde{y} = \frac{2x' + y'}{\sqrt{5}}$ (Fig. 130); 4) hyperbolic equation; represents the degenerate hyperbola $x'^2 - 4y'^2 = 0$ (a pair of intersecting lines), to which form it can be reduced by two consecutive transformations of coordinates: $x = \tilde{x} - 2, y = \tilde{y}$ and $\tilde{x} = \frac{x' + 3y'}{\sqrt{10}}, \tilde{y} = \frac{-3x' + y'}{\sqrt{10}}$ (Fig. 131); 5) elliptic equations; represents no geometric object (the imaginary ellipse $x'^2 + 2y'^2 = -1$, to which form it can be reduced by two consecutive transformations of coordinates):

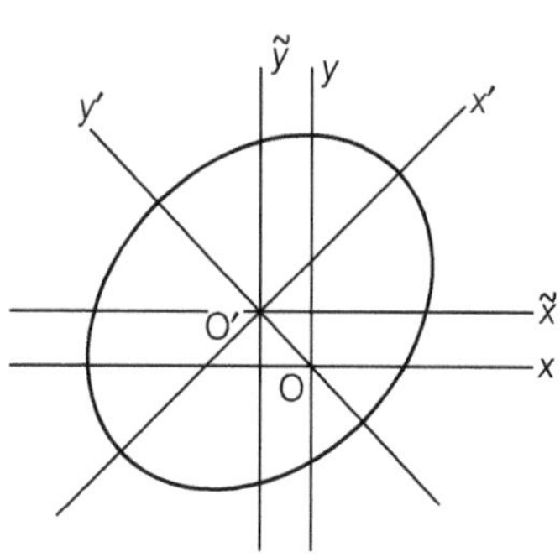

Fig. 129.

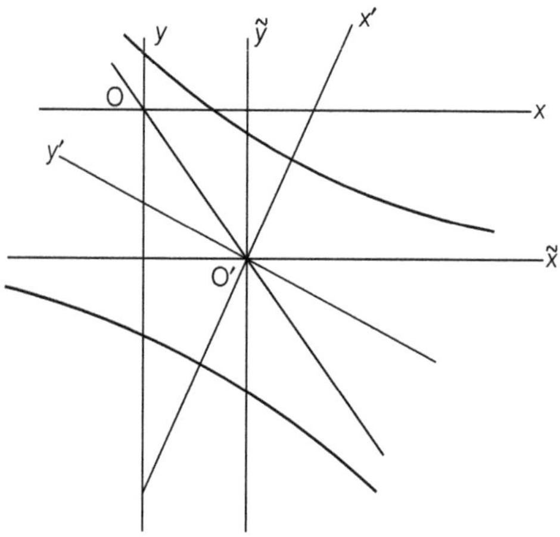

Fig. 130.

$x = \tilde{x} - 1,\ y = \tilde{y}$ and $\tilde{x} = \dfrac{x' + 3y'}{\sqrt{10}},\ \tilde{y} = \dfrac{-3x' + y'}{\sqrt{10}}\Bigg)$;

6) elliptic equation: represents the degenerate ellipse $2x'^2 + 3y'^2 = 0$ (a single point), to which form it can be reduced by two consecutive transformations of coordinates: $x = \tilde{x},\ y = \tilde{y} - 2$ and $\tilde{x} = \dfrac{x' - y'}{\sqrt{2}},\ \tilde{y} = \dfrac{x' + y'}{\sqrt{2}}$.

677. 1) The ellipse $\dfrac{x^2}{30} + \dfrac{y^2}{5} = 1$; 2) the hyperbola $9x^2 - 16y^2 = 5$;

3) the degenerate hyperbola $x^2 - 4y^2 = 0$ (the pair of intersecting lines whose equations are $x - 2y = 0, x + 2y = 0$); 4) the imaginary ellipse $2x^2 + 3y^2 = -1$; the equation represents no geometric object;

5) the degenerate ellipse $x^2 + 2y^2 = 0$; the equation represents a single point, namely, the origin of coordinates;

6) the ellipse $\frac{x^2}{9} + \frac{y^2}{4} = 1;$

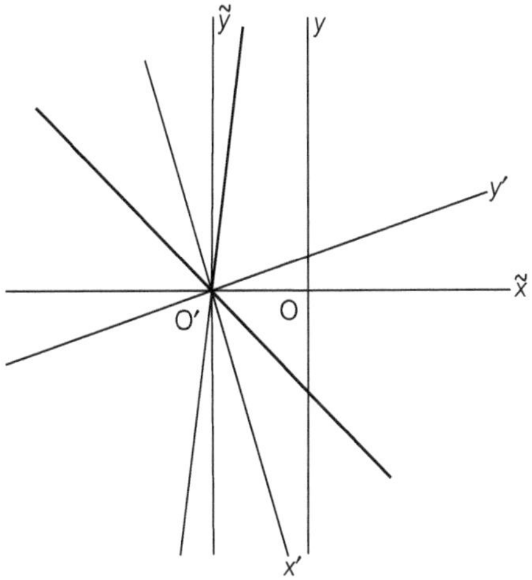

Fig. 131.

7) the hyperbola $\frac{x^2}{4} - y^2 = 1;$

8) the ellipse $\frac{x^2}{9} + y^2 = 1.$

678. 1) 3 and 1; 2) 3 and 2;

3) 1 and $\frac{1}{2}$; 4) 3 and 2.

679. 1) $x = 2, y = 3$; 2) $x = 3, y = -3$; 3) $x = 1, y = -1$; 4) $x = -2, y = 1$.

680. 1) 2 and 1; 2) 5 and 1;

3) 4 and 2; 4) 1 and $\frac{1}{2}$.

681. 1) $x + y - 1 = 0, 3x + y + 1 = 0;$

2) $x - 4y - 2 = 0, x - 2y + 2 = 0;$

3) $x - y = 0, x - 3y = 0;$

4) $x + y - 3 = 0, x + 3y - 3 = 0.$

682. 1) An ellipse;

2) a hyperbola;

3) a pair of intersecting lines (a degenerate hyperbola);

4) the equation represents no geometric object (an imaginary ellipse);

5) a point (a degenerate ellipse).

689. 1) Parabolic equation; represents the parabola $y''^2 = 2x''$, to which form it can be reduced by two consecutive transformations of coordinates: $x = \frac{-4x' + 3y'}{5}$, $y = \frac{-3x' - 4y'}{5}$ and

$$x' = x'' - 3,\ y' = y'' + 2 \text{ (Fig. 132)};$$

2) parabolic equation; represents the degenerate hyperbola $x''^2 = 1$ (a pair of parallel lines), to which form it can be reduced by two consecutive transformations of coordinates:

$x = \frac{3x' - 2y'}{\sqrt{13}}$, $y = \frac{2x' + 3y'}{\sqrt{13}}$ and

$x' = x'' + \frac{4}{\sqrt{13}}$, $y' = y''$ (Fig. 133);

3) parabolic equation; represents no geometric object and is reduced to the form $y''^2 + 1 = 0$ by two consecutive transformations of coordinates: $x = \frac{(3x' - 4y')}{5}$, $y = \frac{4x' + 3y'}{5}$ and $x' = x''$, $y' = y'' - 4$.

690. 1) The parabola $y^2 = 6x$; 2) the degenerate parabola $y^2 = 25$ (the pair of parallel lines whose equations are $y - 5 = 0$, $y + 5 = 0$);

3) the degenerate parabola $y^2 = 0$ (the pair of coincident lines which coincide with the x– axis).

693. 1) $(x + 2u)^2 + 4x + y - 15 = 0$;

2) $(3x - y)^2 - x + 2y - 14 = 0$;

3) $(5x - 2y)^2 + 3x - y + 11 = 0$;

4) $(4x + 2y)^2 - 5x + 7y = 0$;

5) $(3x - 7y)^2 + 3x - 2y - 24 = 0$;

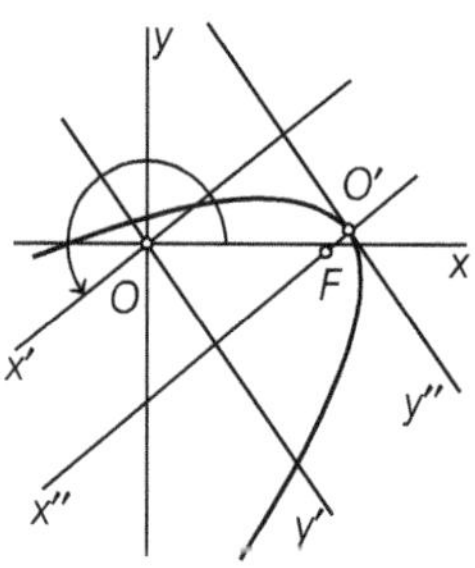

Fig. 132.

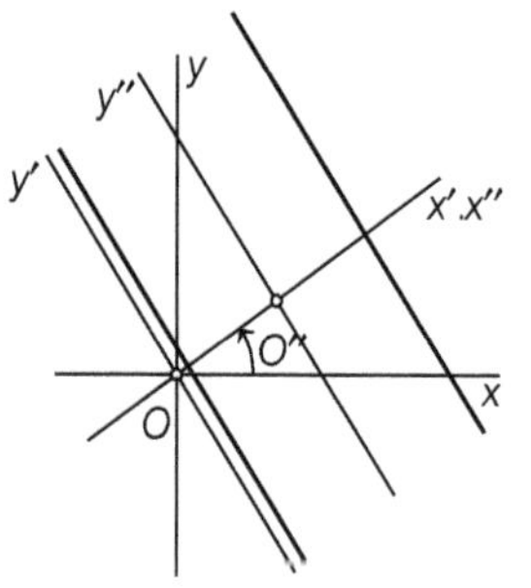

Fig. 133.

697. 1) 3; 2) 3; 3) $\sqrt{2}$; 4) $\frac{1}{2}\sqrt{10}$.

699. 1) $2x + y - 5 = 0, 2x + y - 1 = 0$; 2) $2x - 3y - 1 = 0, 2x - 3y + 11 = 0$; 3) $5x - y - 3 = 0, 5x - y + 5 = 0$.

700. 1) $x-3y+2=0$;

2) $3x+5y+7=0$;

3) $4x-2y-9=0$.

701. $(x^2+y^2)^2-2c^2(x^2-y^2)=a^4-c^4$.

702. $(x^2+y^2)^2=2a^2(x^2-y^2); \rho^2=2a^2\cos 2\theta$.

703. $\rho^2=S\sin 2\theta; (x^2+y^2)^2=2Sxy$.

705. $\rho=\frac{v}{\omega}\theta$ and $\rho=-\frac{v}{\omega}\theta$.

706. $(2r-x)y^2=x^3$.

707. $x(a^2+y^2)=a^3$.

708. $\rho=\frac{a}{\cos\theta}\pm b; x^2y^2+(x+a)^2(x^2-b^2)=0$.

709. $\rho=\frac{a}{\cos\theta}\pm a\tan\theta; x^2[(x+a)^2+y^2]=a^2y^2$.

710. $\rho=2a\cos\theta\pm b; (x^2+y^2-2ax)^2=b^2(x^2+y^2)$.

711. $\rho=a|\sin 2\theta|; (x^2+y^2)^3=4a^2x^2y^2$.

712. $x=a\cos^3 t,\ y=a\sin^3 t; x^{\frac{2}{3}}+y^{\frac{2}{3}}=a^{\frac{2}{3}}$.

713. $\rho=a\cos^3\theta, (x^2+y^2)^2=ax^3$.

714. $x=a(\cos t+t\sin t),\ y=a(\sin t-t\cos t)$.

715. $x=a(t-\sin t);\ y=a(1-\cos t); x+\sqrt{y(2a-y)}=a\,\mathrm{arc}\cos\frac{a-y}{a}$.

716. $x=a(2\cos t-\cos 2t),\ y=a(2\sin t-\sin 2t); \rho=2a(1-\cos\theta)$.

717. $x=(a+b)\cos t-a\cos\frac{a+b}{a}t,\ y=(a+b)\sin t-a\sin\frac{a+b}{a}t$.

718. $x=(b-a)\cos t+a\cos\frac{b-a}{a}t,\ y(b-a)\sin t-a\sin\frac{b-a}{a}t$.

Part Two

720. 1) (4, 3, 0), (−3,2,0), the point C lies in the plane Oxy and hence the projection of C on Oxy coincides with C, (0,0,0);

2) (4,0,5), (−3,0,1), (2,0,0), the point D lies in the plane Oxz and hence the projection of D on Oxz coincides with D;

3) (0,3,5), (0,2,1) (0,−3,0), the point D lies in the plane Oyz and hence the projection of D on Oyz coincides with D;

4) (4, 0, 0), (−3,0,0), (2,0,0), (0,0,0); 5) (0,3,0), (0,2,0), (0,−3,0); (0,0,0);

6) (0,0,5), (0,0,1), (0,0,0), the point D lies on the $z-$ axis and hence the projection of D on the $z-$ axis coincides with D.

721. 1) $(2,3,-1), (5,-3,-2), (-3,2,1), (a,b,-c)$;
2) $(2,-3,1), (5,3,2), (-3,-2,-1), (a,-b,c)$;
3) $(-2,3,1), (-5,-3,2), (3,2,-1), (-a,b,c)$;
4) $(2,-3,-1), (5,3,-2), (-3,-2,1), (a,-b,-c)$;
5) $(-2,3,-1), (-5,-3,-2), (3,2,1), (-a,b,-c)$;
6) $(-2,-3,1), (-5,3,2), (3,-2,-1), (-a,-b,c)$;
7) $(-2,-3,-1), (-5,3,-2), (3,-2,1), (-a,-b,-c)$.

722. $(a,a,-a), (a,-a,a), (-a,a,a), (-a,-a,a)$.

723. 1) The first, third, fifth and seventh octants;
2) the second, fourth, sixth and eighth octants;
3) the first, third, sixth and seventh octants;
4) the second, fourth, fifth and eighth octants;
5) the third, fourth, sixth and seventh octants.

724. 1) The first, third, fifth and seventh octants;
2) the second, third, fifth and eighth octants;
3) the first, second, seventh and eighth octants;
4) the first, third, sixth and eighth octants;
5) the second, fourth, fifth and seventh octants.

725. 1) $(-3,3,3)$; 2) $(3,3,-3)$;
3) $(-3,3,-3)$ 4) $(-3,-3,-3)$;
5) $(3,-3,-3)$.

726. 1) 7; 2) 13; 3) 5.

727. $OA = 6; OB = 14; OC = 13; OD = 25$.

730. $\measuredangle M_1M_3M_2$ is obtuse.

732. $(5,0,0)$ and $(-11,0,0)$.

733. $(0,2,0)$ **734.** $C(3,-3,-3), R=3$.

735. $(2,-1,-1), (-1,-2,2), (0,1,-2)$.

736. 7 **737.** $x=4, y=-1, z=3$

738. $C(6,1,19)$ and $D(9,-5,12)$.

739. $D(9,-5,6)$.

740. the fourth vertex of the parallelogram may coincide with one of the following points: $D_1(-3,4,-4)$, $D_2(1,-2,8)$, $D_3(5,0,4)$.

741. $C(1,5,2), D(3,2,1), E(5,-1,0), F(7,-4,-1)$.

742. $A(-1,2,4), B(8,-4,-2)$.

743. $\frac{2}{3}\sqrt{74}$. **744.** $\frac{3}{2}\sqrt{14}$.

745. $x = \frac{x_1+x_2+x_3+x_4}{4}, y = \frac{y_1+y_2+y_3+y_4}{4}, z = \frac{z_1+z_2+z_3+z_4}{4}$

746. $x = \dfrac{m_1x_1 + m_2x_2 + m_3x_3 + m_4x_4}{m_1 + m_2 + m_3 + m_4}$,

$y = \dfrac{m_1y_1 + m_2y_2 + m_3y_3 + m_4y_4}{m_1 + m_2 + m_3 + m_4}$,

$z = \dfrac{m_1z_1 + m_2z_2 + m_3z_3 + m_4z_4}{m_1 + m_2 + m_3 + m_4}$,

747. $(2,-3,0), (1,0,2), (0,3,4)$.

748. $|a|=7$.

749. $z=\pm 3$.

750. $\overline{AB}=\{-4,3,-1\}, \overline{BA}=\{4,-3,1\}$.

751. $N\ (4,1,1)$.

752. $(-1,2,3)$.

753. $X=\sqrt{2}, Y=1, Z=-1$.

754. $\cos\alpha = \dfrac{12}{25}, \cos\beta = -\dfrac{3}{5}, \cos\gamma = -\dfrac{16}{25}$.

755. $\cos\alpha = \dfrac{3}{13}, \cos\beta = \dfrac{4}{13}, \cos\gamma = \dfrac{12}{13}$.

756. 1) Possible; 2) impossible; 3) possible.

757. 1) Impossible; 2) possible; 3) impossible.

758. 60° or 120°.

759. $\mathbf{a}=\{1,-1,\sqrt{2}\}$ or $\mathbf{a}=\{1,-1,-\sqrt{2}\}$.

760. $M_1(\sqrt{3},\sqrt{3},\sqrt{3}), M_2(-\sqrt{3},-\sqrt{3},-\sqrt{3})$.

761. See fig. 134.

762. $|a-b|=22$.

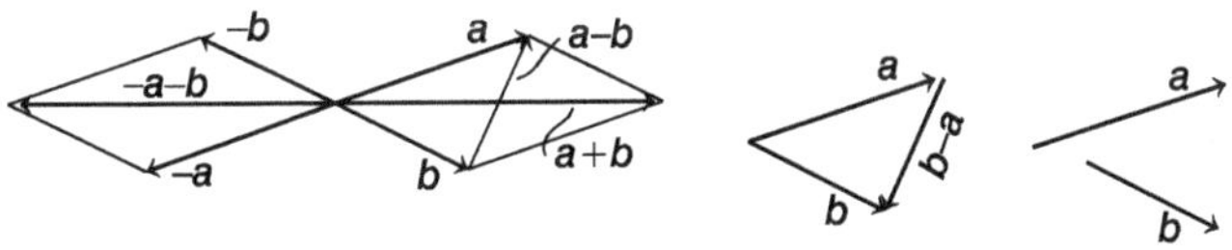

Fig. 134.

763. $|\boldsymbol{a}+\boldsymbol{b}|=20$.

764. $|\boldsymbol{a}+\boldsymbol{b}|=|\boldsymbol{a}-\boldsymbol{b}|=13$

765. $|\boldsymbol{a}+\boldsymbol{b}|=\sqrt{129}\approx 11.4, |\boldsymbol{a}-\boldsymbol{b}|=7$.

766. $|\boldsymbol{a}+\boldsymbol{b}|=\sqrt{19}\approx 4.4, |\boldsymbol{a}-\boldsymbol{b}|=7$.

767. 1) Vectors $\boldsymbol{a}$ and $\boldsymbol{b}$ must be mutually perpendicular;
2) the angle between vectors $\boldsymbol{a}$ and $\boldsymbol{b}$ must be acute;
3) the angle between vectors $\boldsymbol{a}$ and $\boldsymbol{b}$ must be obtuse.

768. $|\boldsymbol{a}|=|\boldsymbol{b}|$.

769. See Fig. 135.

774. $|\boldsymbol{R}|=15$.

775. 1) $\{1,-1,6\}$;

2) $\{5,-3,6\}$; 3) $\{6,-4, 12\}$;

4) $\left\{1,-\frac{1}{2},0\right\}$; 5) $\{0,-1,12\}$;

6) $\left\{3,-\frac{5}{3},2\right\}$.

776. The vector $\boldsymbol{b}$ is three times as long as the vector $\boldsymbol{a}$; the vectors are oppositely directed.

777. $\alpha = 4, \beta = -1$.

779. The vector $\overline{AB}$ is two times as long as the vector $\overline{CD}$; the vectors are similarly directed.

780. $a^\circ = \left\{\frac{6}{7},-\frac{2}{7},-\frac{3}{7}\right\}$. **781.** $a^\circ = \left\{\frac{3}{13},\frac{4}{13},-\frac{12}{13}\right\}$.

782. $|\boldsymbol{a}+\boldsymbol{b}|=6, |\boldsymbol{a}-\boldsymbol{b}|=14$. **783.** $\boldsymbol{d} = -48\boldsymbol{i} + 45\boldsymbol{j} - 36\boldsymbol{k}$.

784. $\mathbf{c}=\{-3,15,12\}$.

785. $\overline{AM} = \{3,4,-3\}, \overline{BN} = \{0,-5,3\}, \overline{CP} = \{-3,1,0\}$.

787. $\boldsymbol{a} = 2\boldsymbol{p} + 5\boldsymbol{q}$.

788. $\boldsymbol{a} = 2\boldsymbol{b} + \mathbf{c}, \boldsymbol{b} = \frac{1}{2}\boldsymbol{a} - \frac{1}{2}\mathbf{c}, \mathbf{c} = \boldsymbol{a} - 2\boldsymbol{b}$.

789. $\boldsymbol{p} = 2\boldsymbol{a} - 3\boldsymbol{b}$.

790. $\overline{AM} = \frac{1}{2}\boldsymbol{b} + \frac{1}{2}\mathbf{c}, \overline{BN} = \frac{1}{2}\mathbf{c} - \boldsymbol{b}, \overline{CP} = \frac{1}{2}\boldsymbol{b} - \mathbf{c}$, where M, N and P are the midpoints of the sides of the triangle ABC.

791. $\overline{AD} = 11\overline{AB} - 7\overline{AC}, \overline{BD} = 10\overline{AB} - 7\overline{AC}, \overline{CD} = 11\overline{AB} - 8\overline{AC}$, $\overline{AD} + \overline{BD} + \overline{CD} = 32\overline{AB} - 22\overline{AC}$. **793.** $\boldsymbol{c} = 2\boldsymbol{p} - 3\boldsymbol{q} + \boldsymbol{r}$.

794. $\boldsymbol{d} = 2\boldsymbol{a} - 3\boldsymbol{b} + \mathbf{c}, \mathbf{c} = -2\boldsymbol{a} + 3\boldsymbol{b} + \boldsymbol{d}, \boldsymbol{b} = \frac{2}{3}\boldsymbol{a} + \frac{1}{3}\mathbf{c} - \frac{1}{3}\boldsymbol{d}$,

$\boldsymbol{a} = \frac{3}{2}\boldsymbol{b} - \frac{1}{2}\mathbf{c} + \frac{1}{2}\boldsymbol{d}$.

795. 1) –6; 2) 9; 3) 16; 4) 13; 5) – 61; 6) 37; 7) 73.

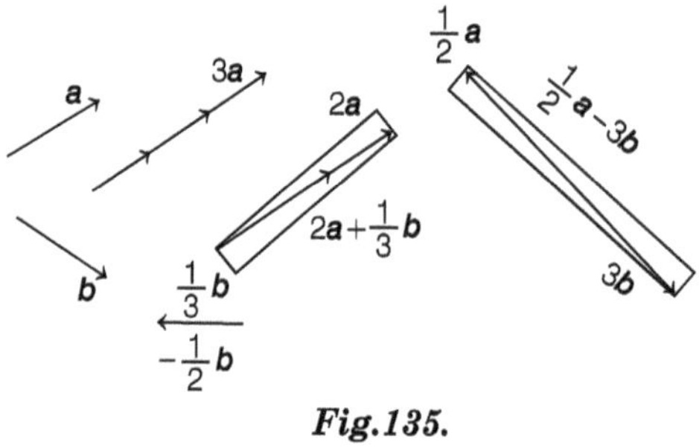

Fig.135.

796. 1) – 62; 2) 162; 3) 373.

797. The sum of the squares of the diagonals of a parallelogram is equal to the sum of the squares of its sides.

798. $-ab = \boldsymbol{ab}$ when vectors $\boldsymbol{a}$ and $\boldsymbol{b}$ are collinear and oppositely directed; $ab = \boldsymbol{ab}$ when vectors $\boldsymbol{a}$ and $\boldsymbol{b}$ are collinear and similarly directed.

799. When $\boldsymbol{b}$ is perpendicular to $\boldsymbol{a}$ and $\boldsymbol{c}$, also when $\boldsymbol{a}$ and $\boldsymbol{c}$ are collinear vectors.

800. $\boldsymbol{ab} + \boldsymbol{bc} + \boldsymbol{ca} = -\frac{3}{2}$.

801. $\boldsymbol{ab} + \boldsymbol{bc} + \boldsymbol{ca} = -13$.

802. $|\boldsymbol{p}| = 10$.

803. $\alpha = \pm\frac{3}{5}$.

804. $|\boldsymbol{a}| = |\boldsymbol{b}|$.

807. $\overline{BD} = \frac{\boldsymbol{bc}}{\boldsymbol{c}^2}\boldsymbol{c} - \boldsymbol{b}$.

808. $\alpha = \arccos\frac{2}{\sqrt{7}}$.

809. $\varphi = \arccos\left(-\frac{4}{5}\right)$.

810. The plane perpendicular to the axis of the vector $\boldsymbol{a}$ and cutting off an intercept equal to $\frac{\boldsymbol{a}}{|\boldsymbol{a}|}$ on this axis.

811. The line of intersection of the planes which are perpendicular to the axes of the vectors $\boldsymbol{a}$ and $\boldsymbol{b}$ and cut off intercepts equal to $\frac{\alpha}{|\boldsymbol{a}|}$ and $\frac{\beta}{|\boldsymbol{b}|}$, respectively, on these axes.

812. 1) 22; 2) 6; 3) 7; 4) –200; 5) 129; 6) 41. **813.** 17.

814. 1) – 524; 2) 13; 3) 3; 4) $(\overline{AB}\cdot\overline{AC})\cdot\overline{BC} = \{-70, 70, -350\}$ and $\overline{AB}\,(\overline{AC}\cdot\overline{BC}) = \{-78, 104, -312\}$.

815. 31.

816. 13.

818. $\alpha = -6$.

819. $\cos\varphi = \frac{5}{12}$.

820. 45°. 821. $\arccos\left(-\frac{4}{9}\right)$.

823. $x = (-24, 32, 30)$.

824. $x = \left\{1, \frac{1}{2}, -\frac{1}{2}\right\}$.

825. $x = -4\boldsymbol{i} - 6\boldsymbol{j} + 12\boldsymbol{k}$.

826. $x = \{-3, 3, 3\}$.

827. $x = \{2, -3, 0\}$.

828. $x = 2\boldsymbol{i} + 3\boldsymbol{j} - 2\boldsymbol{k}$.

829. $\sqrt{3}$.

830. -3.

831. -5.

832. 6.

833. -4.

834. 5.

835. -11.

836. $X = -\frac{14}{3}, Y - \frac{14}{3}, Z = -\frac{7}{3}$.

837. 3.

838. $-6\frac{5}{7}$.

839. $|[\boldsymbol{ab}]| = 15$.

840. $|[\boldsymbol{ab}]| = 16$.

841. $\boldsymbol{ab} = \pm 30$.

842. 1) 24; 2) 60.

843. 1) 3; 2) 27; 3) 300.

844. Vectors $\boldsymbol{a}$ and $\boldsymbol{b}$ must be collinear.

846. When vectors $\boldsymbol{a}$ and $\boldsymbol{b}$ are perpendicular.

850. 1) {5, 1, 7}; 2) {10, 2, 14}); 3) {20, 4, 28}.

851. 1) {6, – 4, – 6}; 2) {–12, 8, 12}.

852. {2, 11, 7}.

853. {–4, 3, 4).

854. 15; $\cos\alpha = \frac{2}{3}, \cos\beta = -\frac{2}{15}, \cos\gamma = \frac{11}{15}$.

855. 28; $\cos\alpha = -\frac{3}{7}, \cos\beta = -\frac{6}{7}, \cos\gamma = \frac{2}{7}$.

856. $\sqrt{66}$; $\cos\alpha = \frac{1}{\sqrt{66}}, \cos\beta = -\frac{4}{\sqrt{66}}, \cos\gamma = -\frac{7}{\sqrt{66}}$.

857. 14 square units.

858. 5.

859. $\sin\varphi = \frac{5\sqrt{17}}{21}$.

860. {–6, –24, 8}.

861. $m = \{45, 24, 0\}$.

862. $\boldsymbol{x} = \{7, 5, 1\}$.

864. $[[\boldsymbol{ab}]\boldsymbol{c}] = \{-7, 14, -7\}$; $[\boldsymbol{a}[\boldsymbol{bc}]] = \{10, 13, 19\}$,

865. 1) Right-handed; 2) left-handed; 3) left-handed; 4) right-handed; 5) the vectors are coplanar; 6) left-handed.

866. $\boldsymbol{abc} = 24$.

867. $\boldsymbol{abc} = \pm 27$; the plus sign is taken if the triad $\boldsymbol{a}$, $\boldsymbol{b}$, $\boldsymbol{c}$ is right-handed; the minus sign, if left-handed.

868. When vectors $\boldsymbol{a}$, $\boldsymbol{b}$, $\boldsymbol{c}$ are mutually perpendicular.

873. $\boldsymbol{abc} = -7$.

874. 1) Coplanar; 2) non-coplanar; 3) coplanar.

876. 3 cubic units.

877. 11.

878. D_1 (0,8,0); D_2 (0, –7.0).

881. $X = -6, Y = -8, Z = -6$.

882. Vectors $\boldsymbol{a}$ and $\boldsymbol{c}$ must be collinear, or vector $\boldsymbol{b}$ must be perpendicular to vectors $\boldsymbol{a}$ and $\boldsymbol{c}$.

885. The points M_1, M_2, M_4 lie on the surface; the points M_3, M_5, M_6 do not lie on the surface. The equation represents a sphere with centre at the origin and radius 7.

886. 1) (1, 2, 2) and (1, 2, –2), 2) there is no such point on the given surface; 3) (2, 1, 2) and (2, –1, 2); 4) there is no such point on the given surface.

887. 1) The plane Oyz; 2) the plane Oxz; 3) the plane Oxy; 4) the plane parallel to the plane Oyz and situated in the near half-space at a distance of two units from Oyz; 5) the plane parallel to the plane Oxz and situated in the left half-space at a distance of two units from Oxz; 6) the plane parallel to the plane Oxy and situated in the lower half-space at a distance of five units from Oxy; 7) the sphere with centre at the origin and radius 5; 8) the sphere with centre at (2, –3, 5) and radius 7; (9) the equation represents a single point, namely, the origin; 10) the equation represents no geometric object in space; 11) the plane which bisects the dihedral angle between the planes Oxz, Oyz and is situated in the 1st, 3rd, 5th and 7th octants; 12) the plane which bisects the dihedral angle between the planes Oyz, Oxy and is situated in the 2nd, 3rd, 5th and 8th octants; 13) the plane which bisects the dihedral angle between the planes Oxy. Oxz and is situated in the 1st, 2nd, 7th and 8th octants, 14) the planes Oxz and Oyz; 15) the planes Oxy and Oyz; 16) the planes Oxy and Oxz, 17) all the three coordinate planes; 18) the plane Oyz and the plane parallel to the plane Oyz and situated in the near half-space at a distance of four units from Oyz; 19) the plane Oxz and the plane which bisects the dihedral angle between the planes Oxz, Oyz and is situated in the 1st, 3rd, 5th and 7th octants; 20) the plane Oxy and the plane which bisects the dihedral angle between the planes Oxz, Oxy and is situated in the 3rd, 4th, 5th and 6th octants.

889. $x^2+y^2+z^2=r^2$.

890. $(x-\alpha)^2+(y-\beta)^2+(z-\gamma)^2=r^2$.

891. $y-3=0$. **892.** $2z-7=0$.

893. $2x+3=0$. **894.** $20y+53=0$.

895. $x^2+y^2+z^2=a^2$. **896.** $x^2+y^2+z^2=a^2$.

897. $x+2z=0$. **898.** $\frac{x^2}{9}+\frac{y^2}{9}+\frac{z^2}{25}=1$.

899. $\frac{x^2}{16}-\frac{y^2}{9}+\frac{z^2}{16}=-1$.

900. The points M_1, M_3 lie on the given curve; the points M_2, M_4 do not lie on the curve.

901. Curves 1) and 3) pass through the origin.

902. 1) (3, 2, 6) and (3, –2, 6); 2) (3, 2, 6) and (–3, 2, 6); 3) the given curve contains no such point.

903. 1) The z-axis; 2) the y-axis; 3) the x-axis; 4) the straight line passing through the point (2, 0, 0) parallel to the axis Oz; 5) the straight line passing through the point (–2, 3, 0) parallel to the axis Oz; 6) the straight line passing through the point (5, 0, –2) parallel to the axis Oy; 7) the straight line passing through the point (0, –2, 5) parallel to the axis Ox; 8) the circle (lying in the plane Oxy) with centre at the origin and radius 3; 9) the circle (lying in the plane Oxz) with centre at the origin and radius 7; 10) the circle (lying in the plane Oyz) with centre at the origin and radius 5; 11) the circle (lying in the plane $z-2=0$) with centre at the point (0, 0, 2) and radius 4.

904. $\begin{cases} x^2+y^2+z^2=9, \\ y=0. \end{cases}$

905. $\begin{cases} x^2+y^2+z^2=25, \\ y+2=0. \end{cases}$

906. $\begin{cases} (x-5)^2+(y+2)^2+(z-1)^2=169, \\ x=0. \end{cases}$

907. $\begin{cases} x^2+y^2+z^2=36, \\ (x-1)^2+(y+2)^2+(z-2)^2=25. \end{cases}$

908. (2; 3; –6), (–2; 3; –6).

909. (1, 2, 2), (–1, 2, 2).

910. 1) The cylindrical surface whose elements are parallel to the axis Oy and whose directing curve is the circle represented (in the plane Oxz) by the equation $x^2+z^2=25$; 2) the cylindrical surface whose elements are parallel to the axis Ox and whose directing curve is the ellipse represented (in the plane Oyz) by the equation $\frac{y^2}{25}+\frac{z^2}{16}=1$; 3) the cylindrical surface whose elements are parallel to the axis Oz and whose directing curve is the hyperbola represented (in the plane Oxy) by the equation $\frac{x^2}{16}-\frac{y^2}{9}=1$; 4) the cylindrical surface whose elements are parallel to the axis Oy and whose directing curve is the parabola represented (in the plane Oxz) by the equation $x^2=6z$; 5) the cylindrical surface whose elements are parallel to the axis Oz and whose directing curve is the pair of lines represented (in the plane Oxy) by the equations $x=0$, $x-y=0$; this cylindrical surface consists of two planes; 6) the cylindrical surface whose elements are parallel to the axis Oy and whose directing curve is the pair of lines represented (in the plane

Oxz) by the equations $x-z=0$, $x+z=0$; this cylindrical surface consists of two planes; 7) the x-axis; 8) the equation represents no geometric object in space; 9) the cylindrical surface whose elements are parallel to the axis Oy and whose directing curve is the circle represented (in the plane Oxz) by the equation $x^2+(z-1)^2=1$; 10) the cylindrical surface whose elements are parallel to the axis Ox and whose directing curve is represented (in the plane Oyz) by the equation $y^2+\left(x+\frac{1}{2}\right)^2=\frac{1}{4}$.

911. 1) $x^2+5y^2-8y-12=0$; 2) $4x^2+5z^2+4z-60=0$; 3) $2y-z-2=0$.

912. 1) $\begin{cases}(8x^2+4y^2-36x+16y-3=0,\\ z=0;\end{cases}$ 2) $\begin{cases}2x-2z-7=0.\\ y=0;\end{cases}$

3) $\begin{cases}4y^2+8z^2+16y+20z-31=0,\\ x=0.\end{cases}$

913. $x-2y+3z+3=0$.

914. $5x-3z=0$. **915.** $2x-y-z-6=0$.

916. $x-y-3z+2=0$. **917.** $x+4y+7z+16=0$.

919. $9x-y+7z-40=0$. **921.** $3x+3y+z-8=0$.

923. 1) $\boldsymbol{n}=\{2,-1,-2\}$, $\boldsymbol{n}=\{2\lambda,-\lambda,-2\lambda\}$; 2) $\boldsymbol{n}=\{1,5,-1\}$, $\boldsymbol{n}=\{\lambda,5\lambda,-\lambda\}$; 3) $\boldsymbol{n}=\{3,-2,0\}$; $\boldsymbol{n}=\{3\lambda,-2\lambda,0\}$; 4) $n=\{0,5\ -3\},\{0,5\lambda,-3\lambda\}$; 5) $\boldsymbol{n}=\{1,0,0\}$, $\boldsymbol{n}=\{\lambda,0,0\}$; 6) $\boldsymbol{n}=\{0,1,0\}$; $n=\{0,\lambda,0\}$, where λ is any number other than zero.

924. 1) and 3) represent parallel planes.

925. 1) and 2) represent perpendicular planes.

926. 1) $l=3, m=-4$; 2) $l=3, m=-\frac{2}{3}$; 3) $l=-3\frac{1}{3}$; $m=-1\frac{1}{5}$.

927. 1) 6; 2) – 19; 3) $-\frac{1}{7}$.

928. 1) $\frac{1}{3}\pi$ and $\frac{2}{3}\pi$; 2) $\frac{1}{4}\pi$ and $\frac{3}{4}\pi$; 3) $\frac{\pi}{2}$; 4) $\arccos\frac{2}{15}$ and $\pi-\arccos\frac{2}{15}$.

929. $5x-3y+2z=0$. **930.** $2x-3z-27=0$.

931. $7x-y-5z=0$. **932.** $x+2z-4=0$.

934. $4x-y-2z-9=0$. **936.** $x=1$, $y=-2$, $z=2$.

939. 1) $a\neq 7$; 2) $a=7$, $b=3$; 3) $a=7$, $b\neq 3$.

940. 1) $z-3=0$; 2) $y+2=0$; 3) $x+5=0$.

941. 1) $2y+z=0$; 2) $3x+z=0$; 3) $4x+3y=0$.

942. 1) $y+4z+10=0$; 2) $x-z-1=0$; 3) $5x+y-13=0$.

943. (12, 0, 0), (0, –8, 0), (0, 0, – 6).

944. $\frac{x}{6}+\frac{y}{3}+\frac{z}{-2}=1.$ **945.** $a=-4, b=3, c=\frac{1}{2}.$

946. 240 square units. **947.** 8 cubic units.

948. $\frac{x}{-3}+\frac{y}{-4}+\frac{z}{2}=1.$ **949.** $\frac{x}{-3}+\frac{y}{3}+\frac{z}{-3/2}=1.$

950. $x+y+z+5=0.$

951. $2x-21y+2z+88=0,$
$2x-3y-2z+12=0.$

952. $x+y+z-9=0, x-y-z+1=0, x-y+z-3=0, x+y-z-5=0.$

953. $2x-y-3z-15=0.$

954. $2x-3y+z-6=0.$

955. $x-3y-2z+2=0.$

956. Planes 1), 4), 5), 7), 9), 11) and 12) are represented by normal equations.

957. 1) $\frac{2}{3}x-\frac{2}{3}y+\frac{1}{3}z-6=0$; 2) $-\frac{3}{7}x+\frac{6}{7}y-\frac{2}{7}z-3=0$;
3) $\frac{2}{7}x-\frac{3}{7}y-\frac{6}{7}z-\frac{11}{14}=0$; 4) $\frac{2}{3}x+\frac{2}{3}y-\frac{1}{3}z-\frac{1}{6}=0$;
5) $-\frac{5}{13}y+\frac{12}{13}z-2=0$; 6) $\frac{3}{5}x-\frac{4}{5}y-\frac{1}{5}=0$;
7) $-y-2=0$; 8) $x-5=0$; 9) $z-3=0$; 10) $z-\frac{1}{2}=0.$

958. 1) $\alpha=60°, \beta=45°, \gamma=60°, p=5$; 2) $\alpha=120°, \beta=60°, \gamma=45°, p=8$; 3) $\alpha=45°, \beta=90°, \gamma=45°, p=3\sqrt{2}$; 4) $\alpha=90°, \beta=135°, \gamma=45°, p=\sqrt{2}$; 5) $\alpha=150°$ $\beta=120°, \gamma=90°, p=5$; 6) $\alpha=90°, \beta=90°, \gamma=0°, p=2$; 7) $\alpha=180°, \beta=90°, \gamma=90°, p=\frac{1}{2}$; 8) $\alpha=90°, \beta=180°, \gamma=90°, p=\frac{1}{2}$; 9) $\alpha=\arccos\frac{1}{3}, \beta=\pi\text{-}\arccos\frac{2}{3}, \gamma=\arccos\frac{2}{3}, p=2$; 10) $\alpha=\pi-\arccos\frac{2}{7}, \beta=\pi-\arccos\frac{3}{7}, \gamma=\arccos\frac{6}{7}, p=\frac{4}{7}.$

959. 1) $\delta=-3, d=3$; 2) $\delta=1, d=1$; 3) $\delta=0, d=0$ – the point M_3 lies in the given plane; 4) $\delta=-2, d=2$; 5) $\delta=-3, d=3.$

960. $d=4.$

961. 1) On the same side; 2) on the same side; 3) on opposite sides; 4) on the same side; 5) on opposite sides; 6) on opposite sides.

964. 1) $d=2$; 2) $d=3.5$; 3) $d=6.5$; 4) $d=1$; 5) $d=0.5$; 6) $d=\frac{5}{6}.$

965. 8 cubic units.

966. The conditions of the problem are satisfied by two points: (0, 7, 0) and (0, –5, 0).

967. The conditions of the problem are satisfied by two points: (0, 0, –2) and $\left(0,0,-6\frac{4}{13}\right)$.

968. The conditions of the problem are satisfied by two points: (2, 0, 0) and $\left(\frac{11}{43},0,0\right)$.

969. $4x-4y-2z+15=0.$

970. $6x+3y+2z+11=0.$

971. $2x-2y-z-18=0,\ 2x-2y-z+12=0.$

972. 1) $4x-y-2z-4=0$; 2) $3x+2y-z+1=0$; 3) $20x-12y+4z+13=0$.

973. 1) $4x-5y+z-2=0, 2x+y-3z+8=0$; 2) $x-3y-1=0,\ 3x+y-2z-1=0$; 3) $3x-6y+7z+2=0,\ x+4y+3z+4=0$.

974. 1) The point M and the origin lie in two complementary angles; 2) the point M and the origin lie inside the same angle; 3) the point M and the origin lie in two vertical angles.

975. 1) The point M and N lie in two complementary angles; 2) the points M and N lie in two vertical angles.

976. The origin lies inside the acute angle. **977.** The point M lies inside the acute angle.

978. $8x-4y-4z+5=0.$

979. $23x-y-4z-24=0.$

980. $x-y-z-1=0.$

981. $x+y+2z=0.$

982. $\begin{cases}5x-7y-3=0,\\ z=0;\end{cases}$

$\begin{cases}5x+2z-3=0,\\ y=0;\end{cases}$

$\begin{cases}7y-2z+3=0.\\ x=0.\end{cases}$

983. $\begin{cases}3x-y-7z+9=0,\\ 5y+2z=0.\end{cases}$

984. (2, –1, 0); $\left(1\frac{1}{3},0,-\frac{1}{3}\right)$; (0, 2, –1).

986. 1) $D=-4$; 2) $D=9$; 3) $D=3$.

987. 1) $A_1 = A_2 = 0$ and at least one of the numbers D_1, D_2 is different from zero; 2) $B_1 = B_2 = 0$ and at least one of the numbers D_1, D_2 is different from zero; 3) $C_1 = C_2 = 0$ and at least one of the numbers D_1, D_2 is different from zero.

988. 1) $\frac{A_1}{A_2} = \frac{D_1}{D_2}$; 2) $\frac{B_1}{B_2} = \frac{D_1}{D_2}$; 3) $\frac{C_1}{C_2} = \frac{D_1}{D_2}$; 4) $A_1 = D_1 = 0,\ A_2 = D_2 = 0$; 5) $B_1 = D_1 = 0,\ B_2 = D_2 = 0$; 6) $C_1 = D_1 = 0,\ C_2 = D_2 = 0$.

989. 1) $2x + 15y + 7z + 7 = 0$; 2) $9y + 3z + 5 = 0$; 3) $3x + 3z - 2 = 0$; 4) $3x - 9y - 7 = 0$.

990. (1) $23x - 2y + 21z - 33 = 0$; 2) $y + z - 18 = 0$; 3) $x + z - 3 = 0$; (4) $x - y + 15 = 0$.

991. $5x + 5z - 8 = 0$.

992. $\alpha(5x - 2y - z - 3) + \beta(x + 3y - 2z + 5) = 0$.

Hint. The line of intersection of the planes $5x - 2y - z - 3 = 0$, $x + 3y - 2z + 5 = 0$ is parallel to the vector $\boldsymbol{l} = (7, 9, 17)$; hence, the conditions of the problem will be satisfied by all planes which belong to the pencil of planes passing through this line.

993. $11x - 2y - 15z - 3 = 0$.

994. $\alpha(5x - y - 2z - 3) + \beta(3x - 2y - 5z + 2) = 0$.

Hint. The line of intersection of the planes $5x - y - 2z - 3 = 0$, $3x - 2y - 5z + 2 = 0$ is perpendicular to the plane $x + 19y - 7z - 11 = 0$; hence, the conditions of the problem will be satisfied by all planes belonging to the pencil of planes through this line.

995. $9x + 7y + 8z + 7 = 0$.

996. $x - 2y + z - 2 = 0,\ x - 5y + 4z - 20 = 0$.

997. The plane belongs to the pencil.

998. The plane does not belong to the pencil.

999. $l = -5,\ m = -11$.

1000. $3x - 2y + 6z + 21 = 0,\ 189x + 28y + 48z - 591 = 0$.

1001. $2x - 3y - 6z + 19 = 0,\ 6x - 2y - 3z + 18 = 0$.

1002. $4x - 3y + 6z - 12 = 0,\ 12x - 49y + 38z + 84 = 0$.

1003. $4x + 3y - 5 = 0,\ 5x + 3z - 7 = 0,\ 5y - 4z + 1 = 0$.

1004. $\begin{cases} 7x - y + 1 = 0, \\ z = 0; \end{cases} \begin{cases} 5x - z - 1 = 0, \\ y = 0; \end{cases} \begin{cases} 5y - 7x - 12 = 0, \\ x = 0. \end{cases}$

1005. $x - 8y + 5z - 3 = 0$.

1006. $\begin{cases} 2x - 4y - 8z + 1 = 0, \\ 2x - y + z - 1 = 0. \end{cases}$

1007. 1) $\frac{x-2}{2}=\frac{y}{-3}=\frac{z+3}{5}$; 2) $\frac{x-2}{5}=\frac{y}{2}=\frac{z+3}{-1}$; 3) $\frac{x-2}{1}=\frac{y}{0}=\frac{z+3}{0}$;
4) $\frac{x-2}{0}=\frac{y}{1}=\frac{z+3}{0}$; 5) $\frac{x-2}{0}=\frac{y}{0}=\frac{z+3}{1}$.

1008. 1) $\frac{x-1}{2}=\frac{y+2}{3}=\frac{z-1}{-2}$; 2) $\frac{x-3}{2}=\frac{y+1}{-1}=\frac{z}{3}$; 3) $\frac{x}{3}=\frac{y+2}{0}=\frac{z-3}{-2}$;
4) $\frac{x+1}{1}=\frac{y-2}{0}=\frac{z+4}{0}$.

1009. 1) $x=2t+1,\ y=-3t-1,\ z=4t-3$; 2) $x=2t+1,\ y=5t-1,\ z=-3$;
3) $x=3t+1,\ y=-2t-1,\ z=5t-3$.

1010. 1) $x=t+2,\ y=-2t+1,\ z=t+1$; 2) $x=t+3,\ y=-t-1,\ z=t$;
3) $x=0,\ y=t,\ z=-3t+1$.

1011. $(9,-4,0), (3,0,-2), (0,2,-3)$.

1012. $x=5t+4,\ y=-11t-7,\ z=-2$.

1013. $\frac{x-1}{1}=\frac{y-2}{-3}=\frac{z+7}{-8}$.

1014. $\frac{x-2}{6}=\frac{y+1}{-1}=\frac{z+3}{-7}$.

1015. $x=3t+3,\ y=15t+1,\ z=19t-3$.

1016. $a=\{1,1,3\}$, $\boldsymbol{a}=\{\lambda,\lambda,3\lambda\}$, where λ is any number other than zero.

1017. $\boldsymbol{a}=-2\boldsymbol{i}+11\boldsymbol{j}+5\boldsymbol{k}$. $\boldsymbol{a}=-2\lambda\boldsymbol{i}+11\lambda\boldsymbol{j}+5\lambda\boldsymbol{k}$, where λ is any number other than zero.

1018. $\frac{x-2}{2}=\frac{y-3}{-4}=\frac{z+5}{-5}$.

1019. 1) $\frac{x-2}{2}=\frac{y+1}{7}=\frac{z}{4}$.

Solution: Setting, for example, $z_0=0$, we find from the given system:
$x_0=2$, $y_0=-1$; thus, we have already found one point, $M_0(2,-1,0)$, of the line. Next, let us find its direction vector. We have
$\boldsymbol{n}_1=\{1,-2,3\}$, $\boldsymbol{n}_2=\{3,2,-5\}$; hence, $\boldsymbol{a}=[\boldsymbol{n}_1\boldsymbol{n}_2]=\{4,14,8\}$, that is, $l=4$, $m=14$, $n=8$. Substituting the values of x_0, y_0, z_0 and l, m, n just found in the relations $\frac{x-x_0}{l}=\frac{y-y_0}{m}=\frac{z-z_0}{n}$, we obtain the canonical equations of the given line: $\frac{x-2}{4}=\frac{y+1}{14}=\frac{z}{8}$,
or $\frac{x-2}{2}=\frac{y+1}{7}=\frac{z}{4}$; 2) $\frac{x}{-5}=\frac{y+1}{12}=\frac{z-1}{13}$; 3) $\frac{x-3}{1}=\frac{y-2}{2}=\frac{z}{1}$.

1020. 1) $x=t+1,\ y=-7t,\ z=-19t-2$; 2) $x=-t+1,\ y=3t+2,$
$z=5t-1$.

1023. 60°. **1024.** 135°.

1025. $\cos\varphi = \pm\frac{4}{21}$.

1027. $l = 3$.

1029. $\frac{x+1}{2} = \frac{y-2}{-3} = \frac{z+3}{6}$.

1030. $\frac{x+4}{3} = \frac{y+5}{2} = \frac{z-3}{-1}$.

1031. $x = 2t - 5,\ y = -3t + 1, z = -4t$.

1032. $v = 13$.

1033. $d = 21$.

1034. $x = 3 - 6t,\ y = -1 + 18t, z = -5 + 9t$.

1035. $x = -7 + 4t, y = 12 - 4t, z = 5 - 2t$.

1036. $x = 20 - 6t,\ y = -18 + 8t, z = -32 + 24t$; (2, 6, 40).

1037. The equations of motion of M are $x = -5 + 6t$, $y = 4 - 12t, z = -5 + 4t$; the equations of motion of N are $x = -5 + 4t$, $y = 16 - 12t,\ z = -6 + 3t$; 1) $P(7, -20, 3)$; 2) this time is 2; 3) this time is 3; 4) $M_0P = 28$; $N_0P = 39$.

1040. 1) (2, –3, 6); 2) the line is parallel to the plane; 3) the line lies in the plane.

1041. $\frac{x-2}{2} = \frac{y+4}{5} = \frac{z+1}{3}$.

1042. $\frac{x-2}{6} = \frac{y+3}{-3} = \frac{z+5}{-5}$.

1043. $2x - 3y + 4z - 1 = 0$.

1044. $x + 2y + 3z = 0$.

1045. $m = -3$.

1046. $C = -2$.

1047. $A = 3, D = -23$.

1048. $A = -3, B = 4\frac{1}{2}$.

1049. $l = -6,\ C = \frac{3}{2}$.

1050. (3, –2, 4). Solution. To find the desired point, solve the equations of the given line simultaneously with the equation of the plane passing through the point P and perpendicular to this line. First of all, note that the direction vector (3, 5, 2) of the given line will be a normal vector to the required plane. The equation of the plane through $P(2, -1, 3)$ and having $\boldsymbol{n} = \{3, 5, 2\}$ as its normal vector will be of the form $3(x-2) + 5(y+1) + 2(z-3) = 0$, or $3x + 5y + 2z - 7 = 0$. By solving simultaneously the equations $\begin{cases} x = 3t,\ y = 5t - 7,\ z = 2t + 2 \\ 3x + 5y + 2z - 7 = 0, \end{cases}$, we find the coordinates of the desired projection to be $x = 3,\ y = -2,\ z = 4$.

1051. $Q(2, -3, 2)$.

1052. $Q(4, 1, -3)$.

1053. (1, 4, –7). Solution. To find the desired point, solve the equation of the given plane simultaneously with the equations of the straight line drawn from the point P perpendicular to this plane. First of all, note that the normal vector {2, –1, 3} to the given plane will be the direction vector of the required line.

The parametric equations of the straight line passing through $P(5, 2, -1)$ and having $\boldsymbol{a} = \{2, -1, 3\}$ as its direction vector will be of the form $x = 2t + 5$, $y = -t + 2$, $z = 3t - 1$. By solving simultaneously the equations

$$\begin{cases} 2x - y + 3z + 23 = 0, \\ x = 2t + 5,\ y = -t + 2,\ z = 3t - 1, \end{cases}$$

we find the coordinates of the desired projection to be $x = 1$, $y = 4$, $z = -7$.

1054. $Q(-5, 1, 0)$.

1055. $P(3, -4, 0)$. *Hint.* The problem can be solved by the following procedure: (1) show that the points A and B lie on the same side of the plane Oxy; (2) find a point symmetric to one of the given points with respect to the plane Oxy: for example, find the point B_1 symmetric *to* B; (3) write the equations of the straight line passing through the points A and B_1; (4) by solving the equations of the line just found simultaneously with the equation of the plane Oxy, we obtain the coordinates of the desired point.

1056. $P(-2, 0, 3)$.

1057. $P(-2, -2, 5)$.

1058. $P(-1, 3, -2)$.

1059. 1) $P(-25, 16, 4)$; 2) this time is equal to 5; 3) $M_0P = 60$.

1060. $x = 28 - 7.5t$, $y = -30 + 8t$, $z = -27 + 6t$; 1) $P(-2, 2, -3)$; 2) the interval of time from $t_1 = 0$ to $t_2 = 4$; 3) $M_0P = 50$.

1061. This time is equal to 3.

1062. $d = 7$. Solution. On the line $\frac{x+3}{3} = \frac{y+2}{2} = \frac{z-8}{-2}$, let us choose a point, say $M_1(-3, -2, 8)$, and assume that the direction vector $\boldsymbol{a} = \{3, 2, -2\}$ of the line is drawn from the point M_1. The modulus of the vector product of the vectors $\boldsymbol{a}$ and $\overline{M_1P}$ will give the area of the parallelogram constructed on these vectors; the altitude drawn from the vertex P of this parallelogram will be the required distance d. Hence, the formula for calculating the distance d is $d = \frac{\left|[\boldsymbol{a}\overline{M_1P}]\right|}{|\boldsymbol{a}|}$. Now find the coordinates of the vector $\overline{M_1P}$ from the coordinates of its terminal and initial points:

$\overline{M_1P} = \{4, 1, -10)$. Evaluate the vector product of the vectors $\boldsymbol{a}$ and $\overline{M_1P}$: $[\boldsymbol{a}\overline{M_1P}]$

$$= \begin{vmatrix} \boldsymbol{i} & \boldsymbol{j} & \boldsymbol{k} \\ 3 & 2 & -2 \\ 4 & 1 & -10 \end{vmatrix} = -18\boldsymbol{i} + 22\boldsymbol{j} - 5\boldsymbol{k}.$$ Find its modulus

$|[\boldsymbol{a}\,\overline{M_1P}]| = \sqrt{18^2 + 22^2 + 5^2} = \sqrt{833} = 7\sqrt{17}$. Calculate the modulus of the vector $\boldsymbol{a}: |\boldsymbol{a}| = \sqrt{9+4+4} = \sqrt{17}$.

Finally, determine the required distance $d = \dfrac{7\sqrt{17}}{\sqrt{17}} = 7$.

1063. 1) 21; 2) 6; 3) 15.

1064. $d = 25$.

1065. $9x + 11y + 5z - 16 = 0$.

1068. $4x + 6y + 5z - 1 = 0$.

1070. $2x - 16y - 13z + 31 = 0$.

1072. $6x - 20y - 11z + 1 = 0$.

1074. $(2, -3, -5)$.

1075. $Q(1, 2, -2)$.

1076. $Q\ (1, -6, 3)$.

1077. $13x - 14y + 11z + 51 = 0$.

1079. $x - 8y - 13z + 9 = 0$.

1081. $\dfrac{x-3}{5} = \dfrac{y+2}{-6} = \dfrac{z+4}{9}$.

1082. $x = 8t - 3,\ y = -3t - 1,\ z = -4t + 2$.

1083. 1); 13; 2) 3; 3) 7.

1084. 1) $x^2 + y^2 + z^2 = 81$; 2) $(x-5)^2 + (y+3)^2 + (z-7)^2 = 4$; 3) $(x-4)^2 + (y+4)^2 + (z+2)^2 = 36$; 4) $(x-3)^2 + (y+2)^2 + (z-1)^2 = 18$; 5) $(x-3)^2 + (y+1)^2 + (z-1)^2 = 21$; 6) $x^2 + y^2 + z^2 = 9$; 7) $(x-3)^2 + (y+5)^2 + (z+2)^2 = 56$; 8) $(x-1)^2 + (y+2)^2 + (z-3)^2 = 49$; 9) $(x+2)^2 + (y-4)^2 + (z-5)^2 = 81$.

1085. $(x-2)^2 + (y-3)^2 + (z+1)^2 = 9$ and $x^2 + (y+1)^2 + (z+5)^2 = 9$.

1086. $R = 5$.

1087. $(x+1)^2 + (y-3)^2 + (z-3)^2 = 4$.

1088. $(x+1)^2 + (y-2)^2 + (z-1)^2 = 49$.

1089. $(x-2)^2 + (y-3)^2 + (z+1)^2 = 289$.

1090. 1) $C\ (3, -2, 5),\ r = 4$; 2) $C(-1, 3, 0),\ r = 3$; 3) $C(2, 1, -1),\ r = 5$; 4) $C(0, 0, 3);\ r = 3$; 5) $C\ (0, -10, 0),\ r = 10$.

1091. $x = 5t - 1,\ y = -t + 3,\ z = 2t - 0.5$.

1092. $\dfrac{x-\frac{1}{2}}{2} = \dfrac{y+\frac{3}{2}}{-3} = \dfrac{z+\frac{1}{2}}{4}$.

1093. 1) Outside the sphere; 2) and 5) on the sphere; 3) and 4) inside the sphere.

1094. 1) 5, 2) 21, 3) 7.

1095. 1) The plane cuts the sphere; 2) the plane touches the sphere; 3) the plane passes outside the sphere.

1096. 1) The line intersects the sphere; 2) The line passes outside the sphere; 3) the line touches the sphere.

1097. $M_1(-2, -2, 7)$, $d = 3$.

1098. $C(-1, 2, 3)$, $R = 8$.

1099. $\begin{cases} (x-1)^2 + (y-2)^2 + (z-1)^2 = 36, \\ 2x - z - 1 = 0. \end{cases}$

1100. $\begin{cases} (x-1)^2 + (y+1)^2 + (z+2)^2 = 65, \\ 18x - 22y + 5z - 30 = 0. \end{cases}$

1101. $\begin{cases} (x-2)^2 + y^2 + (z-3)^2 = 27, \\ x + y - 2 = 0. \end{cases}$

1103. $5x - 8y + 5z - 7 = 0$.

1104. $x^2 + y^2 + z^2 - 10x + 15y - 25z = 0$.

1105. $x^2 + y^2 + z^2 + 13x - 9y + 9z - 14 = 0$.

1106. $x^2 + (y+2)^2 + z^2 = 41$.

1107. $6x - 3y - 2z - 49 = 0$.

1108. $(2, -6, 3)$.

1109. $a = 6$.

1110. $2x - y - z + 5 = 0$.

1111. $x_1 x + y_1 y + z_1 z = r^2$.

1112. $A^2R^2 + B^2R^2 + C^2R^2 = D^2$.

1113. $(x_1 - \alpha)(x - \alpha) + (y_1 - \beta)(y - \beta) + (z_1 - \gamma)(z - \gamma) = r^2$.

1114. $3x - 2y + 6z - 11 = 0,\ 6x + 3y + 2z - 30 = 0$.

1115. $x + 2y - 2z - 9 = 0, x + 2y - 2z + 9 = 0$.

1116. $4x + 3z - 40 = 0, 4x + 3z + 10 = 0$.

1117. $4x + 6y + 5z - 103 = 0, 4x + 6y + 5z + 205 = 0$.

1118. $2x - 3y + 4z - 10 = 0, 3x - 4y + 2z - 10 = 0$

1120. $x - y - z - 2 = 0$.

1122. $Ax + By + Cz + D = 0$.

1123. $x\cos\alpha + y\cos\beta + z\cos\gamma - p = 0$.

1124. $d = |\mathbf{r}_1\mathbf{n} - p|$; $d = |x_1 \cos\alpha + y_1 \cos\beta + z_1 \cos\gamma - p|$.

1125. $(\mathbf{r}_2 - \mathbf{r}_1)(\mathbf{r} - \mathbf{r}_1) = 0$,

$(x_2 - x_1)(x - x_1) + (y_2 - y_1)(y - y_1) + (z_2 - z_1)(z - z_1) = 0$.

1126. $\mathbf{a}_1\mathbf{a}_2(\mathbf{r} - \mathbf{r}_0) = 0; \begin{vmatrix} x - x_0 & y - y_0 & z - z_0 \\ l_1 & m_1 & n_1 \\ l_2 & m_2 & n_2 \end{vmatrix} = 0,$

1127. $(\boldsymbol{r}_2-\boldsymbol{r}_1)\times(\boldsymbol{r}_3-\boldsymbol{r}_1)\times(\boldsymbol{r}-\boldsymbol{r}_1)=0;\ \begin{vmatrix} x-x_1 & y-y_1 & z-z_1 \\ x_2-x_1 & y_2-y_1 & z_2-z_1 \\ x_3-x_1 & y_3-y_1 & z_3-z_1 \end{vmatrix}=0.$

1128. $\boldsymbol{n}_1\boldsymbol{n}_2(\boldsymbol{r}-\boldsymbol{r}_0)=0;\begin{vmatrix} x-x_0 & y-y_0 & z-z_0 \\ A_1 & B_1 & C_1 \\ A_2 & B_2 & C_2 \end{vmatrix}=0.$

1131. $\dfrac{x-x_0}{l}=\dfrac{y-y_0}{m}=\dfrac{z-z_0}{n}.$

1132. $[(\boldsymbol{r}-\boldsymbol{r}_1)(\boldsymbol{r}_2 0 \boldsymbol{r}_1)]=0;[\boldsymbol{r}(\boldsymbol{r}_2-\boldsymbol{r}_1)]=[\boldsymbol{r}_1\boldsymbol{r}_2],\boldsymbol{r}=\boldsymbol{r}_1+(\boldsymbol{r}_2-\boldsymbol{r}_1)t.$

1133. $\boldsymbol{a}\,(\boldsymbol{r}-\boldsymbol{r}_1)=0;\ l(x-x_1)+m\,(y-y_1)+n(z-z_1)=0.$

1134. $\boldsymbol{a}_1\boldsymbol{a}_2(\boldsymbol{r}-\boldsymbol{r}_0)=0.$

1135. $\boldsymbol{n}_1\boldsymbol{n}_2(\boldsymbol{r}-\boldsymbol{r}_0)=0.$

1136. $\boldsymbol{r}=\boldsymbol{r}_0+\boldsymbol{n}t,\dfrac{x-x_0}{A}=\dfrac{y-y_0}{B}=\dfrac{z-z_0}{C}.$

1137. $\boldsymbol{r}=\boldsymbol{r}_0+[\boldsymbol{n}_1\boldsymbol{n}_2]t,\dfrac{x-x_0}{\begin{vmatrix} B_1 & C_1 \\ B_2 & C_2 \end{vmatrix}}=\dfrac{y-y_0}{\begin{vmatrix} C_1 & A_1 \\ C_2 & A_2 \end{vmatrix}}=\dfrac{z-z_0}{\begin{vmatrix} A_1 & B_1 \\ A_2 & B_2 \end{vmatrix}}.$

1138. $\begin{cases} \boldsymbol{r}_0\,\boldsymbol{n}+D=0, \\ \boldsymbol{a}\boldsymbol{n}=0; \end{cases}\begin{cases} Ax_0+By_0+Cz_0+D=0, \\ Al+Bm+Cn=0. \end{cases}$

1139. $\boldsymbol{a}_1\boldsymbol{a}_2(\boldsymbol{r}-\boldsymbol{r}_0)=0.$

1140. $\boldsymbol{a}_1\boldsymbol{a}_2(\boldsymbol{r}_2-\boldsymbol{r}_1)=0$

1141. $\boldsymbol{r}_0=\dfrac{\boldsymbol{r}_0\boldsymbol{n}+D}{\boldsymbol{a}\boldsymbol{n}}\boldsymbol{a};\ x=x_0-\dfrac{Ax_0+By_0+Cz_0+D}{Al+Bm+Cn}l,$

$$y=y_0-\frac{Ax_0+By_0+Cz_0+D}{Al+Bm+Cn}\mathrm{m}$$

$$z=z_0-\frac{Ax_0+By_0+Cz_0+D}{Al+Bm+Cn}n.$$

1142. $\boldsymbol{r}_1-\dfrac{\boldsymbol{r}_1\boldsymbol{n}+D}{\boldsymbol{n}^2}\boldsymbol{n},\ x=x_1+\dfrac{Ax_1+By_1+Cz_1+D}{A^2+B^2+C^2}A,$

$$y=y_1-\frac{Ax_1+By_1+Cz_1+D}{A^2+B^2+C^2}B,\ z=z_1-\frac{Ax_1+By_1+Cz_1+D}{A^2+B^2+C^2}C.$$

1143. $\boldsymbol{r}_0+\dfrac{(\boldsymbol{r}_1-\boldsymbol{r}_0)\boldsymbol{a}}{\boldsymbol{a}^2}\boldsymbol{a},\ x=x_0+\dfrac{(x_1-x_0)l+(y_1-y_0)m+(z_1-z_0)n}{l^2+m^2+n^2}l,$

$$y=y_0+\frac{(x_1-x_0)l+(y_1-y_0)m+(z_1-z_0)n}{l^2+m^2+n^2}m,$$

$$z=z_0+\frac{(x_1-x_0)l+(y_1-y_0)m+(z_1-z_0)n}{l^2+m^2+n^2}$$

1144. $d = \dfrac{\sqrt{[(\boldsymbol{r}_1 - \boldsymbol{r}_0)\boldsymbol{a}]^2}}{\sqrt{\boldsymbol{a}^2}}$,

$$d = \frac{\sqrt{\begin{vmatrix} \dfrac{y_1 - y_0}{m} & \dfrac{z_1 - z_0}{n} \end{vmatrix}^2 + \begin{vmatrix} \dfrac{z_1 - z_0}{n} & \dfrac{x_1 - x_0}{l} \end{vmatrix}^2 + \begin{vmatrix} \dfrac{x_1 - x_0}{l} & \dfrac{y_1 - y_0}{m} \end{vmatrix}^2}}{\sqrt{l^2 + m^2 + n^2}}.$$

1145. $d = \dfrac{|\boldsymbol{a}_1\boldsymbol{a}_2(\boldsymbol{r}_2 - \boldsymbol{r}_1)|}{\sqrt{[\boldsymbol{a}_1\boldsymbol{a}_2]^2}}$;

$$d = \frac{\text{absolute value of} \begin{vmatrix} l_1 & l_2 & x_2 - x_1 \\ m_1 & m_2 & y_2 - y_1 \\ n_1 & n_2 & z_2 - z_1 \end{vmatrix}}{\sqrt{\begin{vmatrix} m_1 & n_1 \\ m_2 & n_2 \end{vmatrix}^2 + \begin{vmatrix} n_1 & l_1 \\ n_2 & l_2 \end{vmatrix}^2 + \begin{vmatrix} l_1 & m_1 \\ l_2 & m_2 \end{vmatrix}^2}}.$$

1147. $\dfrac{R}{|\boldsymbol{a}|}\boldsymbol{a}$ and $-\dfrac{R}{|\boldsymbol{a}|}\boldsymbol{a}$; $x_1 = \dfrac{Rl}{\sqrt{l^2 + m^2 + n^2}}$, $y_1 = \dfrac{Rm}{\sqrt{l^2 + m^2 + n^2}}$,

$z_1 = \dfrac{Rn}{\sqrt{l^2 + m^2 + n^2}}$ and $x_2 = -\dfrac{Rl}{\sqrt{l^2 + m^2 + n^2}}$, $y_2 = -\dfrac{Rm}{\sqrt{l^2 + m^2 + n^2}}$,

$z_2 = -\dfrac{Rn}{\sqrt{l^2 + m^2 + n^2}}$.

1148. $\boldsymbol{r}_0 + \dfrac{R}{|\boldsymbol{a}|}\boldsymbol{a}$ and $\boldsymbol{r}_0 - \dfrac{R}{|\boldsymbol{a}|}\boldsymbol{a}$;

$x_1 = x_0 + \dfrac{Rl}{\sqrt{l^2 + m^2 + n^2}}$, $y_1 = y_0 + \dfrac{Rm}{\sqrt{l^2 + m^2 + n^2}}$,

$z_1 = z_0 + \dfrac{Rn}{\sqrt{l^2 + m^2 + n^2}}$ and $x_2 = x_0 - \dfrac{Rl}{\sqrt{l^2 + m^2 + n^2}}$,

$y_2 = y_0 - \dfrac{Rm}{\sqrt{l^2 + m^2 + n^2}}$, $z_2 = z_0 - \dfrac{Rn}{\sqrt{l^2 + m^2 + n^2}}$.

1149. $(\boldsymbol{r}_1 - \boldsymbol{r}_0)(\boldsymbol{r} - \boldsymbol{r}_0) = R^2$.

1150. $(\boldsymbol{r} - \boldsymbol{r}_1)^2 = \dfrac{(\boldsymbol{r}_1\boldsymbol{n} + D)^2}{\boldsymbol{n}^2}$;

$$(x - x_1)^2 + (y - y_1)^2 + (z - z_1)^2 = \frac{(Ax_1 + By_1 + Cz_1 + D)^2}{A^2 + B^2 + C^2}.$$

1151. $\dfrac{\boldsymbol{nr}}{|\boldsymbol{n}|} - R = 0$, $\dfrac{\boldsymbol{nr}}{|\boldsymbol{n}|} + R = 0$; $\dfrac{Ax + By + Cz}{\sqrt{A^2 + B^2 + C^2}} - R = 0$,

$\dfrac{Ax + By + Cz}{\sqrt{A^2 + B^2 + C^2}} + R = 0$.

1152. $\dfrac{\boldsymbol{a}(\boldsymbol{r}-\boldsymbol{r}_0)}{|\boldsymbol{a}|}-R=0, \dfrac{\boldsymbol{a}(\boldsymbol{r}-\boldsymbol{r}_0)}{|\boldsymbol{a}|}+R=0;$

$$\frac{l(x-x_0)+m(y-y_0)+n(z-z_0)}{\sqrt{l^2+m^2+n^2}}-R=0.$$

$$\frac{l(x-x_0)+m(y-y_0)+n(z-z_0)}{\sqrt{l^2+m^2+n^2}}+R=0.$$

1153. $3, \sqrt{3}$; $(2, 3, 0), (2, -3, 0), (2, 0, \sqrt{3}), (2, 0, -\sqrt{3})$.

1154. 4.3; $(4, 0, -1), (-4, 0, -1)$.

1155. $15; \left(0, -6, -\dfrac{3}{2}\right)$.

1156. The equations of the projections on the planes *Oxy*, *Oxz* and *Oyz* are, respectively:

(a) $\begin{cases} x^2+4xy+5y^2-x=0, \\ z=0; \end{cases}$

(b) $\begin{cases} x^2-2xz+5z^2-4x=0, \\ y=0; \end{cases}$

(c) $\begin{cases} y^2+z^2+2y-z=0, \\ x=0. \end{cases}$

1157. An ellipse with centre at $(2, -1, 1)$.
Hint. The centre of the section is projected into the centre of the projection.

1158. A hyperbola with centre at $(1, -1, -2)$.

1159. 1) An ellipse with centre at $(-1, 1, 3)$; 2) a parabola; has no centre; 3) a hyperbola with centre at $(2, -3, -4)$.

1160. 1) $1<|m|<\sqrt{2}$; 2) $|m|<1$.

1161. 1) $m \neq 0$ and $m \geq -\dfrac{1}{4}$; when $m=-\dfrac{1}{4}$, a degenerate ellipse (a point); 2) $m=0$.

1162. $(9, 5, -2)$.

1163. $(3, 0, -10)$.

1164. $(6, -2, 2)$.

1165. $m=\pm 18$.

1166. $2x-y-2z-4=0$.

1167. $x-2y+2z-1=0, x-2y+2z+1=0; \dfrac{2}{3}$.

1168. $\dfrac{x^2}{9}+\dfrac{y^2+z^2}{25}=1$.

1169. $\dfrac{x^2}{36}+\dfrac{y^2}{16}+\dfrac{z^2}{9}=1$.

1170. $q_1=\dfrac{2}{5}, q_2=\dfrac{4}{5}$.

1172. $\dfrac{x^2}{a^2}+\dfrac{y^2+z^2}{b^2}=1$.

1173. $\dfrac{x^2+y^2}{a^2}-\dfrac{z^2}{c^2}=1$.

1178. $\frac{x^2}{p} - \frac{y^2}{q} = 2z.$

1180. 1) (3, 4, –2) and (6, – 2, 2); 2) (4, –3, 2)—the line touches the surface; 3) the line and the surface have no points in common; 4) the line lies on the surface.

1181. $\begin{cases} 2x - 12y - z + 16 = 0, \\ x - 2y + 4 = 0; \end{cases} \begin{cases} 2x - 12y - z + 16 = 0, \\ x + 2y - 8 = 0. \end{cases}$

1182. $\begin{cases} y + 2z = 0, \\ x - 5 = 0; \end{cases} \begin{cases} 2x - 5z = 0, \\ y + 4 = 0. \end{cases}$

1183. $\frac{x}{1} = \frac{y+1}{4} = \frac{z-1}{-2}, \frac{x}{1} = \frac{y+9}{12} = \frac{z+3}{2}.$

1184. $\frac{x}{1} = \frac{y-3}{0} = \frac{z}{-2}, \frac{x-2}{0} = \frac{y}{3} = \frac{z}{-4}.$

1185. $\arccos \frac{1}{17}.$

1186. 1) $\frac{x^2}{a^2} + \frac{y^2}{b^2} - \frac{z^2}{c^2} = 0$; 2) $\frac{x^2}{a^2} - \frac{y^2}{b^2} + \frac{z^2}{c^2} = 0$; 3) $-\frac{x^2}{a^2} + \frac{y^2}{b^2} + \frac{z^2}{c^2} = 0.$

1188. $x^2 + y^2 - z^2 = 0.$

1189. $\frac{x^2}{a^2} + \frac{y^2}{b^2} - \frac{(z-c)^2}{c^2} = 0.$

1190. $3x^2 - 5y^2 + 7z^2 - 6xy + 10xz - 2yz - 4x + 4y - 4z + 4 = 0.$

1191. $\frac{x^2}{25} + \frac{y^2}{25} - \frac{z^2}{49} = 0.$ **1192.** $x^2 - 3y^2 + z^2 = 0.$

1193. $35x^2 + 35y^2 - 52z^2 - 232xy - 116xz + 116yz + 232x - 70y - 116z + 35 = 0.$

1194. $xy + xz + yz = 0$ — the axis of the cone is situated in the first and seventh octants; $xy + xz - yz = 0$—the axis of the cone is situated in the second and eighth octants; $xy - xz - yz = 0$— the axis of the cone is situated in the third and fifth octants; $xy - xz + yz = 0$—the axis of the cone is situated in the fourth and sixth octants.

1195. $9x^2 - 16y^2 - 16z^2 - 90x + 225 = 0.$

1196. $x^2 + 4y^2 - 4z^2 + 4xy + 12xz - 6yz = 0.$

1197. $4x^2 - 15y^2 - 6z^2 - 12xz - 36x + 24z + 66 = 0.$

1198. $16x^2 + 16y^2 + 13z^2 - 16xz + 24yz + 16x - 24y - 26z - 23 = 0.$

1199. $x^2 - y^2 - 2xz + 2yz + x + y - 2z = 0.$

1200. $5x^2 + 5y^2 + 2z^2 - 2xy + 4xz + 4yz - 6 = 0.$

1201. $45x^2 + 72y^2 + 45z^2 + 36xy + 72xz - 36yz + 54x + 216y - 54z - 567 = 0.$

1202. $5x^2 + 10y^2 + 13z^2 + 12xy - 6xz + 4yz + 26x + 20y - 38z + 3 = 0.$

1203. $x^2 + 4y^2 + 5z^2 - 4xy - 125 = 0.$

1204. 1) 18; 2) 10; 3) 0; 4) – 50; 5) 0; 6) $x_2 - x_1$; 7) 0; 8) 1.

1205. 1) = 12; 2) $x = 2$; 3) $x_1 = -1$, $x_2 = -4$; 4) $x_1 = -1/6; x_2 = 3/2$; 5) $x_{1,2} = \pm 2i$; 6) $x_1 = 2x_{2,3} = -2 \pm i$; 7) $x = (-1)^n \frac{\pi}{12} + \frac{\pi}{2} n$, where n is any integer; 8) $x = \frac{\pi}{6}(2n+1)$, where n is any integer.

1206. 1) $x > 3$; 2) $x > -10$; 3) $x < -3$; 4) $-1 < x < 7$.

1207. 1) $x = 16$, $y = 7$; 2) $x = 2$, $y = 3$; 3) the system has no solutions; 4) the system has an infinite number of different solutions, each of which may be determined from the formula $y = \frac{x-1}{3}$ by assigning arbitrary values to x and computing the corresponding values of y; 5) $x = \frac{ac+bd}{a^2+b^2}$, $y = \frac{bc-ad}{a^2-b^2}$; 6) the system has no solutions.

1208. 1) $a \neq -2$; 2) $a = -2, b \neq 2$, 3) $a = -2$; $b = 2$.

1209. $a = 10/13$.

1210. 1) $x = -2t$, $y = 7t$, $z = 4t$; 2) $x = 2t$, $y = 3t$, $z = 0$; 3) $x = 0$, $y = t$, $z = 3t$; 4) $x = 0$, $y = t$, $z = 2t$; 5) $x = 2t$, $y = 5t$, $z = 4t$: 6) $x = 4t$, $y = 2t$, $z = 3t$; 7) $x = t$, $y = 5t$, $z = 11t$; 8) $x = 3t$, $y = 4t$, $x = 11t$; 9) $x = 0$, $y = t$, $z = 3t$; 10) $x = (a+1)t$, $y = (1-a^2)t$, $z = -(a+1)t$ provided that $a \neq -1$. (If $a = -1$, then any solution of the system will consist of three numbers x, y, z, where x, y may be chosen at will and $z = x + y$); 11) $x = (b-6)t$, $y = (3a-2)t$, $z = (ab-4)t$ provided that $a \neq \frac{2}{3}$ or $b \neq 6$ (If $\left(\text{If } a = \frac{2}{3} \text{and } b = 6, \text{ then} x, y \text{ may be chosen at willand} z = \frac{2}{3}x + 2y\right)$

; 12) $x = 3(1-2a)t$, $y = (ab+1)t$, $z = 3(b+2)t$ provided that $a \neq -\frac{1}{2}$ or $b \neq -2$.

[If $a = -\frac{1}{2}$ and $b = -2$, then x, y may be chosen at will and $z = 2(3y - x)$.]

1211. –12.

1212. 29.

1213. 87.

1214. 0.

1215. –29.

1216. $2a^3$.

1223. –4.

1224. 180.

1225. 87.

1226. 0.

1227. $(x-y)(y-z)(z-x)$.

1229. $2a^2b$.

1230. $\sin 2\alpha$

1231. $xyz(x-y)(y-z)(z-x)$.

1232. $(a+b+c)(a^2+b^2+c^2-ab-ac-bc)$.

1234. 1) $x=-3$; 2) $x_1=-10$, $x_2=2$.

1235. 1) $x>7/2$; 2) $-6<x<-4$.

1236. $x=24\frac{1}{2}$, $y=21\frac{1}{2}$, $z=10$.

1237. $x=1$, $y=1$, $z=1$.

1238. $x=2$, $y=3$, $z=4$.

1239. $x=1$, $y=3$, $z=5$.

1240. $x=13\frac{1}{4}$, $y=8\frac{1}{4}$, $z=14\frac{1}{2}$.

1241. $x=2$, $y=-1$, $z=1$.

1242. $x=\frac{b+c}{2}$, $y=\frac{a-b}{2}$, $z=\frac{a-c}{2}$.

1243. $x=\frac{a+b}{2}$, $y=\frac{b+c}{2}$, $z=\frac{a+c}{2}$.

1244. The system has an infinite number of solutions, each of which can be found from the formulas $x=2z-1$, $y=z+1$ by assigning arbitrary values to z and calculating the corresponding values of x, y.

1245. The system has no solutions.

1246. The system has no solutions.

1247. 1) $a\neq-3$; 2) $a=-3$, $b\neq\frac{1}{3}$; 3) $a=-3$, $b=\frac{1}{3}$.

1249. The system has the unique solution $x=y=z=0$.

1250. The system has an infinite number of solutions, each of which can be found from the formulas $x=2t$, $y=-3t$, $z=5t$ by assigning arbitrary values to t and calculating the corresponding values of x, y, z.

1251. $a=5$.

1252. 30.

1253. -20.

1254. 0.

1255. 48.

1256. 1800.

1257. $(b+c+d)(b-c-d)(b-c+d)(b+c-d)$.

1258. $(a+b+c+d)\times(a+b-c-d)\ (a-b+c-d)\ (a-b-c+d)$.

1259. $(a+b+c+d)\times(a-b+c-d)\ [(a-c)^2+(b-d)^2)]$.

1260. $(be-cd)^2$.

www.ingramcontent.com/pod-product-compliance
Ingram Content Group UK Ltd.
Pitfield, Milton Keynes, MK11 3LW, UK
UKHW020050200726
13845UKWH00055B/616

9 789324 191755